34

# LABORATORY
## SAFETY
### Theory and Practice

# LABORATORY SAFETY
## Theory and Practice

EDITED BY

## ANTHONY A. FUSCALDO
*Director, Division of Tumor Biology*
*Department of Medical Oncology and Hematology*
*Hahnemann Medical College and Hospital*
*Philadelphia, Pennsylvania*

## BARRY J. ERLICK
*Division of Virology*
*United States Army Medical Research*
*Institute of Infectious Diseases*
*Fort Detrick,*
*Frederick, Maryland*

## BARBARA HINDMAN
*Franklin Research Center*
*Franklin Institute*
*Philadelphia, Pennsylvania*

1980

**ACADEMIC PRESS**
*A Subsidiary of Harcourt Brace Jovanovich, Publishers*
New York   London   Toronto   Sydney   San Francisco

This book is a guide to provide general information concerning
its subject matter; it is not a procedural manual. Laboratory
safety is a rapidly changing field. The reader should consult
current procedural manuals for state–of–the–art instructions
and applicable governmental safety regulations. The Publisher
does not accept responsibility for any misuse of this book,
including its use as a procedural manual or as a source of
specific instructions.

ACADEMIC PRESS, INC.
111 Fifth Avenue, New York, New York 10003

*United Kingdom Edition published by*
ACADEMIC PRESS, INC. (LONDON) LTD.
24/28 Oval Road, London NW1 7DX

**Library of Congress Cataloging in Publication Data**
Main entry under title:

Laboratory safety.

Includes index.
1. Biological laboratories––Safety measures.
2. Medical laboratories––Safety measures. 3.
Laboratories––Safety measures. I. Fuscaldo,
Anthony A. II. Erlick, Barry J. III. Hindman,
Barbara. [DNLM: 1. Laboratories. 2. Accident
prevention. QY 23 L123]
QH323.2.L3      574'.028'9      80–24955
ISBN 0–12–269980–7

PRINTED IN THE UNITED STATES OF AMERICA

81 82 83     9 8 7 6 5 4 3 2

# Contents

# PART TWO: BIOLOGICAL LABORATORY SAFETY

Chapter Four
## VIRUSES AND CANCER
Robert Gallagher, Riccardo Dalla Favera, and Robert Gallo

Chapter Five
## RECOMBINANT DNA RESEARCH
Robert B. Helling

Chapter Six
## IDENTIFICATION, ANALYSIS, AND CONTROL
## OF BIOHAZARDS IN VIRAL CANCER RESEARCH
D. L. West, D. R. Twardzik, R. W. McKinney, W. E. Barkley, and A. Hellman

Chapter Seven
## BIOHAZARDS ASSOCIATED WITH LABORATORY ANIMALS
Peter J. Gerone

# PART THREE: MEDICAL AND PSYCHOLOGICAL FACTORS

Chapter Eight
## SELECTED MEDICAL PROBLEMS OFTEN ASSOCIATED WITH LABORATORY PERSONNEL
Linda Haegele

Chapter Nine
## MEDICAL ASPECTS OF OCCUPATIONAL HEALTH IN A LABORATORY SETTING
James H. Anderson, Jr.

Chapter Ten
## GENETIC MONITORING
Kathryn E. Fuscaldo

Chapter Eleven
**BEHAVIORAL FACTORS IN LABORATORY SAFETY:
PERSONNEL CHARACTERISTICS AND THE MODIFICATION
OF UNSAFE ACTS**
Joan C. Martin

# List of Contributors

*Numbers in parentheses indicate the pages on which authors' contributions begin.*

**James H. Anderson, Jr.*** (281), United States Army Medical Research Institute of Infectious Diseases, Fort Detrick, Frederick, Maryland 21701

**W. E. Barkley** (167), Division of Safety, National Institutes of Health, Bethesda, Maryland 20205

**Riccardo Dalla Favera** (115), Laboratory of Tumor Cell Biology, National Cancer Institute, National Institutes of Health, Bethesda, Maryland 20014

**Kathryn E. Fuscaldo** (297), Director, Division of Medical Genetics, Department of Hematology and Medical Oncology, Hahnemann Medical College and Hospital, Philadelphia, Pennsylvania 19102

**Robert Gallagher** (115), Laboratory of Tumor Cell Biology, National Cancer Institute, National Institutes of Health, Bethesda, Maryland 20014

**Robert Gallo** (115), Laboratory of Tumor Cell Biology, National Cancer Institute, National Institutes of Health, Bethesda, Maryland 20014

**Peter J. Gerone** (225), Delta Regional Primate Research Center, Tulane University, Covington, Louisiana 70433

**Linda Haegele** (251), Department of Medical Oncology and Hematology, Hahnemann Medical College and Hospital, Philadelphia, Pennsylvania 19102

**Robert B. Helling** (145), Division of Biological Sciences, University of Michigan, Ann Arbor, Michigan 48109

**A. Hellman** (167), Office of Biohazards Safety, National Cancer Institute, National Institutes of Health, Bethesda, Maryland 20014

**Mark Hite** (29), Merck Institute for Therapeutic Research, West Point, Pennsylvania 19486

**R. W. McKinney** (167), Enviro Control, Inc., Rockville, Maryland 20852

---

* Present Address: Brooke Army Medical Center, Ft. Sam Houston, Texas 78231.

**Joan C. Martin** (319), Department of Psychiatry and Behavioral Sciences, University of Washington, Seattle, Washington 98195

**Michael Slobodien** (59), Nuclear Regulatory Commission, Region I, King of Prussia, Pennsylvania 19406

**Norman V. Steere** (3), Laboratory Safety and Design Consultant, Norman V. Steere and Associates, Minneapolis, Minnesota 55414

**D. R. Twardzik** (167), National Cancer Institute, National Institutes of Health, Bethesda, Maryland 20014

**D. L. West** (167), National Cancer Institute, National Institutes of Health, Bethesda, Maryland 20014

# Preface

There is an ironic relationship between the actual or potential hazards of certain acts or situations and the fear which they produce in the individuals involved. For example, the health hazards produced by habitual ingestion of alcohol or frequent cigarette smoking has been well-publicized for many years, yet the use of these substances is still almost mandatory in social situations. Furthermore, legislation to limit their use has continually met strong resistance. Similarly, neither are drivers forestalled from using their automobiles by the carnage on the highways, nor are a majority likely to observe the speed limits. Conversely, the very sight of a snake or shark, even on film, makes most people recoil in fear. This fear occurs despite the fact that the actual number of deaths caused by shark and snake attacks is a very small fraction of those caused by automobile accidents, cigarette smoking, or alcoholism. It is obvious, therefore, that the level of anxiety evoked often bears little relationship to the actual danger.

This disparity can be better understood if the question of what instills fear is examined. People generally fear what they do not understand or what they cannot control. Whether one smokes, drinks, or drives a car is self-controlled; whether a shark attacks is notoriously unpredictable and is controlled by the shark rather than the person.

A similar ironic relationship between hazard and anxiety is evident in a consideration of the scientific laboratory. There was, at one time, an enormous degree of anxiety in both public and scientific sectors over the hazards involved in genetic engineering. Although this fear still exists in a portion of the general public's mind, most scientists now feel that the actual danger is minimal.

For many years microbiologists have been engaged in research with virulent pathogens and toxins. Although these agents strike fear in the hearts of the nonscientific community, both the scientists and the technicians who work with them daily tend to view them with about as much fear as the general public views a cigarette or a speeding automobile. This over-familiarity has, in several unfortunate instances, been one of the causes of accidents wherein laboratory workers were harmed. Because of the design of these laboratories and the fortuitous limitations of biological systems, an uncontrolled epidemic has never resulted from a laboratory accident. However, an injury to an individual laboratory worker or an epidemic resulting from man overriding nature's limitations should be avoided at all costs.

This book is intended to bridge the gap between fear of the unknown and familiarity bred of daily work. In the chapters that follow, the authors will present information on the hazards encountered in the laboratory by students, technicians, and scientists. The theoretical aspects of the hazards have been emphasized in order to increase the readers' understanding of the practices described, and to teach, by example, methods of risk assessment that can be applied to new technologies as they are translated from the scientist's mind to the laboratory bench.

The book is divided into three sections: (1) General Laboratory Safety; (2) Biological Laboratory Safety; and (3) Medical and Psychological Factors. The first section is subdivided into three chapters. Chapter 1 is a description of the hazards found in almost all laboratories, regardless of their specific functions. This includes the pertinent safety theories and practices. Chapter 2 is concerned with those almost ubiquitous compounds that are either toxic or carcinogenic. This area is becoming increasingly important as more and more laboratory reagents are recognized as carcinogenic. This chapter also stresses the insidious nature of these substances and provides guidelines for their use. Finally, Chapter 3 deals with radiation hazards which, because of the wider use of radioisotopic tracers and radioimmunoassays, have become a concern in many laboratories.

The second section of this book (Chapters 4–7) concentrates on safety in the biological laboratory. Most laboratory personnel are aware of the dangers in working with organisms such as smallpox, rabies, yellow fever, and even such exotic agents as crimean hemorrhagic fever or machupo virus. There are other areas of microbiological research, however, where the laboratory workers are not so fully cognizant of the dangers. One of these areas is the study of oncogenic viruses. Although a cause-and-effect relationship between viruses and cancer in humans has not been unequivocally established, the fact that this relationship exists in the rest of the animal kingdom must make laboratory personnel in this area aware of the problem. Therefore, Chapter 4 includes a discussion of this relatively complex group of viruses. Chapter 5 presents the readers with an approach to recombinant DNA research. It includes a general introduction to this new field of study and alerts investigators to the possible hazards associated with this research. Chapter 6 is a comprehensive approach to the design and function of biohazard laboratories. Such laboratories began almost simultaneously with the study of pathogenic microorganisms and acquired greater urgency because of the increased emphasis on oncogenic virus research of the late 1960s and early 1970s and the beginning of recombinant DNA research in the middle 1970s. Laboratory safety became a prime concern of the National Cancer Institute (NCI), which formed groups to conduct research on these problems and to disseminate information to workers in the field. Various agencies such as the Public Health Services (PHS), the National Cancer Institute, and the Office of Recombinant DNA Research (ORDR) used different names for laboratory containment levels, but these eventually evolved to four very similar levels used by all groups. Chapter 6 was written in relation to a virus oncology laboratory, but the levels of containment discussed can easily be translated to the nomenclature used by either the Center for Disease Control (CDC) or the ORDR. The specific levels of containment needed for various microbial pathogens are given in the Appendix at the end of the book. This is the latest listing by the CDC. The levels needed for various experiments in recombinant DNA research are not included in this book because the ORDR is in the process of reevaluating its current levels. As this book goes to press, it appears that work using the *E. coli* K12 system cloning the recombinant DNA will be lowered to the P1 containment level. Because of the wide variety of potential combinations that can be envisioned by researchers in the field, most of the innovative work in recombinant DNA research has to be judged on a case-by-case basis. An additional safety factor for recombinant DNA research is a requirement by ORDR that every institution engaging in this type of work have an Institutional Biosafety Committee (IBC) which reviews work in this area. Therefore, all laboratories should have an institutional committee that has the latest information and should be able to monitor individual safety practices. The final chapter in this section discusses the hazards relating to laboratory animals. The description of the often-overlooked possibility of contracting human diseases from uninfected experimental animals provides information for laboratory workers about a serious potential hazard. Other aspects common to animal research laboratories including hazards associated with experimentally infected animals, safe animal room practices, and animal-risk assessment are included.

The third section of this book contains a discussion of medical surveillance of persons at risk and the psychological factors involved in accident control. This section does not attempt to provide a detailed treatment of medical surveillance in the laboratory, but will provide an overview of the screening techniques available for the early detection of disease and personnel risks. Furthermore, this book does not discuss the techniques of cardiopulmonary resuscitation (CPR), since techniques such as this are most effectively acquired by taking an appropriate course. Such courses are readily available through the Red Cross, as well as many hospitals, medical institutions, and fire departments. Chapters 8 and 9 are directed to some of the topics discussed in earlier sections but approach hazards such as physical agents, toxins, carcinogens, and infectious agents from the viewpoint of the treating physician. Also included is a comprehensive list of chemical agents, their sources, their subsequent physical effects, and the accepted mode of medical surveillance. This list should provide the reader with a rapid means of ascertaining medically important information about commonly encountered laboratory chemicals.

With the introduction of recent genetic monitoring techniques, a greater awareness of the long term and subtle physical damage from mutagens has been brought to light. In Chapter 10, the reader will find an interesting discussion of various genetic screening tests available and their potential use for the evaluation of presumptive and actual mutagens. It is hoped that an insight into the realistic use, limitation, and potential of genetic screening may be obtained by the reader. The final chapter of the book discusses human behavior as related to laboratory safety. The author provides a brief review and discussion designed to identify and isolate important behavioral factors that can cause accidents. This chapter is included as an acknowledgment of the important basic concept that even the best equipped and designed laboratory with the most stringent hypothetical safety standards is only as "safe" as the people who work in that laboratory make it. The human factor ("safe practices") cannot be overstated. In the final analysis, if the principles discussed in *this* chapter are disregarded, the information contained in the rest of the book will, in time, be forgotten and laboratory safety will remain an administrator's theory rather than a laboratory reality.

The editors would like to thank Ms. Angela Venuto for her time and energy, attentiveness to detail, and persistence. Without her help, the transition from a proposed book to reality would never have been completed.

Anthony A. Fuscaldo*
Barry J. Erlick
Barbara Hindman

* Present Address: Ecology and Environment, Inc., Philadelphia, Pennsylvania 19103.

# PART ONE
# GENERAL LABORATORY SAFETY

# Chapter One

# Physical, Chemical, and Fire Safety

## NORMAN V. STEERE

LABORATORY SAFETY: THEORY AND PRACTICE
Copyright © 1980 by Academic Press, Inc.
ISBN 0-12-269980-7

## I. INTRODUCTION

Epidemiology provides a framework for developing a rational approach to safety and accident prevention. Just as the disease process is an interaction of susceptible host, environment, and disease agent, the causation of injury is an interaction of several factors and should not usually be blamed primarily on a single factor, such as an unsafe condition or an unsafe act.

The parallel between disease causation and injury causation was first elucidated in 1949 by the epidemiologist John Gordon. He stated that injury, like disease, requires a susceptible host, an environmental reservoir, and a third factor corresponding to the disease agent. In an injury, a dose of energy in excess of the damage threshold of the susceptible host can be likened to the infective dose of a pathogen. The amount of energy necessary to cause injury will vary with the damage threshold of the host, just as the infective dose varies in disease causation. An example of injury caused by direct energy transfer is the electrically induced injury that occurs when a person's body receives electrical energy in excess of the "let-go" current level for that person. Haddon elaborated on the concept of injury resulting from direct transfer of energy and added a classification for damage caused indirectly through interference with the body's ability to transfer energy within body systems (Haddon *et al.,* 1964). For example, sulfuric acid causes tissue destruction by direct transfer of energy, but carbon monoxide causes a chemical asphyxiation indirectly by interfering with the ability of hemoglobin to transfer oxygen to body tissues. Extreme cold can cause injury directly by frostbite and freezing of body tissues or indirectly by hypothermia, which is a failure of the body's thermoregulation system.

The types of energy or hazards that can cause damage directly or indirectly have been classified (Steere, 1974) as follows:

| Form of energy or hazard | Modes of action |
| --- | --- |
| Mechanical energy | Direct |
| Electrical energy | Direct |
| Ionizing radiation | Direct and indirect |
| Nonionizing radiation | Direct and indirect |
| Biological organisms | Indirect |
| Chemicals | Direct and indirect |
| Atmospheric pressure differentials | Indirect |
| Thermal | Direct and indirect |
| Fire | Direct and indirect |

Although the major emphasis of this book is on biological hazards, radiation hazards, and hazards from toxic chemicals, most laboratories also have other types of hazards that can cause injury, damage, or other interference with research and testing. This chapter will discuss accident causation, including mechanical, electrical, and thermal energy hazards; chemical and fire hazards; and several types of control measures.

## II. HUMAN FACTORS

It is important to realize that there are many human factors that affect an individual's susceptibility to injury. Knowledge or lack of knowledge of safe procedures, awareness or ignorance of hazards, and physiological and psychological factors must be considered in assessing an incident or in preventing one from happening.

In any working population there will be significant variations in susceptibility because of physiological factors. Some of these are relatively consistent from situation to situation and others vary greatly with circumstances. For example, women in general are more susceptible to electrical shock than men because women tend to have thinner skin, which has less electrical resistance (Dalziel, 1971). However, resistance to electrical shock will vary for *any* individual, depending on the path between electrical contacts, pressure and area of contact, and whether the skin is dry or wet.

In setting up and reviewing laboratory procedures it is important to recognize physiological variations in the personnel who are doing the procedures. Some of the physiological factors that can affect the susceptibility of an individual either to injury or to actions that may result in injury include hearing ability, visual acuity, strength, and mobility. Many of these factors will vary in "normal" individuals and in employees with some unrecognized physical handicap. The use of standard laboratory furniture by people who may range in height from 4 ft 11 in. to 6 ft 4 in. and whose reach varies proportionately may lead to unsafe adaptations.

Temporary variations in physiological and psychological factors can include the effect of fatigue, illness, and therapeutic and nontherapeutic chemicals on reaction time, attention span, and perception of hazards. Individuals who are fatigued, under the influence of chemicals, or in the prodromal or frank stages of illness will find their reactions to normal and unusual working conditions affected by their physical condition. Their altered reactions may not be fast enough, accurate enough, or appropriate.

Another human factor basic to prevention of injuries and damaging incidents is specific knowledge. Technology and research procedures change rapidly, and job training, education, and supervision often lag. Frequently, employees have not been taught laboratory safety procedures during their formal education or in their previous work experience. It is dangerous to assume that any employee new to a procedure has had adequate previous education, training, or experience to carry out the procedure safely.

Training should be provided for all new employees and for all employees who will be using new or revised procedures. Habits and reactions appropriate to one set of procedures may not be safe for another. Chemical injuries should be expected if laboratory personnel have not been provided with instructions on the proper use of emergency procedures or have not been given adequate practice using the emergency equipment, such as showers, spray hoses, or eye wash stations.

There may be a particular problem in communication of important information and effective training in safe procedures to employees who have some degree of learning disabilities. Since functional illiteracy is not uncommon but is often disguised to avoid embarrassment and loss of employment, subtle and considerate methods are needed to identify these problems and to adapt instructions for those who have difficulties in one mode of communication. There may be a need to modify or supplement teaching and communication techniques to assure that necessary learning and understanding take place.

For a discussion of attitudinal and motivational factors see Chapter 11.

## III. ENERGY FACTORS AND CONTROL STRATEGIES

If we think of injury as due primarily to transfer of energy in excess of an individual's damage threshold, we should consider the amount, rate, and concentration of energy and whether the transfer may cause injury. Control of one of these three factors can prevent injury.

Since the severity of injury is proportional to the total amount of excess energy transferred to the host, any method that reduces or blocks the transfer will reduce the severity of injury. If the total amount transferred can be reduced sufficiently, no injury will occur. It may be possible to reduce the amount of energy available by limiting such factors as pressure, speed, noise, radiation, or concentration of chemicals in a particular area. For example, laboratory exhaust ventilation can be an effective means of limiting the amount of chemicals (or infectious organisms) that enter the breathing zone of the user.

Reducing the rate of transfer of energy may also be a very effective way of preventing injury or injury-causing incidents. Just as a padded dashboard reduces collision injury by absorbing energy and changing the rate of impact, plastic-coated bottles or bottle carriers can prevent bottles from breaking or can limit the spillage of chemicals.

Reducing the concentration of energy by eliminating or covering sharp edges and points may prevent injury by spreading the contact over enough area to reduce the concentration of impact.

Haddon has proposed a hierarchy of measures that can be used to control energy factors and limit the effect of laboratory incidents or injuries. Control measures are listed in order of decreasing effectiveness. Two or more control measures may be desirable if the consequences of energy transfer are great. Although this sytem of selecting control measures can be extended to protection of laboratory equipment, animals, and buildings, the following table and examples will be limited here to selecting control measures to prevent or limit injuries to laboratory personnel.

## IV. HADDON'S MEASURES FOR THE PREVENTION OR LIMITATION OF LABORATORY INCIDENTS AND INJURIES

### A. Prevention of Injuries

#### 1. Prevent Accumulation of Hazardous Amounts of Energy

Measures can include laboratory ventilation, limits on quantities of chemicals used or stored in the laboratory, and electrical devices that will limit voltages or currents to safe levels for humans or protect them from ig-. nition of explosive atmospheres that may be present.

#### 2. Prevent or Modify Release of Energy

Measure can include (a) bottle carriers or plastic-coated bottles; (b) secondary containers for protection of bottles, ampules, or culture plates that could cause serious problems if broken and spilled; (c) animal cages to contain laboratory animals; and (d) electrical circuit breakers and rupture discs to control the release of any unexpected excess of energy.

#### 3. Provide Time or Space Separations between Energy and the Damageable Structure

Measures include the monitoring of radiation exposure, use of remote handling equipment to limit the total dose of radiation in a given time period, a fire alarm system and annual practice evacuation drills as means of providing time and space separation between laboratory personnel and the effect of fire, animal cages, isolation rooms, and enclosures or guards on electrical and mechanical equipment.

## 4. Provide Barrier to Block or Attenuate Energy Transfer

Measures include eye protection, gloves, laboratory coats, and other protective equipment designed to block or limit body contact with chemical splashes, flying particles, or other forms of energy. (For example, bricks for the attenuation of radiation, thermal insulation for the prevention of burns, electrical insulation for the prevention of shock, and ear protectors to reduce hearing loss).

## 5. Raise Damage Threshold to Prevent or Reduce Damage

Immunization is one of the few means available to laboratory personnel to raise damage thresholds, although physical conditioning might be appropriate for persons who expect to do strenuous work or lifting.

## B. Limiting Damage Occurring in an Accident

If the preceding control measures have not been taken or have not been effective and personnel are injured, concern must shift from the control of energy to limiting the extent of the injuries or the duration of the effects. These measures are as follows.

## 1. Provide Optimum Emergency Response

First aid and cardiopulmonary resuscitation training for a number of laboratory personnel are likely to prove useful for many common occurrences, as well as for some unusual ones. Rapid and effective communication with public emergency services is needed to obtain the help they can provide to back up the initial response from laboratory emergency teams. Since special knowledge for correct emergency treatment of chemical splashes or exposures may not be immediately available locally, the laboratory needs to acquire the special information in advance or to find and test the response and quality of special information from sources such as Poison Control Centers, Toxline, Medline, Hazmat, and so forth.

## 2. Provide Services to Reduce Damage Results

Special information may be needed to decontaminate equipment or a portion of a laboratory and such information is best obtained in advance of the need. Special equipment or supplies may be needed for such a decontamination, and it is much easier to obtain before it is needed.

## V. MECHANICAL HAZARDS AND CONTROLS

Lacerations or inoculations caused by impacts with hypodermic needles and sharp or broken glassware may be the most common causes of injury in biological laboratories. Control measures can include impact-resistant storage containers for used needles and used or broken glassware and measures to prevent movement of sharp objects toward the unprotected hand or body. For example, a towel may be used to protect the hand while inserting glass tubing into stoppers, and needle covers can be placed in a holder to prevent injection of the hand when returning the needle to its cover.

Other mechanical hazards in the laboratory are also relatively simple: objects that can interfere with motion, moving objects, and stored mechanical energy.

## A. Objects That Can Interfere with Motion

Obstructions or projections into corridors or other passageways are likely to impede traffic, particularly if the path is too narrow for two-way traffic or for the amount of traffic utilizing the passageway. Fire extinguishers, drinking fountains, delivery carts, supplies, door knobs, and sharp edges or equipment may interfere with motion. Impact with these objects may result in bruises, lacerations, or other injuries.

Use of working surfaces for storage is likely to cause interference with work activities, and this may have serious consequences if the interference causes glassware to break or chemicals to spill. Use of floors in passageways for storage or for placement of wires, hoses, or tubing will present tripping hazards that may cause serious falls, especially if visibility is reduced by smoke or power failure or if laboratory personnel move rapidly because of fire, chemical spills, or other emergencies.

## B. Moving Objects

Drawers, doors, machinery, and laboratory apparatus can transfer mechanical energy and cause serious injuries if parts of the body are caught, pinched, squeezed, or sheared.

Rotating machinery can catch clothing or parts of the body on rotating sharp parts or in gears, chains, or pulleys and belts. The most common examples of this type of hazard in the laboratory are the exposed pulleys and V-belts on vacuum pumps. If clothing, hair, or fingers get caught in the point where the belt runs on to the pulley, the possibility of personal injury is highly likely. The control for this type of hazard is to isolate or guard the rotating parts so that they are completely protected on all sides against contact. Guards can be purchased or fabricated, or machinery can be put in a separate enclosure where inadvertent contact is not possible.

## C. Stored Mechanical Energy

Mechanical energy is stored in the laboratory in objects located above floor level and in vessels that hold pressure. Controls may vary from simple measures, such as ways to keep things from falling off shelves, to more complex measures, such as special inspection of pressure vessels.

Autoclaves, compressed gas cylinders, vacuum traps, and Dewar vessels are common examples of vessels that store pressure inside or outside of the vessel. Sudden opening or failure of the vessel can result in release of mechanical energy and propulsion of objects that can cause serious injuries. Control measures may include operating and inspection procedures for autoclaves, secure storage of cylinders, and pressure-resistant glassware, plus protective outer shields and taping to control implosion. Pressure-reducing regulators should always be used in compressed gas lines.

Safe storage above floor level requires shelves or cabinets that can support the loads that will be placed on them, safe means of access to shelves above arm's reach, storage of breakable containers of liquid corrosives below eye level, and guarding of open edges of shelves if severe or continuing vibration is likely to occur. Shallow lips on open shelves will prevent objects from "walking" off the shelves as a result of machinery vibration or moderate tremors. Corrosive liquids and heavy objects should not be stored on high shelves, but at levels where they can be handled with less risk. Elimination of high shelves and cabinets is one possible control measure, and provision of an adequate and safe ladder is an alternate control measure for the problem of safe access to storage above eye level.

## VI. THERMAL HAZARDS AND CONTROL MEASURES

When laboratory personnel are working with cryogenic fluids such as liquid nitrogen, they should use face shields and other personal protective equipment to prevent splashes from reaching the eyes or skin. Cryogenic lines should be located where there can be no accidental contact, and Dewar flasks should be secured so as to avoid tipping or breaking.

Since liquid nitrogen tends to cause localized oxygen enrichment due to a fractional distillation of air, containers such as formed plastics should not be used even for temporary storage of liquid nitrogen due to the resultant increased combustibility.

High flow rates from cylinders of liquified or compressed gases will have a refrigerating effect that can cool metal parts to temperatures that can freeze human tissue. Insulated gloves may be desirable to prevent injury in case it is necessary to touch cold metal lines or connections.

Controls for gas burners should be located so that gas can be controlled without reaching past hot flames or burners.

High-temperature equipment should be located so that it will not heat or ignite combustible material, and so that chances of inadvertent personal contact are minimized.


## VII. CHEMICAL HAZARDS AND CONTROL MEASURES

Hazards of laboratory chemicals include violent reactions, fires, behavioral effects, corrosive and irritating effects, sensitization, toxic effects, and delayed or long-term effects, such as carcinogenesis, mutagenesis, and teratogenesis. Although many chemicals present several kinds of hazards, there is no uniform system in use for labeling containers to identify all the hazards. Similarly, there is no uniform system for classifying all the degrees of hazards or for communicating such information concisely and clearly, even though chemical hazards can vary from minor to extreme and carcinogens can be weak or potent.

Although toxicity and various forms of long-term effects are covered thoroughly in Chapter 2 of this book, some understanding of terms used in federal labeling requirements will help employees prevent injuries.


### A. Labeling Terms and Systems

Since a number of federal agencies and technical associations have required or recommended several different labeling systems for several different purposes, it is not likely that all potential hazards will be listed on any single label on containers of laboratory chemicals. For example, there are differing federal requirements for warning labels on pesticides, household products, *in vitro* diagnostic products, cancer-suspect agents, and shipping containers of hazardous materials.

Basic precautionary information and labeling terms have been defined for many years by the Manufacturing Chemists Association in their manual "Labeling and Precautionary Information," which served as the basis for labeling requirements of the Federal Hazardous Substances Act. This Act, administered by the Consumer Product Safety Commission, requires precautionary labeling on every hazardous chemical intended for household use or likely to be taken for household use. The terms defined and used for labeling by many chemical manufacturers include TOXIC, HIGHLY TOXIC, FLAMMABLE, EXTREMELY FLAMMABLE, CORROSIVE, and IRRITANT. The term POISON, required for some chemicals by other federal laws, is defined as being the same as HIGHLY TOXIC.

The Department of Transportation (DOT) requires shipping containers of hazardous chemicals to be labeled with the "major" category of hazard and sets definitions for POISON, FLAMMABLE LIQUID, FLAMMABLE GAS, CORROSIVE, IRRITANT, EXPLOSIVES, and RADIO-ACTIVE MATERIALS.

The Occupational Safety and Health Administration has established requirements for labeling a limited number of carcinogens and CANCER-SUSPECT AGENTS, and it is proposing to regulate many more.

The Food and Drug Administration (FDA) requires certain labeling of *in vitro* diagnostic products, but its concern is not focused on protection of the user of the chemical. However, the FDA has accepted use by one distributor of a labeling system that communicates five possible degrees of hazard in each of three categories of hazard. The hazard identification system is based on one developed by the National Fire Protection Association.

The National Fire Protection Association standard defines five degrees of hazard in each of three categories: emergency health hazard, fire hazard, and instability or reactivity hazard. Its system does not deal with ingestion or long-term exposure hazards, nor does it include any precautionary wording. It is thus used primarily to signal degrees of hazard in case of spill or acute exposure. The American National Standards Institute (ANSI) has adopted, as a standard, the manual on labeling and precautionary information developed by the Manufacturing Chemists Association. This ANSI standard does not call for the labeling of chemicals that may cause sensitization or cancer or that may explode incidentally.

## B. Toxic Effects

The definition of the term TOXIC is based on the acute dose that will kill approximately one-half of a specified animal test population, or the $LD_{50}$ (50% lethal dose). The dose is given in milligrams (mg) of chemical per kilogram (kg) of body weight of the test animal, and it is assumed that the effects on humans will be proportional at similar doses. TOXIC is used to refer to a substance that:

(a) When ingested has an $LD_{50}$ in rats in 14 days of more than 50 mg/kg but not more than 500 mg/kg.
(b) When administered by continuous contact with the bare skin of rabbits for 24 hr has an $LD_{50}$ of more than 200 mg/kg but not more than 2000 mg/kg.
(c) When inhaled, has an $LC_{50}$ (lethal concentration) of more than 200

parts per million (ppm) but not more than 20,000 ppm of gas or vapor or more than 2 mg but not more than 200 mg/liter by volume of mist or dust.

HIGHLY TOXIC substances are defined as those that have an $LD_{50}$ in test animals of 50 mg/kg or less by ingestion or 200 mg/kg or less by contact, and an $LC_{50}$ of 200 ppm or less or 2 mg/liter or less by inhalation.

Some examples of chemicals that fit in the category of TOXIC by ingestion are aliphatic and aromatic hydrocarbons, e.g., aspirin, aniline, and acrylonitrile. Some chemicals that are highly toxic by ingestion are arsenic, acrolein, and antimony and azide salts. Phenol and aniline are examples of chemicals that are highly toxic or poison by skin contact.

One of the unexpected toxic effects of some laboratory chemicals is development of an allergic sensitization in some individuals. Examples of such chemicals are formaldehyde, paraphenylenediamine, 1-thioglycerol, diisocyantes, and some of the epoxy resin components that may be used in plastic embedding for electron or light microscopy. Recommendations have been developed for a "no-touch" technique to avoid contact that may result in a disabling sensitization (Leppert, 1974).

## C. Corrosive and Irritant Effects

The term CORROSIVE is used for chemicals that destroy living tissue. Examples of corrosive chemicals are sulfuric acid, nitric acid, and sodium hydroxide.

The term IRRITANT is used for chemicals that cause a local inflammatory reaction or primary irritation but not destruction of tissue or irreversible changes at the site of contact. Many solvents and other laboratory chemicals are irritants.

## D. Behavioral Effects

Narcosis is the most obvious behavioral effect that can result from excessive exposure to solvent vapors. Narcotic effects on the central nervous system may range from subtle loss of coordination to loss of motor control to unconsciousness. Other behavioral effects from exposure to chemicals are likely to occur before frank illness occurs. Any adverse effect should justify limiting exposures to the agent.

## E. Reactive and Explosive Hazards

Some chemicals are shock sensitive or deteriorate in prolonged storage to become shock sensitive. In addition, some can form explosive compounds or react violently with other chemicals. One of the common chemicals that is shock sensitive is benzoyl peroxide. It is sold in paper bags because a threaded bottle cap can generate enough friction to detonate it.

Picric acid or trinitrophenol is a high explosive that becomes shock sensitive, as well as heat sensitive, if it dries out so that it contains less than 10% moisture. Ethers and other peroxidizable compounds also can deteriorate in prolonged storage and form shock-sensitive peroxides. A documented death resulted from an effort to open an old ground glass-stoppered bottle of isopropyl ether in a laboratory stockroom (Manufacturing Chemists Association). All ethers and other compounds that can become explosive in prolonged storage should be dated when they are opened, to allow an inventory control to prevent serious hazards and dangerous disposal problems (Jackson et al., 1974). In general, peroxidizable compounds either should be tested within three months of the date of opening (twelve months for some compounds) to assure that peroxide concentrations are acceptable or removed, or the compounds should be disposed of to prevent development of excessive concentrations that can be hazardous. As chemical prices and shortages increase and waste disposal problems and costs increase, regular testing and removal of peroxides will become important. There are a number of methods for detection of peroxides, such as test papers, as well as several effective techniques of removing peroxides before they become hazardous (Jackson et al., 1974).

Sodium azide used as a preservative or biocidal agent in solutions in water baths, counters, or other apparatus can react with copper or lead to form compounds that are violently explosive from thermal or mechanical shock. Injuries have occurred when such compounds formed in copper or lead drain lines were struck during pipe cleaning. Scraping copper azide deposits off of a thermostat used in a water bath caused an explosion violent enough to blow off one of the fingers of the instrument repairman.

Hot digestion with concentrated perchloric acid causes vaporization and subsequent condensation that is likely to result in formation of explosive perchlorates in a ventilation system, usually in the duct beyond the digestion hood. Although these explosive perchlorates may not specifically endanger the laboratory worker, explosions can occur if there is impact or maintenance work on the ducts or ventilation system. Case histories of perchlorate explosions and recommendations for handling perchloric acid and perchlorates have been published (Everett and Graf, 1971; Wolsey, 1974). If perchlorates may have been deposited in ductwork, it is advisable to determine whether perchlorates are present before dismantling ductwork or doing other work that may generate enough impact to detonate explosive residues. An elegant test for perchlorates consists of collecting some wash water flushed down the ductwork in question and adding to half a test tube of the wash water a few drops of a 0.1% solution of methylene blue in water. If a violet precipitate forms, perchlorates are present. To correct this condition, flush the ductwork with steam (to reach all surfaces) for about 24 hr, test for perchlorates, and continue flushing until the test is negative.

## F. Control Measures

Control measures for chemical hazards include evaluation of the hazards of chemicals in use or to be ordered; determination of methods, equipment, or other precautions necessary for receiving and handling them; clean up of spills; and disposal of wastes. Information on necessary precautions should be contained on labels and in Material Safety Data Sheets and technical information sheets required as a condition of purchasing any chemical. In some cases the information on hazardous properties will suggest substitution of other less hazardous but equally effective chemicals.

Although intentional mixing of incompatible chemicals is not likely to occur in biomedical laboratories, there are two basic precautions that can be taken. One is to determine whether a hazardous chemical reaction has been reported before undertaking experimental mixing of chemicals (National Fire Protection Association, No. 491M-1975b). A second precaution is to prevent unintentional mixing of known incompatible chemicals by segregating or protecting glass bottles of such chemicals. For example, organic acids and solvents should be segregated or protected from storage with such acids as sulfuric, nitric, or perchloric acid and with other strong oxidizers. Large bottles of these substances should be protected with bottle carriers or by special plastic-coated bottles that have considerable impact resistance.

Chemical storage areas should be examined periodically to determine that all labels are legible and securely attached, that containers have not leaked or deteriorated, that reactive and incompatible chemicals are properly protected, and that chemicals have not dangerously deteriorated or peroxidized.

In areas where irritating or corrosive chemicals are present, the laboratory should provide an adequate, piped emergency water supply, which may include eyewash fountains, emergency hoses, and safety showers. Eyewash fountains and emergency hoses should be flushed out weekly or daily to assure that personnel are familiar with their location and operation and that they are fully operational. Safety showers should be tested annually by the maintenance staff, including an individual who is familiar with the location of the shutoff valve, in the event the shower valve fails to close. Adequate testing of showers will require the use of a shower "curtain" to direct the flow into a drum on wheels or some other safe way of minimizing splash and facilitating disposal of large amounts of water.

Although there are no national standards yet, it is recommended that the emergency water source be supplied from a cold water line with a connection to a hot water line through a normally closed valve that can be opened gradually to bring shower water temperatures up to about 50°F (10°C) for most of the necessary 15 or 20 min of flushing needed. Cold water reduces damage from thermal and corrosive burns. Floor drains are recommended for emergency showers installed in new buildings. For adequate flushing of the body in cases where massive splashes are possible, a shower plus a hose is recommended. An emergency spray hose alone seems adequate and more likely to be used than a shower where splashes are not likely to be massive or to involve most of the body. Flushing of chemicals from the eyes can be done best in most cases with eyewash fountains or other emergency water sources that are equipped with valves that remain open without being held.

One should limit exposure to hazardous materials by use of appropriate ventilated enclosures (such as glove boxes or hoods); separate facilities for food, drink, and smoking; handwashing facilities; and techniques to minimize contact, ingestion, injection, or inhalation of such materials. Glove boxes will be needed to provide adequate control for handling many carcinogens and some pathogens. Other biological safety cabinets used for pathogens or sensitive culturing work will require special design, certification, and techniques, in addition to recognition that the reduction of air drawn through the open face of the cabinet reduces the protection for the worker. This will be discussed in detail in Chapter 6.

Satisfactory performance of laboratory hoods depends on design, location, air supply quality, air supply location and velocity, and proper use of techniques that take maximum advantage of hood capabilities. Some of the imporant techniques for obtaining the maximum benefit from a hood include working so that release or dispersal of the contaminant is at least 6 in. inside of the hood (behind the sash), pulling the sash or panels to keep the hood face opening as small as practicable (to boost capture velocity at the face of the hood), and keeping the hood as free of storage and obstructions as possible, to minimize interference with the work and the removal of exhaust air.

Personnel exposure should be monitored by periodic sampling of air in employee breathing zones, appropriate physical examinations, bioassay of body fluids if appropriate (see Chapter 8), and documentation of exposure effects relative to inhalation, ingestion, and contact exposures. Documenting exposures that cause adverse effects will provide more information for improving exposure limits.

Laboratory workers should prepare for emergency control of chemical spills, particularly if chemicals are stored or used in large, unprotected glass bottles. Personnel protection is of primary importance and cleanup of spills is secondary. Respiratory protection is essential for personal protection in the event of spills or leaks that release gases, vapors, or aerosols that are toxic, anesthetic, radioactive, or pathogenic.

## VIII. ELECTRICAL HAZARDS AND CONTROL MEASURES

Electrical hazards in the laboratory include the possibilities of shock, fire, and ignition of flammable vapors and gases. Good design and maintenance usually minimize such hazards, but regular inspection, testing, and other precautions are also necessary.

The precautions against electrical shock are (1) grounding of all metal cases and exposed metal parts of electrical equipment; (2) guarding to prevent contact with any terminals or metal parts that cannot be grounded, such as resistance heaters and electrophoresis equipment; (3) providing electrical interlocks to shut off power when contacts are exposed; (4) making periodic inspection of cords to assure integrity of insulation and contacts; (5) providing ground fault circuit interrupters where such devices are required or useful; and (6) testing regularly to assure that grounding circuits are satisfactory and that leakage currents are below safe minimums for the conditions under which the equipment is being used. Leakage currents from grounded equipment must be considerably less if such equipment is to be connected to human subjects. Laboratories affiliated with hospitals may be able to obtain testing services and information on current standards from hospital-employed biomedical electronic engineers or technicians. Other laboratories using human subjects can contract for such services. As an added precaution in case of electrical shock injuries, laboratories should have a number of personnel trained in cardiopulmonary resuscitation.

Effects of contact with electrical circuits can range from slight sensation to painful shock to life-threatening conditions, such as ventricular fibrillation, respiratory inhibition, cardiac arrest, damage to the central nervous system, or deep burns. The resistance of the human body decreases and the severity of injuries increases with increases in voltage, current flow, and contact time. Since even commercial alternating current at voltages of 110 V can cause serious or fatal injuries, it is important not only to follow all the preceding precautions but to be sure that all "tingles" or slight shocks from electrical equipment are reported by laboratory personnel and prompt corrective action taken. Table I lists the quantitative effects of 60-Hz current on men and women. The range of currents and the differences between men and women should be noted (Dalziel, 1971).

Dalziel (1971) points out that minute electrical shocks "constitute an

**TABLE I**

Quantitative Effects of Electric Current, 60 Hz[a]

|  | Values (in mA) | |
| --- | --- | --- |
| Effect | Women | Men |
| Minimum perception threshold—slight sensation on hand | 0.3 | 0.4 |
| Median perception threshold | 0.7 | 1.1 |
| Shock—not painful, muscular control not lost | 1.2 | 1.8 |
| Painful shock—muscular control lost by ½% of subjects | 6 | 9 |
| Painful shock—muscular control lost by 50% of subjects ("let-go threshold)[b] | 10.5 | 16 |
| Painful and severe shock—muscular control lost by 99½%, breathing difficult | 15 | 23 |

[a] From Dalziel, 1971.

[b] "Let-go" currents represent values at which test subjects could not control muscles to let go of electrical contacts.

ominous warning of the presence of potentially hazardous conditions" and that "currents in excess of one's let-go current may produce collapse, unconsciousness and death." A "let-go" current is defined as that current level at which test subjects could not control muscles to let go of contacts.

Precautions against electrical overheating and possible ignition of electrical insulation or other combustible material include safe design of equipment and circuits and regular inspection and maintenance. If equipment performs a heating function, there should be fail-safe protection against overheating, as well as adequate clearance from combustible materials. Maintenance service should be provided routinely and in case of equipment malfunction. Plug connections should be inspected at regular intervals (such as annually) to assure adequate contacts and to prevent localized overheating. Insulation on cords should also be inspected to see that it has not deteriorated or been damaged.

If flammable solvents such as ethyl ether or isopentane are stored in refrigerators or freezers that are not specifically designed to prevent ignition of flammable vapors, violent explosions should be expected. Opened containers of ether, isopentane, and similar solvents with high vapor pressure and extremely low flash-point temperatures CANNOT be stored safely in any enclosure that has within it any electrical switches, motors, thermostats, lights, or heaters. (Electrical equipment safe for use with flammable concentrations of ether would have to be rated and labeled as explosion-proof for Class I and Group C atmospheres.)

Extremely flammable solvents with flash-point temperatures below refrigeration temperatures can safely be stored in limited quantities in (1) laboratories on open shelves with normal ventilation; (2) flammable liquid storage cabinets; (3) flammable liquid storage rooms; and (4) special refrigerators/freezers, which are identified on the manufacturer's label as "laboratory safe" or "flammable material storage refrigerator" (having all electrical contacts removed from the storage compartment), or which are explosion-proof for Class I, Group C and D atmospheres and bear the label of Underwriters Laboratories.

Explosion-proof electrical equipment is required only when an entire area is classified as a hazardous location because it is exposed to high concentrations of flammable vapors or gases. Except in the case of usage of large quantities of extremely flammable solvents, laboratories will not be classified as hazardous locations and will not require all electrical equipment to be classified as explosion-proof for the atmospheres present (National Fire Protection Association, No. 45-1975a). Another way of describing the National Fire Code requirements for laboratories is that a completely explosion-proof refrigerator/freezer would not be required in a laboratory unless the flammable vapor/gas hazards were so great that all electrical equipment in the laboratory was required to be explosion-proof. (If an explosion-proof refrigerator or freezer is used in a laboratory as a means of assuring that the interior of the refrigeration compartment is explosion-proof, then the expense of wiring the refrigerator/freezer to the power supply with explosion-proof wiring is not justified if the refrigerator/freezer is the only explosion-proof device in the laboratory.)

## IX. FIRE HAZARDS AND CONTROL MEASURES

Laboratory fire hazards range from simple and relatively slow combustion of paper (charts, notes, books, files) to the fast-burning and rapidly spreading combustion of organic solvents and occasionally to the explosive combustion of flammable gases or solids. The consequences of an uncontrolled fire in a biomedical laboratory may range from little or no damage to very costly damages. For example, a 30-min fire in a solvent-filled tissue processor in a laboratory without an automatic fire extinguishing system caused $1,000,000 in damage in 1976 and required evacuation of patients from the nine-story hospital in which the laboratory was located.

Even if no injuries were incurred in a fire and the damage repair costs were completely covered by insurance, a major fire would probably destroy priceless research records and interrupt research for long periods of time while laboratory facilities are rebuilt or other facilities are adapted for use.

There are two consensus standards that are intended to provide reasonable requirements for fire protection in laboratories (National Fire Protection Association, 1973, 1975a). The NFPA Safety Standard for Laboratories in Health-Related Institutions, NFPA No. 56C, was adopted in 1973 and the NFPA Fire Protection Standard for Laboratories using Chemicals, NFPA No. 45, was adopted in 1975; both are being revised and new additions are expected in 1980 or 1981. Since these standards are commonly used by insurance companies for recommendations and are adopted as regulations by fire authorities and the Joint Commission on Accreditation of Hospitals (JCAH), laboratory management must be cognizant of the current requirements.

Fire hazard control measures must be tailored to the problems and needs of each laboratory in order to be effective and to provide protection without unduly restricting laboratory activities.

Several types of control measures can be used by laboratory management to provide fire protection appropriate to the hazards encountered. The basic control measures are as follows:

1. Limiting ignition sources.
2. Limiting oxidizer availability.
3. Limiting fuel availability.
4. Detecting fire and smoke.
5. Evacuating occupants promptly.
6. Extinguishing fires.

These control measures will be discussed in the following sections.

## A. Limiting Ignition Sources

Since laboratories are not generally considered to be hazardous locations requiring explosion-proof electrical equipment and other special means for total elimination of possible ignition sources, special attention must be given to selective control of ignition sources. General laboratory ventilation is usually adequate to prevent the accumulation of flammable concentrations of vapors and gases if operations are controlled so that only very small quantities are released. If there are pouring operations or activities that require open containers of flammable liquids, such as staining or cover-slipping, it is good practice to provide local exhaust ventilation (for fire and health protection) *and* to keep all ignition sources out of the immediate vicinity. Motors, switches, and other possible sources of ignition that are in fixed positions several feet away from vapor sources will probably not have flammable concentrations reach the ignition sources. Portable sources of ignition, such as lighted cigarettes and burners and lamps, must be kept out of areas with possible vapor concentrations.

Prohibition of smoking in laboratories is generally justified to keep ignition sources out of possible vapor concentrations, as well as to prevent accidental ignition of other combustible materials, including wastebaskets into which ashtrays are often emptied.

Prohibition of smoking in laboratories is also a good practice to prevent accidental ingestion of chemicals or microorganisms that may contaminate smoking materials set down on various surfaces in the laboratory or that may contaminate the fingers holding the smoking materials. Further reasons for limiting smoking to areas outside the laboratory are to prevent irritation of nonsmoking employees and to prevent contamination of laboratory samples or optical instruments.

Refrigerators, freezers, and environmental chambers should not be used for storage of ethyl ether, isopentane, petroleum ether, or other solvents that can produce flammable concentrations of vapor at refrigerated temperatures. If there is any likelihood that such materials may be stored in refrigerated enclosures, it will be extremely important that all possible ignition sources have been removed from such refrigerated enclosures, preferably by the manufacturer of the refrigeration device and so stated on the manufacturer's nameplate on the device. Dilute ethanolic solutions do not present any more hazard under refrigeration than do beer or wine. Before purchasing a special and very expensive refrigerator or freezer for storing ethyl ether or similar solvent, the laboratory should determine whether there is any real need to provide refrigerated storage or whether it would be feasible simply to store the solvent in a flammable liquids storage cabinet or on an open shelf.

If electrical equipment is needed in areas in which flammable concentrations of vapors or gases may accumulate, there are three possible means to prevent the equipment from providing ignition sources. Explosion-proof equipment is available for some operations, such as refrigerating, blending, mixing, and lighting. Some other equipment is available that has been designed and tested to be intrinsically safe, so that electrical failure cannot generate sufficient heat to ignite flammable gases or vapors. The third alternative is to enclose electrical switches and arcing, sparking and heating elements and to ventilate the enclosures to prevent flammable vapors and gases from reaching any possible source of ignition. The alternative of enclosing and ventilating electrical equipment provides great flexibility at relatively low cost if the electrical equipment must be used where there may be flammable concentrations. As an example, air from the laboratory supply can be connected to a shaker housing to prevent vapors entering from flammable mixtures being shaken.

Static electricity is generated by pouring solvents, and rapid pouring of large quantities can generate enough static electricity to ignite solvent vapors. Buildup of static electricity can be minimized by electrically bonding metal containers between which the solvent is being transferred, by minimizing the distance through which the solvent falls, and by providing humidity of about 50% in areas in which quantities of solvents are transferred. Conductive shoes and special clothing may be required if large quantities of solvents are transferred or if the solvents are as susceptible to development of static electricity as ethyl ether (Stecher, 1968).

## B. Limiting Oxidizer Availability

Where both fuel and ignition sources are present, oxidizers must be excluded to prevent fire or explosion. For example, in anaerobic incubation chambers using hydrogen it is critical to purge the chamber to flush out oxygen before introducing the hydrogen.

Strong oxidizers, such as nitric acid, perchloric acid, and sulfuric acid, should be protected by bottle carriers or by separate storage from acetic acid, solvents, and other combustible and flammable liquids and solids.

## C. Limiting Fuel Availability

Four measures are commonly recommended or required to limit the amount of fuel available in laboratories.

1. Exclude or protect large glass containers of flammable liquids.
2. Store as much as possible of needed flammable liquids in safety cans.
3. Protect as much as possible of needed flammable and combustible liquids by storage in flammable liquid storage cabinets.
4. Limit total quantities of flammable and combustible liquids.

A strong argument for avoiding use of gallon glass bottles for flammable solvents or for protecting them against breakage is that ignition of flammable vapors from a gallon spill can produce ceiling temperatures of nearly 900°F within 1 min after ignition. Occupational Safety and Health Association (OSHA) standards allow gallon size for glass bottles of solvent such as acetone, benzene, ethyl alcohol, methanol, methyl ethyl ketone, and toluene, only if glass containers are necessary to protect the purity of the solvent and if 2 oz or more are needed at one time. Gallon glass containers of solvents that are exempt from OSHA limits are not exempt from breakage and ignition.

Safety cans provide resistance to breakage from impact or fire and commonly contain a flash arrester in the pouring spout so that flame cannot flash back into the can explosively. Safety cans are available in polyethylene, stainless steel, tin plate, and Teflon lining, and in sizes from 1 pt to 5 gal.

Quantities of flammable and combustible liquids needed in the laboratory for convenient operation can be protected from immediate involvement in a fire by storage in insulating flammable liquids storage cabinets made of two layers of metal or one layer of 1-in. exterior grade plywood. Both types of cabinets are available to provide approximately 10 min of protection of the contents from adding fuel to a fire. Plywood cabinets made to the specifications listed in OSHA and NFPA standards are less expensive and provide greater insulation than double-walled metal cabinets. Plywood cabinets can be built conveniently to fit in spaces where working supplies are needed. Ventilation of storage cabinets is not required by NFPA, OSHA, or JCAH standards.

Limitation of quantities of flammable and combustible liquids within laboratory working areas is justified for reduction of fire load and fire hazard and for economical use of valuable laboratory space. Limitation is also required by OSHA and NFPA standards. The limits on quantities of flammable and combustible liquids in laboratories do *not* include quantities in flammable liquids storage cabinets. The NFPA Standard for Laboratories in Health-Related Institutions, used by the Joint Commission for Accreditation of Hospitals, sets a limit of 10 gal within a fire area.* The College of American Pathologists (CAP) has set a limit of 5 gal in a laboratory, not including quantities in approved storage cabinets.

The principle for limiting quantities and numbers of cylinders of flammable gases in laboratories is to use the smallest practicable size for the work and to limit the number to the cylinders in use and a single standby cylinder for each one in use. If gases can safely be piped in from the outside or from a fire-resistant storage room, the laboratory has less fire hazard and more working space, as well as less interference from cylinder deliveries.

## D. Detecting Fire and Smoke

Although fire and smoke that may originate from combustion or overheating are most quickly and sensitively detected by humans, unoccupied areas and unattended operations may need to be supervised by automatic fire and smoke detection systems, depending on the values at risk. Smoke detectors and detectors of ionization products of combustion are relatively reliable means of detecting combustion early enough to alert oc-

---

* Fire area is defined as an area completely separated from all other interior areas by one-hour fire resistant construction with all door and window openings in the separation protected by fire-rated assemblies.

cupants for evacuation and emergency response. Selection of a detection system should be based on evaluation of the types of hazards present and the needs for special speed or detection of specific types of hazards.

Although a detection system can alert occupants and summon assistance from a local fire department, most detection systems can do nothing to suppress the fire. The notable exception is the automatic sprinkler system that detects fire by loss of pressure and water flow in a piping system provided with spray openings held closed by fusible links. The heat from a fire melts the nearest fusible link above the fire signaling the presence of a fire by water flow, which actuates an alarm; the water spray cools the fire and surrounding combustibles within the spray pattern of the sprinkler head. The water spray has a density usually less than half a gallon per square foot per minute and only the sprinkler heads heated by the fire open up. Automatic sprinkler systems can be used even for protection of computers and other valuable electronic equipment, as wet equipment is more readily restored than equipment that has had cases distorted and wiring insulation destroyed.

Detection of smoke is very important in ventilation systems which could circulate smoke throughout large areas of a building. If it is necessary to keep doors open to stairwells or hazardous areas, smoke detection systems are required that will release doors automatically in case of smoke and incipient fire.

Laboratory equipment that is very valuable or that presents unattended fire hazards (such as tissue processors using flammable solvents) should be equipped with overtemperature protection or smoke detection to shut down the equipment and summon assistance.

Automatic sprinkler systems and other fire and smoke detection systems should alert occupants and summon the local fire department, either directly or through an alarm detection company.

## E. Evacuation

Rapid and effective evacuation in case of fire or similar emergency depends on having an alarm system that can be heard in all areas that may be affected. Also, regular evacuation drills should be held at least annually so that occupants have practiced how to evacuate and are aware of a clear and unobstructed means of egress. If the building is occupied during hours of darkness or has rooms and corridors without windows, emergency lighting is required.

An evacuation alarm system should be heard clearly in all occupied and working areas and the signal should be unmistakable and not confused with alarm signals on refrigerators or other equipment. One of the best ways to test the alarm for audibility is to use the system once a year for a practice evacuation. Initially, it is a good idea to be sure that all occupants know the purpose of evacuation drills and that all know they are expected to leave the building, with rare and justifiable exceptions. Announcing the day of the first drill in advance will help everyone plan their emergency shutdowns and procedures to evacuate and assemble outside the building. If it is necessary to have more frequent drills, as in hospitals and hospital-connected facilities where four drills are required for each shift every year, subsequent drills should be unannounced and evaluated carefully to see that emergency actions are appropriate.

Means of egress from laboratories should be evaluated and inspected regularly to determine whether they are adequate and kept unobstructed. The means of egress include the corridors to the exits, the exits (which are usually stairwells protected at each level by self-closing fire doors), and the passageways from the bottom of stairwells to the street or other public way. If corridors are so wide that they invite storage or cannot be kept clear, it is important to have a clear path along one wall of the corridor for evacuation and for access by the fire department or emergency team. Corridors in clinical laboratories are generally required by code to be 60 in. wide.

Doors to stairwells should generally have vision panels to help prevent personnel collisions, and the doors should have operating latches as well as knobs or handles appropriate to traffic needs. Such doors must not be kept open by wedges, wires, or other devices which would allow smoke and fire to spread up the stairwells.

Stairwells that serve as required exits should not discharge in the middle of a building but should discharge directly to the outside or into a protected passageway, as required by the NFPA Life Safety Code. As a minimum, the direction from the exit to safety should be clear, and the path should be unobstructed.

## F. Extinguishing Fires

Extinguishing laboratory fires may be possible with small portable fire extinguishers used by trained personnel, or it may require many firefighters equipped with self-contained breathing apparatus, protective clothing, and special extinguishing agents. Automatic extinguishing systems provide quick and automatic response, almost always reduce fire damage, and make it possible for more effective use of special portable extinguishers by occupants or the fire department.

Small carbon dioxide fire extinguishers are recommended for protection of electrical equipment, if flammable liquids are present only in very small quantities (less than a pint or 500 ml). Carbon dioxide extinguishers are usually not effective for extinguishing flammable-liquid-spill fires. Dry-chemical or other types of fire extinguishers are required that have ratings of 12B or greater. Dry-chemical fire extinguishers have much more extinguishing capacity per pound than carbon dioxide fire extinguishers and are available in light, portable models with ratings as high as 60B. Dry-chemical extinguishing agents, such as sodium bicarbonate and ammonium phosphate, are very effective in extinguishing solvent fires by interrupting the flame chain reaction, but the extremely small particles are difficult to clean out of fine electrical contacts. Halon extinguishing agents are clean but are expensive to use and hazardous to some people who may have a cardiac sensitivity to the halogenated agent. Special foam extinguishers are available for solvent spill fires where electrical equipment will not be encountered.

Since any solvent fire larger than about 4 ft$^2$ may be too large and hot to be approached without protective clothing, efforts should be made to limit the sizes of spills and possible fires, and employees should be trained to leave rather than fight such dangerous fires. If large spills and spill fires are possible and it is desirable to have control measures begin immediately and effectively, it will be necessary to equip and train a team of laboratory occupants for emergency response. Equipment should include protective clothing (preferably heat reflective) and positive pressure-type, self-contained breathing apparatus (SCBA). If the SCBA is interchangeable with the local fire department, training will be greatly facilitated.

We believe that efforts to control spills of a gallon or more of flammable or toxic materials require respiratory protection and special protective clothing for the safety of the person attempting to control the spill. If spill control materials are provided, amounts should be appropriate for the spills that are possible, and personnel should be well aware of the needs for personal protective equipment.

Although there are a number of special extinguishing agents that can be used in automatic extinguishing systems, the least expensive and generally the most reliable system uses water as the extinguishing agent. Performance of automatic sprinkler systems can be enhanced and water damage reduced by use of special sprinkler heads that turn off automatically when the fire goes out and that turn on again automatically if the fire reignites.

As the final note in this section on fire hazards we should mention that there is a need for prompt and accurate reporting of information on laboratory fires. This knowledge will make hazards known quickly and vividly; more important, it will provide an improved basis for scientifically based fire protection programs and standards. Reports, which do not have to identify the source, should be sent to some national source for information exchange such as the NFPA and the R & D section of the National Safety Council.

## X. SAFETY MANAGEMENT

Management of safety requires anticipation of problems and advance preparation to prevent or control situations that may endanger laboratory personnel, equipment, or facilities. Unfortunately, many safety programs focus only on an after-the-fact response to crises, such as spills, injuries, or fires.

An increasing number of laboratories are beginning to manage safety by developing systems for evaluating proposed research and new procedures to determine what safety and occupational health problems may be anticipated and what safety measures and special training are needed. Existing laboratory activities that are frequently or potentially hazardous may also be evaluated on the same basis to see what preventive measures are needed.

Basic safety practices and rules or codes of practice should be derived from the hazards that may be present. Employees who are expected to follow basic safe practices or specific rules should have the opportunity to participate in the development or approval of the guidelines they will be expected to follow. Safety training and other special job training related to safety and occupational health should be designed carefully. Specific safety training should be given as soon as possible after an employee is hired or transferred to a new job or procedure.

Based on the evaluation of job hazards, inspection procedures can be developed and appropriate emergency procedures can be practiced. Safety programs based on anticipation and control of problems in advance will be more acceptable and effective than programs based on exhortations to "Be Careful."

## REFERENCES

Daziel, C. F. (1971). Deleterious effects of electric shock. *In* "Handbook of Laboratory Safety" (N. V. Steere, ed.), p. 521. CRC Press, Boca Raton, Florida.

Everett, K., and Graf, F. A., Jr. (1971). Handling perchloric acid and perchlorates. *In* Handbook of Laboratory Safety" (N. V. Steere, ed.), p. 265. CRC Press, Boca Raton, Florida.

Gordon, J. E. (1949). The epidemiology of accidents. *Am. J. Public Health* **39**, 504.

Haddon, W., Jr., Suchman, E. A., and Klein, D. (1964). "Accident Research—Methods and Approaches." Harper, New York.

Jackson, H. L., McCormack, W. B., Rondestvedt, C. S., Smeltz, K. C., and Viele, I. E. (1974). Control of peroxidizable compounds. *In* "Safety in the Chemical Laboratory" (N. V. Steere, ed.), Vol. 3, p. 114. Division of Chemical Education of the ACS, Easton, Pennsylvania.

Leppert, C. (1974). Health hazards and precautions for epoxy resin systems in electron microscopy and other laboratory uses. *In* "Safety in the Chemical Laboratory" (N. V. Steer, ed.), Vol. 3. Division of Chemical Education of the ACS, Easton, Pennsylvania.

Manufacturing Chemists Association. "Accident Case History No. 603."

National Fire Protection Association (1973). "Fire Protection Standard for Laboratories in Health-Related Institutions, NFPA 56C-1973. NFPA, Boston, Massachusetts.

National Fire Protection Association (1975a). "Fire Protection Standard for Laboratories Using Chemicals," NFPA 45-1975. NFPA, Boston, Massachusetts.

National Fire Protection Association (1975b). "Hazardous Chemical Reactions," NFPA 491M-1975. NFPA, Boston, Massachusetts.

Stecher, P. (1968). "Merck Index," 8th ed. Merck & Co., Inc., Rahway, New Jersey.

Steere, N. V. (1974). Identifying multiple causes of laboratory accidents and injuries. *In* "Safety in the Chemical Laboratory" (N. V. Steere, ed.), Vol. 3, p. 3. Division of Chemical Education of the ACS, Easton, Pennsylvania.

Wolsey, W. C. (1974). Perchlorate salts, their uses and alternatives. *In* "Safety in the Chemical Laboratory" (N. V. Steere, ed.), Vol. 3, p. 125. Division of Chemical Education of the ACS, Easton, Pennsylvania.

# Chapter Two
# Classes of Toxic Compounds: Procedures and Principles for Evaluating Toxicity

## Mark Hite

## I. INTRODUCTION

Toxicology may be defined as the study of the harmful effects of chemicals on biologic mechanisms. The discipline borrows from a knowledge of chemistry, immunology, genetics, physiology, and pharmacology. A knowledge of statistics and public health is fundamental to the study of toxicology. Pathology is an important part of toxicology, but a harmful effect from a chemical on a cell, tissue, or organ must necessarily manifest itself in the form of gross, microscopic, or submicroscopic deviations from the normal. Thus modern toxicology is a multidisciplinary field, and as such it borrows freely from several disciplines in the basic sciences (Casa-

**29**

LABORATORY SAFETY: THEORY AND PRACTICE
Copyright © 1980 by Academic Press, Inc.
ISBN 0-12-269980-7

rett, 1975a,b; Dow Chemical Company, 1960; Loomis, 1968; National Academy of Sciences, 1977; Wallace, 1978).

Toxicity is the ability of a chemical molecule or compound to produce injury once it reaches a susceptible site in or on the body. The extent of the injury produced by a compound will depend on several factors, such as its physical state, concentration, and duration of exposure. Another basic concept that must be considered is that of hazard. This is the probability that a substance will cause harm. In evaluating hazard, toxicity is but one factor. Others include the chemical and physical properties of a chemical, warning properties such as odor and pain, and the intended use (Doull, 1975; Durham, 1975; Eckardt and Scala, 1978).

## II. CLASSIFICATION OF TOXICITY

### A. Chemical

There are two basic ways in which toxic materials may be classified—chemically and physiologically. With regard to the chemical factors related to an unknown agent that will influence its toxicity one should consider the chemical composition; pH; form of ion; physical characteristics (particle size, method of formulation, etc.); presence of impurities or contaminants; stability and storage characteristics; solubility in biologic fluids; choice of vehicle; and presence of excipients, such as adjuvants, emulsifiers, surfactants, binding agents, coating agents, coloring agents, flavoring agents, preservatives, antioxidants, and other intentional and nonintentional additives.

A brief explanation of these points shows that pH may affect toxicity because of the possible innate corrosive properties of the chemical. Large particles perhaps cannot be inhaled, whereas very small particles (less than .03 $\mu$m) may be inhaled and then promptly exhaled with the next breath. Sometimes the compound may degrade in storage to form either more toxic or less toxic chemical. It is essential to know what happens to a compound in storage. Carcinogenic oils are kept in brown bottles and stored in dark cabinets because ultraviolet light may destroy polynuclear aromatics.

### B. Physiological

In consideration of the physiologic approach, the toxicologist must consider the dose or exposure of the compound. That is, how much material is to be introduced or how much would the person absorb after an exposure (Cohn and Linnecar, 1978; Doull, 1975; Eckardt and Scala, 1978). Consideration of absorption would include excretion; metabolic transformation, which could increase or decrease the toxicity; and storage or deposition of the material in some compartment of the body, such as in the liver or fat.

Specific examples of chemicals affecting a tissue or organ system are listed below (Loomis, 1968).

1. *Central nervous system depressants:* ethanol, methanol, and general anesthetics.
2. *Nerve poisons:* botulism toxin and curare.
3. *Muscle poisons:* those affecting smooth muscle (e.g., barium salts) and those affecting cardiac muscle (e.g., chloroform, cocaine, nicotine, and potassium salts).
4. *Protoplasmic poisons:* those giving rise to local effects on the skin and in the respiratory tract. Such materials as irritants and corrosives would fall in this category. Compounds that cause degenerative lesions of the nervous system include carbon monoxide, cyanide, and magnesium.
5. *Hepatotoxins:* bismuth, mercury, uranium, carbon tetrachloride, heavy metals, mushroom poisons, and chlorinated hydrocarbons.
6. *Nephrotoxins:* arsenic, bismuth, uranium, turpentines, chromium, oxalates, glycols, phenols, and creosols.
7. *Gastrointestinal tract:* various irritants and corrosive materials as well as pharmaceuticals, such as aspirin.
8. *Hematopoietic poisons:* arsenic, benzene, fluoride, iodide, and various nitro compounds.

## 1. Route of Administration

The route of administration is of prime import (McNamara, 1974; Stokinger, 1967). Of the various means of exposure to the body by toxic agents, skin contact is first in the number of afflictions occupationally related. Intake by inhalation ranks second, whereas oral intake is generally of minor importance, except as it becomes a part of the intake by inhalation or when an exceptionally toxic agent is involved. For some materials, as might be inferred, there are multiple routes of entry.

**a. Dermal Contact.** Upon contact of an industrial agent with the skin, four actions are possible:

1. The skin and its associated film of sweat and lipid may act as an effective barrier that the agent cannot disturb, injure, or penetrate.
2. The agent may react with the skin surfaces and cause primary irritation.

3. The chemical may penetrate the skin, conjugate with tissue protein, and effect skin sensitization.
4. The chemical may penetrate the skin through the folliculosebaceous route, enter the bloodstream, and act as a systemic poison.

The skin, however, is normally an effective barrier for protection of underlying body tissues and relatively few substances are absorbed through this barrier in dangerous amounts (Durham, 1975). Fortunately, many compounds are just not absorbed to a significant degree by the skin. For example, botulin toxin is extremely toxic to humans when taken orally, but it is relatively harmless when applied to the skin. The principal reason for this difference is that the toxin molecule is very large compared to most chemicals and is not readily absorbed by the skin. On the other hand, when it enters the digestive tract, the toxin is immediately absorbed and transported to the susceptible sites of action. Thus the chemical and physical properties of the toxicant determine which route of exposure will produce what toxic effects. Serious and even fatal poisonings can occur from short exposures of skin to strong concentrations of extremely toxic substances, such as parathion and related organic phosphates, tetraethyl lead, analine, and hydrocyanic acid. Moreover, the skin as a means of contact may also be important when an extremely toxic agent penetrates body surfaces from flying objects or through skin lacerations or open wounds.

There are some compounds that are more toxic when spilled on the skin than when swallowed (Durham, 1975). Usually these are compounds that are rapidly detoxified by the liver. The detoxification site is important since materials absorbed by the gastrointestinal tract are carried directly to the liver and only subsequently are distributed to the rest of the body. Materials absorbed from the skin or the lungs are distributed by the blood to the rest of the body at the same instant that a portion passes to the liver.

Skin absorption attains its greatest importance in connection with the organic solvents (Cohn and Linnecar, 1978). It is generally recognized that significant quantities of these compounds might enter the body through the skin either as a result of direct accidental contamination or indirectly when the material has been spilled on the clothing. An additional source of exposure is found in the fairly common practice of using industrial solvents for removing grease and dirt from the hands and arms, in other words, for washing purposes. This procedure, incidentally, is a frequent source of dermatitis. This condition could traumatize the skin so that greater amounts of the toxic agent could be absorbed through the skin and cause a systemic reaction.

**b. Ocular Exposure.** Associated with skin contact is *eye exposure* (Dow Chemical Company, 1960). The eye is a very vascular organ and is innervated directly to the brain. Thus compounds might be absorbed in sufficient quantities by this organ so that a systemic effect could be observed. Of course, the mucous membranes of the eye are highly sensitive and would react to small amounts of corrosive or irritant materials to produce local effects.

**c. Inhalation.** The respiratory tract is by far the most important means by which injurious substances enter the body (Durham, 1975; Stokinger, 1967). A great majority of occupational poisonings that affect the internal structures of the body result from breathing airborne substances. These substances, lodging in the lungs or other parts of the respiratory tract, may affect the system, or pass from the lungs to other organ systems by way of the blood, lymph, phagocytic cells, or broncho-tracheal fluid. The type and severity of the action of toxic substances depend on the nature of the substance, the amounts absorbed, the rate of absorption, individual susceptibility, and many other factors. Harmful substances may be suspended in the air in the form of dusts, fumes, mists, or vapors and may be mixed with respiratory air in the case of true gases. Since an individual under conditions of moderate exertion will breathe about 10 m$^3$ of air in the course of an ordinary 8-hr working day, it is readily understood that any poisonous material present in the respired air offers a serious threat to health. The lung surface area of a human adult is enormous (90 m$^2$ total surface, 70 m$^2$ alveolar surface); together with the capillary network surface (140 m$^2$) with its continuous blood flow, it presents to toxic substances an extraordinary leaching system that makes for an extremely rapid rate of absorption of many substances from the lungs. Despite this action there are several occupationally important substances that resist solubilization by the blood or phagocytic removal by combining firmly with the components of lung tissue. Such substances include beryllium, thorium, silica, and toluene-2,4-diisocyanate (TDI). When such resistance occurs, irritation, inflammation, fibrosis, malignant change, and allergic sensitization might result.

In consideration of inhaled material, one must consider the physical state of the chemical agent, that is , whether the material is made of particulates, such as dusts, fumes, mists, and fogs, or is in the form of a gas or vapor.

### (1) *Physical State of Particles in Air*

**(a.) Dust.** A dust is composed of solid particulates generated by grinding, impact, detonation, crushing, or other forms of energy resulting in attrition of organic or inorganic materials, such as metals, rock, coal, wood, and grain. Dusts do not tend to flocculate except under electrostatic forces. If their particle diameter is greater than a few tenths micron, they do not diffuse in the air, but rather settle under the influence of gravity. On the other side of the spectrum, particles that are generally less than 0.3 $\mu$m may be inhaled but then exhaled with the succeeding breath, since the material remains in the air stream. Examples of dusts are silica dust, sawdust, and coal dust.

**(b) Fumes.** Fumes are composed of solid particles generated by condensation from the gaseous state, as from volatilization from molten metals, and are often accompanied by oxidation. A fume tends to aggregate and collect into chains or clumps. The diameter of individual particles is less than 1 $\mu$m, but the aggregate would be considerably larger. Examples of fumes are lead vapor on cooling in the atmosphere and uranium hexafluoride ($UF_6$), which sublimes as a vapor, hydrolyzes, and oxidizes to produce a fume of uranium oxyfluoride ($UO_2F_2$).

**(c) Mists.** A mist is composed of suspended liquid droplets generated by condensation from the gaseous to the liquid state as by atomization, foaming, or splashing. Examples would be oil mists, sprayed paint, and chromium trioxide mist.

**(d) Fogs.** Fog is composed of liquid particles of condensates whose particle size is larger than mist, usually greater than 10 mm. An example of a fog is a supersaturation of water vapor in air. Because of the large particle size, a fog may not present a toxicologic hazard by the inhalation route, but it may produce a hazard to the skin. Of course, if the material has sufficient vapor pressure, then it could in fact be inhaled and present a hazard by the inhalation route.

### (2) *Particle Shape*

The size and surface area of particulates play an important role in occupational lung disease, especially the pneumoconioses. The particle diameter associated with most injury is believed to be less than 1 $\mu$m; larger particles either do not remain suspended in the air sufficiently long to be inhaled or, if inhaled, cannot negotiate the tortuous passages of the upper respiratory tract. Smaller particles tend to be more injurious than larger particles for other reasons. On inhalation, a larger percentage (possibly as much as 10-fold) of the exposure concentration is deposited in the lungs from small particles than from large ones. In addition, smaller particles appear to be less readily removed from the lungs. The additional dosage and residence time act to increase the injurious effect of a chemical particle.

### (3) *Particle Density*

The density of the particle also influences the amount of deposition and retention of particulate matter in the lungs on inhalation. Those with high density behave as larger particles of small density on passage down the respiratory tract by virtue of the fact that their greater mass and consequent inertia tend to impact them on the walls of the upper respiratory tract. Thus a particle of uranium oxide with a density of 11, and 1 $\mu$m in diameter, will behave in the respiratory tract as a particle of several microns in diameter, and thus its pulmonary deposition will be less than that of a low-density particle of the same measured size.

### (4) *Other Factors*

Other factors affecting the toxicity of inhaled particulates are the rate and depth of breathing and the amount of physical activity occurring during breathing. Slow deep breathing tends to result in larger amounts of particulates deposited in the lungs. High physical activity acts in the same direction not only because of greater number and depth of respirations, but also because of increased circulation rate. The ambient temperature also might modify the toxic response of inhaled materials since high temperatures in general tend to worsen an effect, as do temperatures below normal, but the magnitude of the effect is less for the latter.

### (5) *Gases and Vapors*

The gaseous state of a toxic agent is that formless fluid that can be changed to the liquid or solid state by the combined effect of increased pressure and decreased temperature. Examples are carbon monoxide and hydrogen sulfide. An aerosol is a dispersion of a particulate in a gaseous medium, whereas smoke is a gaseous product of combustion, rendered visible by the presence of particulate carbonaceous matter.

A vapor is a gaseous form of a substance that is normally (at standard temperature and pressure) in the liquid or solid state and that can be transformed to these states either by increasing the presence or decreasing the temperature. Examples of vapors include carbon disulfide, naphthalene, and iodine.

The absorption and retention of inhaled gases and vapors by the body are governed by certain factors different from those that apply to particulates. Solubility of the gas in the aqueous environment of the respiratory tract governs the depth to which a gas will penetrate in the respiratory tract. Thus very little, if any, inhaled, highly soluble ammonia or sulfur dioxide will reach the pulmonary alveoli, depending on concentration, whereas relatively little insoluble ozone or carbon dioxide will be absorbed in the upper respiratory tract.

After inhalation of gas or vapor, the amount that is absorbed into the bloodstream depends not only on the nature of the substance but more particularly on the concentration of the inhaled air and the rate of elimination from the body. For a given gas, the limiting concentration in the blood is attained that is never exceeded no matter how long it is inhaled, providing the concentration of the inhaled gas in the air remains constant.

**d. Ingestion.** Poisoning by ingestion of toxic materials in the workplace is far less common than by inhalation or by contact with the skin or eyes. The reason for this is that the frequency and degree of contact with toxic agents on the hands, food, and cigarettes are far less than by the other routes. Also, many industries are restricting the use of food and smoking to specific areas in the plant and encouraging the use of protective clothing such as gloves. Because of this, ingestion is of concern only with the most highly toxic agents.

The oral route passively contributes to the intake of toxic substances by inhalation, since that portion of the inhaled material that lodges in the upper respiratory tract is removed by ciliary action and is subsequently swallowed, thereby contributing to the body intake.

The absorption of toxic substances from the gastrointestinal tract into the bloodstream commonly is far from complete, despite the fact that substances in passing through the stomach are subjected to relatively high acidity and on passing through the intestine are subjected to alkaline media. However, one must consider that the original toxic agent might be altered chemically by the action of either one of these chemical environments.

Other factors that favor low absorption of toxic materials from the gastrointestinal route include dilution of the material with food and water. Also, this mixing may cause a chemical reaction combining the toxic agent with the material contained in the gastrointestinal tract. Second, there is a certain selectivity in absorption through the intestine that tends to prevent absorption of "unnatural" substances or to limit the amount absorbed. Third, following absorption into the bloodstream, the toxic material goes directly to the liver, which metabolically alters, degrades, and detoxifies most substances.

One outstanding example of ingesting highly toxic material by the oral route is the case of the radium dial painters who followed the practice of "pointing" their brushes between their lips, thus ingesting lethal quantities of the radioactive material. This notorious example should serve as a model against such laboratory practices as mouth pipetting and testing materials in the laboratory.

A combined tabulation of toxicity by different routes is presented in Table I.

**TABLE I**

Combined Tabulation of Toxicity Classes: Various Routes of Administration[a]

| Toxicity rating | Oral LD$_{50}$ single-dose rats (mg/kg) | Inhalation 4-hr vapor exposure mortality of 2/6–4/6 rats (ppm) | Skin LD$_{50}$ rabbits (mg/kg) | Oral probable lethal dose for man |
|---|---|---|---|---|
| Extremely | 1 or less | <10 | 5 or less | A taste, a grain |
| Highly | 1–50 | 10–100 | 5–43 | 1 teaspoon, 4 cm$^3$ |
| Moderately | 50–500 | 100–1000 | 44–340 | 1 oz, 30 g |
| Slightly | >0.5–5 g | 1000–10,000 | 0.35–2.8 g/kg | 1 pint, 250 g |
| Practically nontoxic | 5–15 g | 10,000–100,000 | 2.8–23 g/kg | 1 quart |
| Relatively harmless | 15 g and more | 100,000 | 22.6 or more g/kg | >1 quart |

[a] Taken from Hine and Jacobson (1954).

## 2. Storage and Excretion

Toxic substances might be retained or stored in the body for indefinite periods of time, being excreted but slowly over periods of months or years (Klaassen, 1975). Lead, for example, is stored primarily in the bones and mercury principally in the kidneys. Smaller amounts may be stored in other organs or tissues. Inhaled particulate matter can be phagocytized and remain in regional lymph nodes, where it may have little effect, as in the case of coal dust, or may produce pathological changes in the lungs, as in the case of silica and beryllium.

The excretion of toxic materials takes place through the same channels as does absorption, namely, lungs, intestines, and skin, but the kidneys (by means of the urine) are the main excretory organs for many substances. Saliva, sweat, and other body fluids may participate to a lesser extent in the excretory processes. Volatile vapors and gases usually are excreted by means of the lungs and breath.

## 3. Biotransformation, Metabolism, and Detoxication

Many organic compounds are not excreted unchanged, but are altered through a process known as biotransformation (Norton, 1975; Williams, 1959). This is also known as metabolism or detoxication. The resulting new compounds, or metabolites, are usually removed from the body through the urine. However, some metabolites may be more toxic than the parent material.

The detoxication of chemicals that are absorbed by the body may be defined as the metabolic alteration of the substance that the body cannot oxidize readily to carbon dioxide in water, detectable in the urine, and involving conjugation of the foreign substance with some substance derived from the body. Detoxified products are end products of metabolism and may or may not be less toxic than the original substance. Foreign substances undergoing detoxication are usually organic compounds.

The major types of detoxication that occur in humans and laboratory animals are briefly described below and are shown in Table II (Williams, 1959).

**a. Methylation.** This reaction occurs in most laboratory species except for rabbits. This particular form of biotransformation applies to most inorganic compounds, such as arsenic, selenium, and tellurium. It also applies to ring nitrogen compounds and certain complex aromatic phenols. An example of this form of detoxication would be the formation of $(CH_3)_2Se$, which is expired in the breath of persons exposed to selenium and is the causative agent for producing the characteristic garlic odor in workers exposed to this metal. In the case of pyridine the methylated product is considerbly more toxic than pyridine itself, so that the body actually makes conditions worse.

**b. Acetylation.** This form of detoxication occurs in most species, except for dogs. It involves reactions with aromatic amines and amino acids. Some known exceptions to this rule are the aromatic amine carcinogens and aliphatic amines.

**c. Ethereal Sulfate.** This form of conjugation occurs in all species so far studied and involves reactions with phenolic compounds. This is a true detoxication step, since phenol is reacted with the hydrogen sulfate ion, which is difficult to split, and thereby is eliminated in the urine.

**d. Acetyl Mercapturic Acid.** This form of detoxication occurs in most species except for pigs and involves detoxication of aromatic hydrocarbons, halogenated aromatic hydrocarbons, polycyclic hydrocarbons, sulfonated esters, and nitro paraffins. This step basically involves oxidation by hydrogenation of the foreign material to a chemical that is ultimately excreted in the urine.

**e. Thiocyanate Formation.** This particular form of detoxication is found with inorganic cyanide compounds and organic cyanide compounds (nitriles). It includes reactions involving a specific enzyme, rhodanase,

**TABLE II**

**Major Types of Detoxication**

| Type | Foreign substance | Detoxication product examples |
|---|---|---|
| Methylation<br>-CH$_3$ | Inorganic compounds of As, Te<br>Ring N compounds<br>Certain complex aromatic phenols | $(CH_3)_2Se$<br><br>NCH$_3$ (on ring, with OH)<br><br>OCH$_3$ (on ring)<br><br>CHOHCH$_2$NHCH$_3$ |
| Acetylation<br>CH$_3$CO- | Aromatic amines<br>Amino acids<br>(known exceptions: aromatic amine carcinogens, also aliphatic amines) | NHCOCH$_3$ (on ring)<br><br>RCHCOOH<br>\|<br>NHCOCH$_3$<br><br>e.g., benzidine, hydroxylated aliphatic amines, aldehydes |
| Ethereal sulfate<br>-OSO$_3$H | Phenols<br>(cyclohexanol glucuronide) | OSO$_3$H (on ring) |
| Acetyl mercapturic acid<br>-SCH$_2$CHCOOH<br>-NHCOCH$_3$ | Aromatic hydrocarbons<br>Halogenated aromatic HCs<br>Polycyclic HCs<br>Sulfonated esters<br>C$_2$H$_5$SO$_3$-CH$_3$<br>Nitroparaffins (C$_4$H$_9$NO$_2$) | SCH$_2$CHCOOH<br>\|<br>NHCOCH$_3$ (on ring)<br><br>Br<br>C$_2$H$_5$-acetylcysteyl-<br>C$_4$H$_9$-acetylcysteyl- |
| Thiocyanate | Cyanide, inorganic<br>Organic cyanides (nitriles) | RCNS |

*(cont.)*

**TABLE II** *(cont.)*

| Type | Foreign substance | Detoxication product examples |
|---|---|---|
| Glycine -NHCH$_2$COOH | Aromatic acids<br>Aromatic-aliphatic acids<br>Furane carboxylic acids<br>Thiophene carboxylic acids<br>Polycyclic carboxylic acids<br>(bile acids) | CONHCH$_2$COOH<br> |
| Glucuronoside | Aliphatic (1°, 2°, 3°)<br>and aromatic hydroxyl<br>Aromatic carboxyl | OC$_6$H$_9$O$_6$<br><br>(ether)<br><br>O=C-OC$_6$H$_9$O$_6$<br><br>(ether) |
| Glucose hydrazone | Hydrazine<br>Hydrazine derivatives? | NH$_2$N=CHC$_5$H$_8$O$_5$ |

which is involved in the reaction of converting cyanides to thiocyanate, which is usually 50 to 80 times less toxic than cyanide.

**f. Glycine Conjugation.** This form of detoxication occurs in most mammals, but not in birds. It was first found in the urine of horses in the form of hippuric acid. The major foreign substances that undergo this particular type of detoxication include aromatic acids, aromatic–aliphatic acids, furane carboxylic acids, thiophene carboxylic acids, and polycyclic carboxylic acids. After appropriate conjugation with glycine, all these kinds of agents would be excreted like hippuric acid.

**g. Glucuronicide Formation.** This type of conjugation involves reactions with primary, secondary, and tertiary aliphatic and aromatic hydroxyl compounds as well as aromatic carboxyl compounds. Basically this reaction reduces the toxicity at the physiologic pH so that the detoxified material can be eliminated in the urine. It generally requires only a slight alteration of the molecule to allow it to be eliminated.

## 4. Frequency and Duration of Exposure

Nearly all substances show a dose-related effect in the test species (Eckardt and Scala, 1978). The dose response curve is one of the more important considerations given to the effect of a test compound. That is, the response (irritation, decreased activity, death, etc.) is usually more prominent or has an earlier onset with an increase in dose. Of course, another form of the dose response curve is that of the time–effect curve in which the duration of exposure or the number of doses over a period of time is one of the major variables.

## 5. Termination of Toxicity

A toxic effect may terminate through several mechanisms (Eckardt and Scala, 1978; Klaassen, 1975). The chemical may be excreted either unchanged (e.g., many solvents, which are excreted in the expired air) or combined (changed) (e.g., phenol when it is excreted in the urine). As previously mentioned, a compound may undergo metabolic transformation in the body that would serve either to increase or to decrease its inherent toxicity. Vinyl chloride is not carcinogenic per se, but after it is metabolized by the body, the chloroethylene oxide metabolite is carcinogenic. Finally, materials may be removed from the circulation and stored in relatively unreactive regions of the body, such as in the fat or in the bones. As long as the agents remain in these deposition sites, they produce no harm, but if mobilized they may react with a target organ to produce toxicity. This usually occurs when the body burden is reached; that is, the storage site is saturated with the chemical. As it is mobilized, it adds to the additional material that is absorbed under certain exposure conditions so that toxicity can be produced.

## 6. Inherent Factors Related to the Subject (Human or Laboratory Animal)

In this section consideration will be given to such factors as the species and strain; the genetic status; immunologic status; nutritional status; hormonal status; age, sex, body weight, and maturity; factors affecting the central nervous system of the laboratory animals or workers; and presence of disease (Eckardt and Scala, 1978).

**a. Species and Strain.** One should appreciate that animal data must be projected to the human counterpart before it is of much use. Basically, the toxicologist will assume that if a given material elicits a specific effect in test animals, it is probable that it will do likewise in man. The seriousness of this effect will guide the choice of safety factors. In this regard, the experience of the toxicologist is invaluable. Certain species of animals may be totally refractive (resistant) to the effect of a toxic chemical, whereas another species may be unduly sensitive. This is true not only for local effects but also for systemic effects. The manner in which an animal or human being metabolizes or detoxifies a chemical so that it is ultimately removed or excreted from the body will in large part determine its toxicity.

**b. Genetic Factors.** Certain factors inherent in the organism may affect the toxicity of a foreign substance in the body (Eckardt and Scala, 1978). These factors include the accumulation of the chemical, prolongation of the chemical's action, and the individual's hypersensitivity to the toxic effects of the compound. These factors have been lumped into a term called *individual susceptibility.* This has long been used to express the well-known fact that under conditions of similar exposure to potentially harmful substances there is usually a marked variability in the manner in which individuals respond to a given exposure. Some may show no evidence of intoxication whereas others may show signs of mild poisoning, and still others may become severely, even fatally, affected (hypersensitivity). Comparatively little is known about the factors responsible for this variability. For example, it is believed that differences in the anatomical structure of noses and respiratory tracts in different species may be concerned with different degrees of efficiency in filtering out harmful material in the inspired air. Likewise, the presence of disease or specific organ pathology greatly influences the toxicity of a chemical. For example, previous infections in the lungs, particularly tuberculosis, are known to enhance susceptibility to silicosis. Obesity is another important predisposing factor among persons who are subject to occupational exposure to organic solvents and related compounds.

**c. Age and Sex.** Age and sex are also known to play an important part in the determination of the toxicity of a foreign chemical (Doull, 1975; Eckardt and Scala, 1978). In newborns and in very young animals, including humans, the enzyme systems responsible for detoxication may not have been fully developed. If these enzymes are responsible for lowering the toxicity, then exposure to a chemical may actually have a higher toxicity rating in the young than in the adults. On the other hand, if the enzyme systems would in fact metabolize the chemical to something that is more toxic, then the very young would be protected. Likewise, the hormonal differences between males and females may influence the toxicity of a chemical. This is especially true in pregnant females where the hormonal levels are quite different from those in the nonpregnant female.

**d. Environmental Factors.** Various environmental conditions are known to affect the toxicity of a chemical both in laboratory animals and in industrial workers. In humans, different rates of working speed, resulting in variation in respiratory rate, in depth of respiration, and in pulse rate, may play a part in the overall toxicity of a chemical. The action of the cilia also may have some import. The permeability of the lungs may influence absorption and the efficiency of the kidneys may govern the rate at which toxic materials are excreted. This is especially true if there is a concurrent disease affecting either the lungs or kidneys, or both, so that toxic metabolites or conjugated products cannot be removed from the body. Likewise, the liver plays a major role in the detoxication and excretion of harmful substances; subnormal functioning of this organ may lead to increased susceptibility to the toxicity of a chemical.

**e. Nutrition.** The nutritional state of the exposed worker or laboratory animal may have a significant role in the toxicity of a chemical. For example, an inadequate intake of nutrients could lead to a physiologic dysfunction. This in turn could make an individual vulnerable to the effects of chemical exposure at a concentration lower than that of the general population.

Similarly, the overindulgence in alcoholic beverages may increase the possibility of occupational poisonings. This is particularly true with organic solvents, since alcohol is a known central nervous system depressant and may add to the depression induced by exposure to an organic solvent. Furthermore, consideration should be given to a worker exposed to chemicals while concurrently taking medication that could affect a particular organ system or systems to cause a synergistic or antagonistic effect.

## C. Kinds of Toxicity

Another way of classifying toxicity is that by which the exposure is labeled "acute" or "chronic" (Durham, 1975; Gardiner, 1977). An acute exposure is defined as one that either is short in duration (a few hours or less) or comes about from a single dose; usually, in order to be termed "acute," all of the material should be administered within a 24-hr period. Chronic exposure is reported over longer periods of time (e.g., days, months, or years), with the implication that a single exposure would not result in any particular harm, notwithstanding the dose. For example, acute poisoning may result in unconsciousness, shock, or collapse; severe inflammation of the lungs; or even sudden death. An exposure to a moderate or strong concentration of hydrogen sulfide would be acute; repeated exposures to traces of halogenated hydrocarbons such as chloroform or methylene chloride would be chronic. Some chemicals, such as mercury compounds, have a biphasic toxicity, since acute toxicity to mercury affects the kidneys, whereas chronic exposure to these compounds would adversely affect the central nervous system. This latter effect also is known as a "cumulative action" and is usually found with many heavy metals. This means that over a period of time the material that is absorbed is only partially excreted and that increasing amounts accumulate in the body. This is to say that the toxicant is coming into the body faster than it is being excreted. Eventually the quantity in the body becomes great enough to cause physiologic disturbances (toxicity).

## D. Site of Action of Poisons

Several examples already have been given to show that different chemicals act on different parts of the body. Therefore one must define whether a chemical has a "local" effect or a "systemic" effect, or both (Dow Chemical Company, 1960; Durham, 1975). Local toxicity refers to the action of a toxic material on the skin or mucous membranes. Systemic toxicity is concerned with the action of a toxic substance after absorption into the body through one or more of the several routes previously discussed. A systemic effect can be a generalized one, such as anemia.

The fumes and mists arising from strong acids have a direct irritating effect on the eyes, nose, throat, and lower air passages. If the chemical reaches the lungs, a severe inflammatory reaction may occur called chemical pneumonitis.

Systemic or indirect effects occur when a toxic substance has been absorbed into the bloodstream and distributed throughout the body. For example, arsenic, when absorbed in toxic amounts, may cause disturbances in several parts of the body: blood, nervous system, liver, kidneys, and skin. Benzene, on the other hand, may affect only one organ system, namely the hematopoietic system (blood-forming bone marrow). Carbon monoxide causes asphyxia by preventing the hemoglobin of the blood from performing its normal function of transporting oxygen from the lungs to the tissues of the body. Although oxygen starvation occurs equally in all parts of the body, the brain tissue is the most sensitive so that the early manifestations are those caused by damage to this organ.

## III. CARCINOGENICITY

Three specialized branches of toxicology are concerned with those compounds that have carcinogenic, mutagenic, or teratogenic activity. Each of these represent three different effects, although frequently a chemical will have activity in more than one area (Arcos, 1978; Goldberg, 1972; Heidelberger, 1977; Ministry of Health and Welfare, 1973; Miller, 1977; New York Academy of Sciences, 1977). A description of carcinogenicity follows; the subjects of mutagenicity and teratogenicity are discussed elsewhere in this book.

Cancer represents a group of diseases characterized by uncontrolled growth of abnormal and nonfunctional cells. A carcinogen may be defined as an agent that causes malignant disease (according to the National Institute of Occupational Safety and Health) or an agent that statistically increases the risk of cancer, whether by initiating or promoting it (according to the American Cancer Society).

The carcinogenic potential of a chemical is usually discovered in humans through epidemiologic studies. In such investigations the incidence of tumors significantly increases in an exposed group of workers compared to a comparable nonexposed group. Of course, animal studies also have shed light on a number of compounds thought to have this kind of activity under the proper circumstances. Table III shows some of the chemicals that epidemiologic data as well as data from experimental animals have implicated as known or highly suspect carcinogenic agents to man. Naturally, as the number of new chemical entities increases, so will be the number with presumed carcinogenic potential (Arcos, 1978).

The sequence of events leading to the development of cancer following exposure to carcinogens is still poorly defined and poorly understood. However, much is known of the interactions of carcinogens with tissues that subsequently develop tumors. Most known chemical carcinogens require metabolic activation to manifest their activity. In some instances this involves activation of the carcinogenic agents to carbonium ions or electrophilic agents. Further, the identity of several of the nucleophilic targets of the electrophiles has been established (Ministry of Health and Welfare, 1973).

Primarily because of public concern, it is obvious that cancer research involving chemicals will increase dramatically during the next few years. This will require organic and physical chemists, who will be studying the properties of known carcinogens; biologists and microbiologists, who will be seeking rapid screening tests for determining possible chemical carcinogens; and physiologists, pathologists, and biochemists, who will be comparing

**TABLE III**

**Chemicals Known to Cause or Suspected of Causing Cancer in Man**

| Major organ affected | Compound(s) |
| --- | --- |
| Hematopoietic system | Melphalan |
| | Benzene |
| Kidney | Phenacetin |
| Liver | Aflatoxins |
| | Progestational contraceptives |
| | Vinyl chloride |
| | Cycasin |
| | Safrole |
| | Pyrrolizidine alkaloids |
| Lungs | Bis(2-chloroethyl)sulfide |
| | Chloromethyl methyl ether |
| | Asbestos |
| | Chromium |
| | Nickel |
| | Soot, tars, and oils |
| Prostate | Cadmium |
| Skin | Arsenic |
| Urinary bladder | 2-Naphthylamine |
| | 4-Aminobiphenyl |
| | Benzidine |
| | Cyclophosphamide |
| | $N,N$-Bis(2-chloroethyl)-2-naphthylamine |
| Vagina | Diethylstilbestrol |

normal cellular processes with chemically altered metabolic pathways. In essence, each discipline of life science will contribute greatly in future studies of neoplastic transformation. This, needless to say, will necessitate the training of personnel with different backgrounds in the safe handling of chemical carcinogens.

## A. Types of Chemical Carcinogens

The diversity of chemical structures that cause cancer in humans and laboratory animals has made it difficult for laboratory workers to predict the carcinogenicity of newly synthesized compounds. Efforts to classify chemicals into various categories have met with limited success. However, in many scientific groups chemical carcinogens have been categorized as polycyclic aromatic hydrocarbons, aromatic amines and aminoazo dyes, nitroso compounds, alkylating agents, naturally occurring organic compounds, and inorganic compounds (Ames *et al.*, 1975; Arcos, 1978).

## 1. Polycyclic Aromatic Hydrocarbons

This class of compounds was the first group of chemicals to be related to cancer in man and constitutes perhaps the most widespread chemical carcinogens in our environment. Polycyclic aromatic hydrocarbons are produced in large quantities in many combustion processes involving fossil fuels and tobacco (cigarette smoke). Polycyclic aromatic hydrocarbon concentrations have been shown to be particularly high in the air pollution of large cities and industrial areas and have been correlated with the increased incidence of lung cancer and urban environments. More recently, compounds of this class have been detected in soil and in both fresh water and sea water as a result of untreated effluence from industrial plants and exhaust fumes from ships. The extent of polycyclic aromatic hydrocarbon production can be put into perspective if one considers that approximately 2000 tons of benz[a]pyrene are put into the air over the United States each year.

Some of the more commonly used aromatic hydrocarbons in the research laboratory include benz[a]pyrene; 7,12-dimethylbenz[a]anthracene; and 3-methylcholanthrene (Fig. 1). Benz[a]pyrene has been used extensively to induce lung cancer in laboratory animals, and both 7,12-dimethylbenz[a]-anthracene and 3-methylcholanthrene are noted for their ability to induce cancer of the mammary gland in experimental animals. These compounds are also used as reference compounds in the design of short-term tests for carcinogenicity. As shown in Fig. 1, some polycyclic aromatic hydrocarbons are not carcinogenic (e.g., anthracene, phenanthrene, and pyrene). The number of aromatic rings, the position of each aromatic ring in relation to other rings, and the presence of various side groups are important in determining the potential for carcinogenic activity.

## 2. Aromatic Amines and Aminoazo Compounds

The attachment of an amine ($NH_2$) group to a polycyclic aromatic hydrocarbon will often enhance its carcinogenic potential. If, for example, an amine is attached to the noncarcinogenic anthracene molecule, the resulting compound is 2-anthramine (Fig. 2), which is a well-known carcinogen. 2-Acetylaminofluorene (2-AAF) was initially synthesized for use as a potential insecticide. Preliminary testing, however, revealed that it was a potent chemical for inducing cancer of the urinary bladder; fortunately, it was never used commercially. 2-Naphthylamine was used extensively in dye factories at the turn of the century. It was only after the death of many workers from cancer of the urinary bladder that a cause and effect relationship was established. The aromatic amine 3-methyl-4-aminobiphenyl and many of its derivatives are known to be potent inducers of colon cancer.

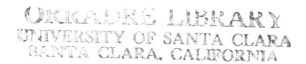

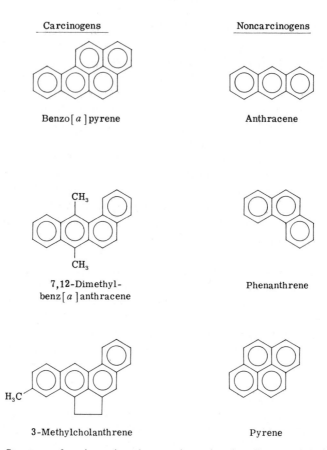

**Fig. 1.**   Structures of carcinogenic and noncarcinogenic polycyclic aromatic hydrocarbons.

The aminoazo compounds contain the group —N = N— and were used frequently as dyestuffs for fabrics and foods. Examples of carcinogenic azo compounds are 4-dimethylaminoazobenzene and 2′,3-dimethyl-4-amino-azobenzene (Fig. 2). The former compound is also known as "butter yellow" and was used to color margarine and cooking oils until it was demonstrated to cause liver cancer in rats.

## 3. *N*-Nitroso Compounds

Compounds in this group include nitrosomines, nitrosomides, and nitro-soureas (Fig. 3). These compounds were found to cause hepatic tumors in rats after a feeding study with dimethylnitrosamine.

Data have accumulated during the last few years that show that nitro-somines are formed from various environmental materials and may well be

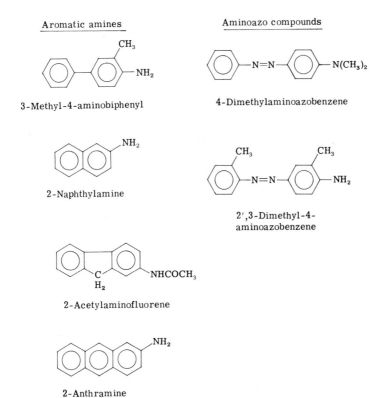

**Aromatic amines**

3-Methyl-4-aminobiphenyl

2-Naphthylamine

2-Acetylaminofluorene

2-Anthramine

**Aminoazo compounds**

4-Dimethylaminoazobenzene

2′,3-Dimethyl-4-
aminoazobenzene

**Fig. 2.**   Chemical structures of carcinogenic aromatic amines and aminoazo compounds.

one of the most important groups of potential carcinogens in humans. Nitrosamines are readily formed by the reaction of amines with nitrous acid; for example, cooking nitrite-cured bacon will convert the nitrite to nitrous acid, which then reacts with amines in the meat, forming these potentially carcinogenic compounds. Substances that have been suggested as potential sources for *N*-nitroso compounds include secondary and tertiary amines, amino acids, and many types of agricultural chemicals, particularly fertilizers and pesticides.

In the research laboratory this group of compounds was studied extensively because of organ specificity. Depending on the chemical structure, route of administration, and species of animal, the compounds can induce cancers in almost any organ of experimental animals. For example, nitrosomethylurea administered by intravenous injection induces cancer of the mammary gland, whereas direct application into the urinary bladder or the respiratory tract induces squamous cell carcinomas.

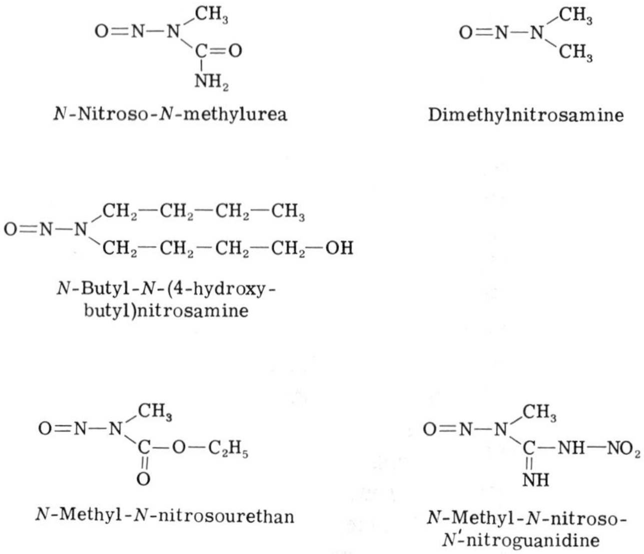

$$O=N-N\diagdown\begin{matrix}CH_3\\ C=O\\ |\\ NH_2\end{matrix}$$

N-Nitroso-N-methylurea

$$O=N-N\diagup\diagdown\begin{matrix}CH_3\\ CH_3\end{matrix}$$

Dimethylnitrosamine

$$O=N-N\diagup\diagdown\begin{matrix}CH_2-CH_2-CH_2-CH_3\\ CH_2-CH_2-CH_2-CH_2-OH\end{matrix}$$

N-Butyl-N-(4-hydroxy-
butyl)nitrosamine

$$O=N-N\diagdown\begin{matrix}CH_3\\ C-O-C_2H_5\\ \|\\ O\end{matrix}$$

N-Methyl-N-nitrosourethan

$$O=N-N\diagdown\begin{matrix}CH_3\\ C-NH-NO_2\\ \|\\ NH\end{matrix}$$

N-Methyl-N-nitroso-
N'-nitroguanidine

**Fig. 3.**    Chemical structures of N-nitroso compounds.

## 4. Alkylating Agents

One of the larger classes of "potential mutagens" present in man's environment is represented by the alkylating agents. These compounds include the nitrogen and sulfur mustards and are among the first chemicals recognized as mutagens in certain organisms.

Alkylating agents can be divided into two classes: (a) those that carry a single reactive alkyl group, that is, the monofunctional alkylating agents, and (b) the bifunctional or polyfunctional agents, which carry two or more reactive alkyl groups. The primary difference between them is related to the latter's ability to form covalent cross links between the two individual strands of a DNA double helix. Since interstrand cross-links prevent the DNA duplex from undergoing complete strand separation necessary for proper replication, bifunctional and polyfunctional agents produce a higher cytotoxicity than their single arm counterparts. It is not surprising, therefore, that in many test systems, polyfunctional and bifunctional agents have been shown to induce a greater number of lethal events per mutational event than their monofunctional analogs.

In addition to cross-linking DNA, bifunctional and polyfunctional alkylating agents may also mimic the action of monofunctional agents by reacting with only one of the two strands of a DNA duplex. The ratio of cross links to single-stranded reactions presumably depends on the nature of the alkylating agent as well as on the DNA involved.

Alkylating agents can be subdivided on the basis of their functional groups, such as aziridines; nitrogen, sulfur, and oxygen mustards; nitrosamines; nitrosamides; epoxides; lactones; aldehydes; dialkyl sulfates; alkane sulfonic esters; and related derivatives.

**a. Ethylenimine.** Specific examples of alkylating agents would include compounds such as ethylenimine (aziridine), which is an extremely reactive compound undergoing two major types of reactions: (1) ring-opening reactions similar to those undergone by ethylene oxide, and (2) ring-preserving reactions in which ethylenimine acts as a secondary amine. The ease with which the ring opening occurs arises from the strained nature of the three-membered ring. Under proper conditions ethylenimine reacts with many organic functional groups containing an active hydrogen to yield an aminoethyl derivative or products derived therefrom. In ring-preserving reactions, ethylenimine can undergo replacement of the hydrogen atom on the nitrogen in various ways or it can form salts and metallic complexes. 2-Triethylenemelamine (TEM) is used in the manufacture of resinous products, as a cross-linking agent in textile technology, in the finishing of rayon fabrics, and in waterproofing of cellophane. In addition, it has been used medically as an antineoplastic agent and as a chemosterilant for houseflies, screwworms, and certain fruit flies.

**b. Triethylene Phosphoramide (TEPA).** TEPA has been the most extensively employed of all the aziridine phosphene oxides. Its industrial applications include the flameproofing of textiles, waterproofing, wash-and-wear fabrics, and crease-resistant fabrics.

**c. Cyclophosphamide.** This alkylating agent has been used as an antitumor agent for several years. Thus it is frequently found as a laboratory tool not only as a cytostatic agent but also as a neoplastic inducing agent.

Other alkylating agents that are commonly used in laboratories include ethylene oxide, propylene oxide, ethylene chlorohydrin, epichlorohydrin, formaldehyde, acetaldehyde, acrolein, chloral hydrate, ethylene sulfide, methyl methanesulfonate, and ethyl methanesulfonate. There is extensive literature covering the biological and chemical properties of all of these, plus many more, alkylating agents. It would be prudent for the laboratory worker to learn as much as possible about those compounds with which he or she will be working prior to initiating any experimental studies.

## 5. Naturally Occurring Carcinogens

The naturally occurring carcinogens constitute one of the most rapidly growing and most exciting areas of the study of chemicals carcinogenesis. The discovery that a number of products of perfectly natural origin contain carcinogenic agents, some of them quite potent, forces a chilling reassessment of the magnitude of the task ahead. In Fig. 4, several examples of well-identified naturally occurring carcinogens are shown. The activities of these range from very weak to highly potent. Safrole is present in sassafrass, capsaicine in chili peppers, parascorbic acid in mountain ash berry, catechin in certain vegetable extracts used in the tanning of leather, cycasin in cycad nuts consumed by natives on the island of Guam, monocrotaline in senecio shrubbery used by the Bantu population in South Africa for preparing medicinal tea, and cotinine in unburned tobacco. Aflatoxin $B_1$, a

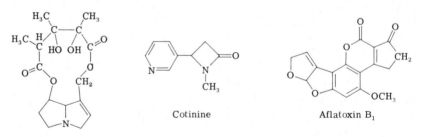

**Fig. 4.** Some chemical carcinogens produced by certain plants and molds.

product of metabolism of the common yellow mold *Aspergillus flavus,* is probably the most potent carcinogen known. One subcutaneous injection of 10 mcg in a mouse is sufficient to induce sarcomas.

## 6. Inorganic Carcinogens—Metals and Minerals

Several inorganic carcinogens are known. With the exception of asbestos and arsenic, the inorganic carcinogens are certain metals or simple salts thereof. These include beryllium salts, which can produce lung cancer and bone tumors. Cadmium, either as its sulfate salt or as powdered metal, produces close to 100% incidence of subcutaneous tumors in laboratory animals. The heavy metals, such as nickel, cobalt, chromium, and lead, either as the pure metal powders or as salts and ores, have substantial potency for causing cancers. Several studies have indicated a high incidence of lung cancer among chromate workers and nickel factory workers. Lung tumors have been induced in laboratory animals by exposure to nickel in the form of metal dust or as nickel tetracarbonile; similar results were obtained by injecting rats and mice with nickel sulfide or nickel oxide. Chronic feeding of tolerable levels of lead (in the form of various salts) induces tumors of the kidney.

The magnitude of carcinogenic hazard due to asbestos is best indicated by the fact that the asbestos industry is probably the one with the highest industrial cancer incidence known. Since asbestos is insoluble, the hazard is due to the microscopic fibers, which become airborne or otherwise detached from a sheet or block. The carcinogenic activity of asbestos is related to the geometry of the mineral fibers; that is, they appear under high magnification in the shape of long, very thin needles. Because of their shape, asbestos fibers readily penetrate and migrate through tissue linings and cell membranes and accumulate in certain parts of cells.

## B. Metabolism Activation and Detoxication

A considerable number of studies have shown that many carcinogenic organic compounds must be biotransformed (metabolized) before they can exert a carcinogenic effect (Ames *et al.,* 1975; Arcos, 1978; Heidelberger, 1977; Weisburger and Weisburger, 1967). If a compound is carcinogenic per se, then it is termed an *ultimate* or *primary carcinogen.* However, if it must undergo biotransformation to chemically reactive metabolites, then it is termed a *proximate carcinogen.* On the other hand, the carcinogen might undergo detoxifying metabolism, which will render it harmless. Biochemically the pathways of activation and detoxication are generally in a competitive situation and the outcome of this competition will ultimately determine whether or not tumors will arise in a particular individual or laboratory animal. Most enzymes that activate and detoxify carcinogenic organic compounds belong in the class of mixed function oxidases, which are enzymes that metabolize various exogenous chemicals and drugs.

## C. Promoting Agents or Factors

Small amounts of a primary chemical carcinogen such as benz[a]pyrene can be applied to the skin of mice and yet rarely lead to cancer. When the treated area is subsequently exposed repeatedly over a period of time to another suitable agent, papillomas and carcinomas appear even though this agent itself, without hydrocarbon pretreatment, does not induce cancer. These materials are termed *promoting agents.*

Initial treatment (initiation) with the carcinogen leads to reduction of transformed latent or dormant tumor cells, which cannot be reorganized morphologically or by any other known technique. They are, however, subsequently stimulated to grow into perceptible tumors by the promoting agent (promotion or co-carcinogen). These two steps represent a much simplified outline of a series of highly complex host interactions with the agents.

The initiating and the promoting agent need not always be applied to the same site. For example, the systemic carcinogens, urethan or 2-fluoroenamine, have been administered orally to mice. Skin tumors were induced by subsequent painting with croton oil. Neither agent alone induced skin cancer.

The effect of some chemical carcinogens is to increase the occurrence of spontaneous types of tumors. Examples are of the pulmonary tumors of strain A mice and the liver and mammary gland tumors in C3H mice. The answer as to whether carcinogenesis can generally be ascribed to a promoting agent hinges on the discovery of the underlying mechanism of the action of carcinogens and their promoting agents.

## D. Handling Suspect and Confirmed Carcinogens in the Laboratory

As new information becomes available regarding the potential carcinogenic properties of a chemical, it is necessary to assess the validity of the evidence and decide if it warrants taking special precautions to protect laboratory workers even though no regulatory mandate has been issued. Naturally, if the compound is listed under the Occupational Safety and Health Administration (OSHA) regulations for handling carcinogens, then the legal steps for handling such materials will have to be followed. Thus it would behoove the laboratory director to review all available material both from a scientific and a regulatory point of view before entering into any studies in which known or suspect carcinogens would be used.

## E. Laboratory Guidelines

The Department of Health, Education, and Welfare (1978) recently issued guidelines to establish work practices and engineering controls for the use of chemical carcinogens in laboratories; these guidelines were for their own use, but certain points bear repeating and would be applicable to almost any laboratory situation.

Three levels of work practices and engineering controls have been established. These levels are identified as Level A, Level B, and Level C. Level A work practices and engineering controls apply to the use of chemical carcinogens for which OSHA has established permissible exposure limits of 1 ppm (or equivalent levels in mg/m$^3$) or greater. Level B work practices and engineering controls apply to the use of chemical carcinogens for which OSHA has established permissible exposure limits less than 1 ppm (or equivalent levels in mg/m$^3$). Level C work practices and engineering controls apply to the use of chemical carcinogens for which OSHA has prohibited occupational exposures. For chemical carcinogens not covered by OSHA regulations but that have been identified by the DHEW committee to coordinate toxicology and related programs, the control level will be established by the committee on a case-by-case basis for each chemical so identified. Carcinogen safety monographs will be developed by the DHEW committee. The monograph will provide specific technical and safety information that may be helpful to the chemical carcinogen user in implementing the guidelines.

Basically the guidelines cover considerations of working in a laboratory fume hood having a sash plane capture velocity of at least 100 ft/min average with no point less than 85 ft/min. Appropriate warning signs should be attached to the hood as well as on the door through which people will enter the laboratory. The warning sign should state "Danger: Contains Suspect Carcinogen." Work should be done on an impervious tray with sufficient volume to contain leaks or spills. Disposable gloves should be used for all work in the hood. These should be removed in the hood and sealed in a plastic bag. Contaminated lab coats should be similarly bagged, then incinerated. Use of disposable coats should be considered. Bags should be sealed and tagged to read "Contents contaminated with cancer suspect agent; do not open prior to incineration." Whenever possible, operations having carcinogenic off gases or where carcinogens would otherwise be released should use scrubbers for the air before it is discharged through the hood. If this is not possible, the hood exhaust should be scrubbed. If a hazard still exists, the roof should be posted and controlled to restrict access while the hazard is present. Material should be decontaminated or reduced to a noncarcinogenic state prior to discharge to a sewer or waste collection system. Equipment in the hood should be decontaminated before it is removed.

## ACKNOWLEDGMENTS

The author is most grateful to Dr. Clinton Grubbs (Illinois Institute of Technology, Chicago) for supplying a considerable amount of material on the subject of carcinogenicity. In addition, the author wishes to thank Mrs. Thelma Demetrius for her assistance in the preparation of this manuscript.

## REFERENCES

Ames, B. N., McCann, J., and Yamasaki, E. (1975). Methods for detecting carcinogens and mutagens with the Salmonella/mammalian-microsome mutagenicity test. *Mutat. Res.* **31**, 347–364.

Arcos, J. C. (1978). Cancer: Chemical factors in the environment. *Am. Lab.* **10**, Part I (June), 65–73; Part II (July), 29–41.

Casarett, L. J. (1975a). Origin and scope of toxicology. *In* "Toxicology—The Basic Science of Poisons" (L. J. Casarett and J. Doull, eds.), pp. 3–10. Macmillan, New York.

Casarett, L. J. (1975b). Toxicologic evaluation. *In* "Toxicology—The Basic Science of Poisons" (L. J. Casarett and J. Doull, eds.), pp. 11–25. Macmillan, New York.

Cohn, P., and Linnecar, D. (1978). Toxic effects of industrial chemicals. *Chem. Ind.* **20**, 445–446.

Department of Health, Education, and Welfare (U.S.A.) (1978). "Guidelines for the Laboratory Use of Chemical Substances Posing a Potential Occupational Carcinogenic Risk" (revised draft). Laboratory Chemical Carcinogen Safety Standards Subcommittee of the DHEW Committee to Coordinate Toxicology and Related Programs. DHEW, Washington, D.C.

Doull, J. (1975). Factors influencing toxicology. *In* "Toxicology—The Basic Science of Poisons" (L. J. Casarett and J. Doull, eds.), pp. 133–147. Macmillan, New York.

Dow Chemical Company (1960). "Symposium on Toxicology" (V. K. Rowe, arranger). Dow Chem. Co., Midland, Michigan.

Durham, W. F. (1975). Toxicology. *In* "Dangerous Properties of Industrial Materials" (N. I. Sax, ed.), 4th ed., pp. 289–298. Van Nostrand-Reinhold, Princeton, New Jersey.

Eckardt, R. E., and Scala, R. A. (1978). Toxicology: Assessing the hazard. *J. Occup. Med.* **20**, 490–493.

Gardiner, J. S. (1977). Health care of people at work. The toxicological screening of industrial chemicals. *J. Soc. Occup. Med.* **27**, 13–19.

Golberg, L. (1972). Safety of environmental chemicals—the need and the challenge. *Food Cosmet. Toxicol.* **10**, 523–529.

Heidelberger, C. (1977). Chemical carcinogenesis. *Cancer* **40**, 430–433.

Hine, C. H., and Jacobson, N. W. (1954). Safe handling procedures for compounds developed by the petro-chemical industry. *Am. Ind. Hyg. Assoc., Q.* **15**, No. 2, 141.

Klaassen, C. D. (1975). Absorption, distribution, and excretion of toxicants. *In* "Toxicology—The Basic Science of Poisons" (L. J. Casarett and J. Doull, eds.), pp. 26–44. Macmillan, New York.

Loomis, T. A. (1968). "Essentials of Toxicology." Lea & Febiger, Philadelphia, Pennsylvania.

McNamara, B. P. (1974). Toxicological test methods. *Assoc. Food Drug Off. U.S., Q. Bull.* **38**, 33–50.

Miller, R. W. (1977). Relationship between human teratogens and carcinogens. *J. Natl. Cancer Inst. 58*, 471–474.

Ministry of Health and Welfare (1973). "The Testing of Chemicals for Carcinogenicity, Mutagenicity, and Teratogenicity." MHW, Ottawa.

National Academy of Sciences (1977). "Principles and Procedures for Evaluating the Toxicity of Household Substances." Committee on Toxicology, Assembly of Life Sciences, National Research Council, Washington, D.C.

New York Academy of Sciences (1977). "Cancer and the Worker." NYAS, New York.

Norton, T. R. (1975). Metabolism of toxic substances. *In* "Toxicology—The Basic Science of Poisons" (L. J. Casarett and J. Doull, eds.), pp. 45–132. Macmillan, New York.

Stokinger, H. E. (1967). Means of contact and entry of toxic agents. *In* "Handbook of Laboratory Safety" (N. V. Steere, ed.), pp. 231–233. Chem. Rubber Publ. Co., Cleveland, Ohio.

Wallace, M. J. (1978). Basic concepts of toxicology. *Chem. Eng. (N.Y.)* **85**, 72–76.

Weisburger, J. H., and Weisburger, E. K. (1967). Tests for chemical carcinogens. *Methods Cancer Res.* **I**, 307–398.

Williams, R. T. (1959). "Detoxication Mechanisms," 2nd ed. Chapman & Hall, London.

# Chapter Three
# Radiation Hazards in the Laboratory*

## Michael Slobodien

* The views and opinions expressed herein are those of the author and are not necessarily those of the United States Nuclear Regulatory Commission or of its contractors. Mention of specific products or manufacturers does not constitute endorsement by the Commission. Such endorsements are those of the author.

# I. INTRODUCTION

## A. Scope

The biological scientist should expect to encounter several types of radiation in his work. There are two broad categories of radiation, which are defined by their ability, or lack thereof, to form charged species (ions) on interaction. Predictably the two categories are termed *ionizing* and *nonionizing*. Many biological scientists are familiar with the use of isotopes or tracers that produce ionizing radiation; however, not so readily apparent is the fact that many laboratories also have sources of nonionizing radiation that may present hazards to personnel. This chapter deals with ionizing radiation primarily, since such sources are in widespread use and, at the present at least, of the two types appear to present a greater threat to personnel safety. This is not meant to minimize the importance of nonionizing radiation safety, but, as will become evident, nonionizing radiation in most biological laboratories is primarily associated with well-protected instrumentation.

## B. Ionizing Radiation

### 1. Definitions

A brief review of the common forms of ionizing radiation and their characteristics is essential to an understanding of the reasons for specific radiation protection procedures. Although there are general precautions to be taken when using any source of ionizing radiation, the specifics of experiment design depend greatly on a number of factors—one of the most important being the type of ionizing radiation in use.

## 2. Alpha Radiation

Alpha radiation or $\alpha$ particles are doubly charged helium nuclei ($He^{2+}$). Sources of $\alpha$ radiation are common in nature and include thorium, uranium, and radium. Alpha-emitting isotopes with one exception ($^{147}Sm$) have atomic numbers greater than 82. These heavy elements are not commonly used in biological research. Alpha radiation is also produced by sources other than the decay of an unstable nucleus such as accelerators. Alpha emitters such as $^{239}Pu$ and $^{241}Am$ are mixed with beryllium in sealed neutron sources. These are used to study neutron effects on materials.

Alpha particles are highly energetic. Nearly all have energies of 4 MeV (million electron volts) or greater. An electron volt is the energy gained by an electron as it traverses a potential difference of 1 V. These energies are up to several orders of magnitude greater than other types of radiation associated with radioisotopes commonly found in biological laboratories. From a radiation safety standpoint $\alpha$ particles are easily stopped by a thin absorber such as a sheet of paper or the skin epidermis. External to the body such sources do not present a great hazard. Inside the body, however, $\alpha$ emitters are highly significant. Because the $\alpha$ particle is doubly charged and relatively massive, it undergoes many interactions with surrounding atoms, depositing all its energy in a very small volume, on the order of $3 \times 10^{-9}$ cm$^3$ in muscle. An energy deposition of this magnitude within a cell will virtually guarantee cell destruction. For this reason great efforts must be taken to prevent sources of $\alpha$ radiation from entering the body. These measures will be detailed in subsequent sections of this chapter.

## 3. Beta Radiation

Beta radiation is an unpaired, singly charged electron possessing kinetic energy. The mass of the $\beta$ particle is about 1/8000 of the mass of an $\alpha$ particle. There are many naturally occurring sources of $\beta$ particles as well as those produced artificially by man. In the biological laboratory $\beta$-emitting isotopes are most common and include $^3H$, $^{14}C$, $^{35}S$, and $^{32}P$. High-energy electrons are also emitted by accelerators and electron microscopes. The ability of a $\beta$ particle to penetrate matter is a function of its energy. A 1 MeV $\beta$ particle will travel about 1 m in air, 1 mm in Lucite, and 1–3 mm in skin.

Low-energy $\beta$ particles (less than 0.2 MeV) are easily absorbed in the outer layer of skin; $\beta$ sources external to the body present a somewhat greater threat of penetration when compared to $\alpha$ particles. As with $\alpha$ radiation, $\beta$ sources inside the body within cells or incorporated into biologically active molecules may give significant doses disabling and killing cells.

## 4. Gamma Radiation and X Rays

Gamma radiation and X rays are electromagnetic radiation traversing matter. They have neither mass nor charge. They create ionization in materials through which they pass by a number of mechanisms that cause excited electrons to be ejected from absorbing or struck atoms. Since X and γ rays are without mass or charge, their interactions per unit volume are fewer than for α or β particles and therefore they are vastly more penetrating than those particles. X rays and γ rays have identical physical characteristics. They differ only in method of production. Gamma rays result from the emission of energy as an excited nucleus drops from one energy level to a lower energy state; X rays result from transitions within the electron cloud of an atom or are emitted when a charged particle is accelerated.

The penetrating ability of X and γ radiation makes them biological hazards whether they are external to or within the body. Additional considerations in experimental design must be made if penetrating γ or X radiation will be present.

## 5. Neutrons

Neutrons are least common in the biological laboratory of the four major types of ionizing radiation. They can be produced with accelerators, nuclear reactors, isotopes that fission spontaneously such as $^{252}$Cf, and as the result of the interaction of α or γ radiation on beryllium or other materials. Neutrons are uncharged particles with a mass 1827 times that of the electron. They are characterized into three broad groups by energy as fast, epithermal, and thermal (high, medium, and low energy). Thermal neutrons have about the same velocity as gas molecules at room temperature and are very readily absorbed by hydrogenous material. Although we can speak of the absorption or penetrating ability of α, β, and elecromagnetic radiations fairly easily, that is not the case for neutrons. Their absorption properties are complex functions of the absorber's atomic weight, neutron to proton ratio, and interaction probabilities with various nuclei. For these reasons alone, it is fortunate that neutrons are encountered infrequently in the biologic laboratory. Neutrons cause ionization indirectly; proton recoils result when atoms are struck by neutrons that sheer away electrons from nearby atoms. In many cases, neutrons are completely absorbed by stable isotopes that then become radioactive. Exposure to neutron radiation is of considerable concern as biological damage from sources external to or within the body is considerably greater than for equivalent amounts of β or γ radiation. In general, where significant sources of neutrons are present, a professional radiation safety expert is nearby. Therefore our discussion of neutrons will be minimal in subsequent sections of this chapter. The general references at the end of the chapter do provide greater detail on radiation protection for situations in which neutrons are encountered.

## C. Sources of Ionizing Radiation in the Biological Laboratory

### 1. Tracer Isotopes

Radioactive isotopes are among the most common sources of ionizing radiation in the laboratory. Radioisotopes are found in many forms including elements, inorganic compounds, and biological molecules and their precursors. They are used to study molecular, cellular, and gross anatomical structure, molecular and chemical pathways, environmental pathways, and biological function, and to assay proteins, hormones, and the like. The amounts of radioactivity used in biological applications cover 12 orders of magnitude from submicrocurie levels to tens of curies. The radioactive materials described above are used in "unsealed" form (also called normal form). This means that the materials may be easily dispersed in air or water or as surface contamination. One of the major goals of this chapter is to provide guidance to minimize unnecessary dispersal and prevent harmful releases of material to the work or public environment.

### 2. Sealed Sources

Radioactive materials encapsulated in strong, leakproof devices such that personnel do not have ready access to the radioactive contents are, in the broad sense, sealed sources. There are a multitude of sealed sources in use in biological laboratories. Nearly all fall into one of the following five categories: electron capture detectors, vacuum gauges, $\gamma$ check sources and standards, mossbauer sources, and $\gamma$ irradiator sources.

Modern electron capture detectors use $^{63}$Ni and $^{3}$H on foils within envelopes to aid in detection and quantification of certain organic compounds in gas chromatography. $^{90}$Sr and $^{226}$Ra sources have been used in electron capture detectors in early models, but, since they may cause measurable external radiation levels outside the device whereas $^{63}$Ni and $^{3}$H do not, their use has been discontinued. The radiation safety requirements involved in using such devices are few and simple. Where $^{63}$Ni and $^{3}$H (both low-energy $\beta$-emitting isotopes) are used, no external radiation is expected. All electron capture detectors except those containing $^{3}$H should be tested for leakage at intervals of six months or at intervals specified by the manufacturer. The $^{3}$H sources do release small quantities of $^{3}$H (in the range of 10 $\mu$Ci per day) under normal use. The exit ports of chromatographs containing $^{3}$H electron capture detectors should be vented to an exhaust hood.

Care should be taken to operate the detectors at the proper temperatures as elevated temperatures will cause greater evolution of $^3$H.

Vacuum gauges also contain radioactive materials ($^{241}$Am, $^{226}$Ra) on electroplated foils. Alpha emitters are used since high ionization density is needed to detect very low gas pressures. There may be small external radiation fields associated with such devices (most $\alpha$ emissions are accompanied by X or $\gamma$ rays) and they should be located with consideration of the exposure potential for nearby workers.

Mossbauer, check and standard sources, and $\gamma$ irradiators use mostly $\gamma$ emitters in sealed sources. The radiation hazards depend on the amount of the radioisotope present and its use. Radiation levels may be significant. The protective measures that may be necessary are covered in Section IV. There are also quite a few plated sources. In these the radioactive material has been electroplated or vapor-deposited on a substrate. The plate may be covered with a thin protective coating or it may be bare. Such sources are found in static eliminators (often $^{210}$Po) and smoke detectors ($^{241}$Am, $^{226}$Ra). They can release the radioisotopes if scratched. In devices containing plated sources the source itself may be protected in a secure container not accessible to personnel. Lab workers should be aware of such devices in use. In most cases the plated sources are found in devices not intended to be opened and if used according to manufacturer's directions do not present a radiation hazard. It is prudent to make a survey as described later in this chapter to ensure that radiation levels around such devices are safe and that the locations of radioactive material are evident to anyone having to work with the devices.

## 3. X-Ray Machines

There are, of course, sources of ionizing radiation that are not associated with the use of radioisotopes. Analytical X-ray machines such as powder diffraction and X-ray fluorescence devices are used in qualitative and quantitative analysis. There are too many types of devices to discuss adequately here but several general rules concerning their use can be given. First, many analytical X-ray units have narrow, very intense beams incident on targets. Fingers may be seriously injured when changing targets if the beam is not turned off or shielded. Eyes may be exposed when aligning targets if X rays are being emitted. Radiation scattered from targets and shielding may be present all around such units. Therefore, it is imperative and cannot be too highly stressed that all users of such devices turn off the X-ray source and close all shutters when changing targets, making gross changes to alignment of detectors, or making any adjustments. Second, units should be surveyed with a calibrated survey instrument to establish radiation levels around the units under actual use conditions. If shielding is needed to make radiation levels acceptable, thin (2.5-mm) portable sheets of lead supported on plywood or a frame may be moved into place to reduce radiation levels. Records of such surveys should be maintained and surveys should be updated at least annually or whenever the machine is used in a different setup, whichever comes first. These machines should be used only by "authorized" individuals who have had safety training. Finally, operators of the units should wear personnel dosimeters on wrists or preferably fingers.

## 4. Electron Microscopes

Another of the nonisotopic sources of ionizing radiation commonly found in life sciences laboratories is the electron microscope. Under normal use high-energy electrons strike targets, screens, and occasionally walls of the microscope. In areas where X rays are expected, shielding is provided. Personnel exposures may result if the electron beam is off target, striking unshielded areas. One of the first indications of such a problem is an inability to obtain an adequate image. It is normal practice to have a survey of the area around the microscope performed at installation and following major adjustments or servicing to determine whether any external radiation is present. An ionization chamber is best suited for this type of survey.

## D. Nonionizing Radiation

### 1. Definition

For the purpose of occupational radiation protection, nonionizing radiation includes microwaves, lasers, ultraviolet light, and ultrasonic sources. The basis for nonionizing radiation safety is the Threshold Limit Value established by the American Conference of Government Industrial Hygienists (ACGIH). A brief description of the hazards associated with each type of nonionizing radiation follows.

### 2. Microwaves

Microwave sources in the biological laboratory include cooking ovens, drying chambers, and transmission antennas. The critical organ is usually the eye, although in medical diathermy devices the critical organ is subcutaneous muscle. Allowed doses and dose rates have been established from experimental evidence with exposures of rabbit eyes (rabbit eyes are anatomically similar to those of humans). For continuous 8-hr exposures, average power densities of 10 mW/cm$^2$ have been determined to cause no adverse effects. Short exposures of up to 10 min in any 1 hr may be in the range of 10–25 mW/cm$^2$ without any adverse effects. Exposures in excess of 25 mW/cm$^2$ are not permissible under any circumstances according to the ACGIH.

Microwave power densities are measured by tuned, calibrated antenna thermocouples. Most commercially available microwave devices must meet radiation level limits set by the U.S. Food and Drug Administration's Bureau of Radiological Health (BRH). BRH regulations set microwave radiation limits well below 10 mW/cm$^2$. In practice, ovens and drying devices—the most common microwave sources in the biological lab—emit little or no stray microwave radiation. It is wise to have an annual radiation survey conducted on all commercial microwave devices to ensure satisfactory performance. Research devices should be checked during construction and quarterly or whenever use conditions change.

## 3. Lasers

Sources of monochromatic, coherent light are lasers. Lasers may emit radiation in the infrared, visible, and ultraviolet regions. As with microwaves, the eye is the critical organ on which protection standards are based. Lasers may be of the continuous or pulsed types. Doses are set in milliwatts or millijoules per square centimeter. Allowable doses are based on the wavelength of the laser. The ACGIH publication "Threshold Limit Values for Chemical Substances and Physical Agents in the Workroom Environment with Intended Changes for 1977" provides several fine data tables for laser dose limits. Since limits are based on many parameters, including continuous versus pulsed emission mode, wavelength, solid angle intercepted by the beam, beam divergence, and reflected versus direct beams, it is not easy to set forth simple principles for laser protection. The reader is advised to consult the instrument manual and the ACGHI publication for specific guidance.

Surveys of laser equipment should be made on initial installation and annually thereafter or whenever changes in the setup are made, whichever comes first.

## 4. Ultraviolet Light and Ultrasonic Sources

Ultraviolet light sources are common in the biologic laboratory as antibacterial and antifungal devices. They should be used so that direct exposure to workers' eyes and skin is minimized. Exposure limits for such sources have not been set (low-intensity sources); however, caution is advised because of the possibility of eye or skin irritation due to chronic exposures to even low-intensity lamps. Ultrasonic sources include cleaners and cell disrupters. These present a hazard only upon direct contact with a portion of the body. The simple protective measure, then, is to avoid contacting the operating device.

## II. UNITS OF RADIATION MEASUREMENT

### A. Activity

Since the discovery by Wilhelm Roentgen in 1895 that energetic electrons impinging on a target of high atomic number produce rays that easily penetrate matter and can expose photographic film—X rays, the scientific community has adopted special units to describe the amount and nature of ionizing radiation. The International Commission on Radiological Units (ICRU) was formed to develop a well-founded system of units and nomenclature specific to the needs of physicians and other persons working with not only X rays but other types of radiation found in nature or produced by man. The units that have been developed were named after pioneers in the field (Roentgen, Curie) or began as descriptive terms that turned into acronyms then into units (rem—"roentgen equivalent man"). The ICRU designated units on the basis of observed quantities. Thus the special unit of activity, the curie, was equal to the number of disintegrations taking place per unit time from 1 g of radium. The curie was later redefined as the activity of that quantity of radioactive material in which the number of disintegrations per second is $3.7 \times 10^{10}$ (a number nearly the same as the number of disintegrations per second from 1 g of radium). We have since learned that a curie of any radioisotope is a very appreciable amount, too great for most laboratory applications, so we commonly find activity expressed as millicurie (mCi, $1 \times 10^{-3}$ curie) or microcurie ($\mu$Ci, $1 \times 10^{-6}$ curie). The symbol for curie is Ci. It is essential that one not confuse the symbol for micro with that for milli. The 1000-fold error that results may mean the difference between an almost inconsequential radiation problem and a major radiation hazard! A useful number to remember is $2.22 \times 10^6$ disintegrations per minute per microcurie. Most tracer applications require microcurie quantities, although it is not unusual to find millicurie quantities of $^3$H, $^{32}$P, and $^{125}$I in many laboratories. Recently, the International System of Units has been adopted by the ICRU for all radiation quantities and units. In this system a new definition of activity, the becquerel (symbolized by Bq) has been adopted. The Bq is defined as the quantity of radioactive material in which the number of disintegrations per second is one. No doubt, both Ci and Bq will be found in the literature in the future.

### B. Exposure

The ICRU defined the special unit of exposure in air to be the roentgen (symbolized by R).

$$R = 2.58 \times 10^{-4} \text{ coulomb/kg air}$$

This unit is special in that it is defined only for X or $\gamma$ radiation in air. Thus the roentgen is not applicable to alphas, betas, or neutrons. Still for many applications the roentgen and more commonly used milliroentgen ($10^{-3}$ R, symbolized by mR) are useful. Many survey instruments provide output data in terms of mR/hr. The roentgen is not always useful for making accurate evaluations of energy absorbed due to radiation impinging on material. It is the absorbed energy that is a true index of biological damage. If one knows how well a certain material can absorb radiation as compared with air, the energy absorbed by that material when exposed to 1 R can be calculated. This procedure turns out to be easy and practical. It is very easy to measure ionization in air with inexpensive equipment, so the roentgen can be measured directly. It is not so easy to measure the energy absorbed in material directly.

## C. Absorbed Dose

The rad is the special unit of absorbed energy. It is defined as that amount of ionizing radiation that deposits 100 ergs/g of material. The rad is applicable to all types of ionizing radiation. Because it is often most difficult to measure directly, normally ionization in the air or another gas is measured and the absorbed dose in a particular material calculated. One roentgen results in 86.9 ergs being absorbed in 1 g of air; if tissue is placed in the same radiation beam, 1 R in air corresponds to about 93 ergs/g. For most applications of X and $\gamma$ rays and $\beta$ radiation it is reasonable to assume that 1 R = 1 rad. One roentgen is a large exposure, therefore, we more often see the term millirad ($1 \times 10^{-3}$ rad, symbolized by mrad). There is no abbreviation for rad; it is always written out.

## D. Dose Equivalent

In the interest of completeness one additional unit, the rem (mrem), must be described. Certain types of radiation such as alphas and neutrons do more biological damage per unit of energy absorbed. This is due in part to the extremely high ionization density associated with such particles as they are absorbed. As mentioned earlier, the higher the ionization density, the more extensive the biological damage per unit of energy deposited. Thus even though two equal masses of tissue may absorb 1 rad of X radiation and 1 rad of $\alpha$ radiation, the very much greater ionization density due to the $\alpha$ particles will inactivate or kill more cells. Lower ionization density means that the critical sites within cells are less likely to be "hit" so a cell has a greater chance of survival. The rem "corrects" for the difference in the cell-killing effectiveness of certain types of radiation. Note that for X and $\gamma$ rays and most $\beta$ radiation, 1 R $\cong$ 1 rad, 1 rad = 1 rem.

**TABLE I**

**Measurement of Radioactivity and Radiation**

| Unit definition | Function |
|---|---|
| Activity (A) | |
| Curie (Ci) $3.7 \times 10^{10}$ d/sec | Describes the amount of radioactive |
| Millicurie (mCi) $3.7 \times 10^{7}$ d/sec | material present by specifying the rate |
| Microcurie ($\mu$Ci) $3.7 \times 10^{4}$ d/sec | of disintegration |
| Exposure (E) | |
| Roentgen (R) $2.58 \times 10^{-4}$ coulomb/kg air | Describes the number of ions liberated in a given mass of air following inter- |
| Milliroentgen (MR) $2.58 \times 10^{-7}$ coulomb/kg air | action of X or $\gamma$ rays. One ion pair $= 34$ eV/ion |
| Absorbed dose (H) | |
| Rad (100 erg/g) | Describes the energy absorbed per unit |
| Millirad (mrad) | mass due to any type of radiation. 1 R in air results in 87.6 erg/g air, about 93 erg/g in soft muscle |
| Dose equivalent | |
| Rem (rad) (Q) | Describes the index of biological damage |
| Millirem (rem) | in erg/g, where Q is the Quality Factor, a term related to the relative damage capacity for a given type of radiation. The Q value for X, $\gamma$, and most $\beta$s is 1; for alpha particles Q is 10; for neutrons Q has a range of values from 10–20 depending on neutron energy |

Table I summarizes the units peculiar to radiation protection. In the following sections in this chapter we will frequently refer to these units to establish levels of acceptable radiation dose, dispersed material, and contamination.

## III. RADIATION EFFECTS ON BIOLOGICAL SYSTEMS

### A. Scope

The effects of ionizing radiation on the body are far too complex to describe in detail in this chapter. Radiation does damage cells, cause mutations, and, if absorbed in large enough doses, cause clinically observable effects in humans including skin burns, loss of hair, and damage to blood-forming tissue. It is quite evident that radiation exposure causes cancer. Taken out of context one might view these statements as a total indictment of the use of any source of ionizing radiation. This must be put in perspective by separating low-dose effects from high-dose effects and then evaluating what doses are typically received in the biological laboratory. It is essential to minimize radiation doses to minimize the possibility of undesirable effects.

Radiation effects are a function of both integrated dose and dose rate. Radiation interactions in the body may be viewed as causing damage with some probability. This probability is a function of the type of radiation and the cell system exposed. Different organs have different sensitivities and certain types of radiation are more likely to cause damage than others. There is evidence that some radiation damage is repaired by normal biological process. When the dose rate (injury rate) exceeds the body's ability to repair damage, the net effect is negative.

## B. Sources of Data

Let us look at the sources of data for radiation effects in humans and the types of damage associated with different degrees of radiation exposure and then examine the types and levels of doses received in the biological laboratory and in our daily routine.

Clinically observable radiation effects in humans have been documented in several "populations." These include cancer patients treated with externally administered radiation, atomic bomb survivors, individuals treated with X rays for diseases of the spine and skin, and persons occupationally exposed such as laboratory workers. These populations cover a broad spectrum of doses from sources of radiation external to the body. Populations with internal radiation sources include radium watch dial painters and certain industrial accident victims. Extensive animal experiments have been done with rats, dogs, and primates for both external and internal sources of radiation. The animal data have been extrapolated to humans with success in some cases.

Most of this data represents exposures at high doses. Low doses give rise to observable effects with extremely low probability and thus in small populations it is difficult to distinguish radiation effects from natural incidences. It should be noted that radiation exposure produces no effects that are unique. The effects of exposure to both low and high doses can also be caused by other forms of stress, including bacteria, viruses, and chemical or physical agents. The radiation-induced monster arising from the deep is a figment of imagination.

## C. Levels of Exposure and Associated Effects

The sources of radiation exposure in nature include cosmic rays and naturally occurring radioisotopes. Doses and dose rates comparable with

**TABLE II**

**Average Annual Doses to General Public in the United States**[a,b]

| Source | Annual dose (mrem) |
|---|---|
| Cosmic rays | 44 |
| External sources of radionuclides | 40 |
| Internally deposited radionuclides | 18 |
| Total | 102 |

[a] Source: National Academy of Sciences report, "The Effects on Populations of Exposure to Low Levels of Ionizing Radiation" (BEIR Report).

[b] Due to naturally occurring radioisotopes and radioisotopes from nuclear weapons testing.

those natural sources are considered to be low. Tables II and III give data for natural radiation sources and also for some man-made sources such as diagnostic X rays. These are low-dose sources.

Table IV indicates the expected biological effect from a single exposure (given in a day or less) for doses comparable to natural background doses and higher. The data are for external radiation sources. Data for low doses are based on calculation and extrapolation from higher doses. The doses typically found in the biological laboratory are rarely in excess of 10–20 mrem per month. Clinical effects would not be observed. It must be stressed that the probability of cancer induction is extremely remote; the risk of developing cancer has been estimated at 1/1,000,000,000 per millirem per year. When radiation exposures are protracted over several days or longer, the effects per unit dose tend to lessen. Thus for individuals receiving doses at or near background (10–12 mrem/month), the effects

**TABLE III**

**Typical Radiation Doses for Medical Diagnostic Exposures**

| Procedure | Dose |
|---|---|
| Chest X ray | 70–130 mrem/film |
| Fluoroscopy lower gastrointestinal | 1000 mrem |
| Complete gastrointestinal series | 700–7000 mrem |
| Nuclear medicine lung scan | 2000–3000 mrem to lung |

[a] Sources: "Radiation Levels and Effects," United Nations Scientific Report by Committee on Effects of Atomic Radiation, 1972, and Medical Internal Radiation Dose Committee Publication 11 © 1975 Society of Nuclear Medicine, New York.

TABLE IV

Expected Biological Effects from Single Whole Body Dose without Treatment

| Dose (rem) | Expected effect |
|---|---|
| 0–0.1 | No clinical symptoms. One additional cancer per 10 million population in 10–30 years. |
| 0.1–1 | No clinical symptoms. One additional cancer per million population in 10–30 years. |
| 1–10 | No clinical symptoms. Ten additional cancers per million population in 10–30 years. |
| 10–50 | Chromosome breaks, rings and dicentrics in cultured blood cells. No other symptoms. Effects are reversible. |
| 50–100 | Chromosome breaks. Slight depression in white blood cell count at higher end of range. No external symptoms. Effects are reversible. |
| 100–250 | Chromosome breaks. White blood cell depression. Some loss of electrolytes, diarrhea or vomiting. Skin changes similar to sunburn seen after 2–5 weeks. |
| 250–600 | All symptoms above plus prolonged loss of blood-forming organ function of several weeks to months. More severe skin changes in shorter time. At higher end of range up to 30–50% deaths in 30–60 days to infection and loss of gastrointestinal function. |
| 600–1000 | Death due to gastrointestinal failure with 3–6 weeks for typical individual plus blood-forming organ damage indicated above. Most skin changes will not have opportunity to evidence themselves. |

should be essentially those experienced by the general population. That is, there should be no discernible difference in cancer rates or ill health among such persons. This will be the case with most workers with radiation sources in the biological laboratory. The only way to minimize the possible adverse effects, however, is by following sound radiation safety practices to minimize doses.

## D. Internally Deposited Radionuclides

The preceding discussion relates to exposure to sources of radiation external to the body. Sources of radiation within the body are often more hazardous, not because the effects of exposure are more pronounced at a given dose but because a dose equivalent to that from a external source dose may be caused by a very small amount of radioactive material in intimate contact with cells. Furthermore, once an external source of radiation is turned off, shielded, or removed, the irradiation of the body ceases. With internally deposited radionuclides the exposure of the body goes on until the material is excreted or decays. There are some types of radiation such as $\alpha$ and low-energy $\beta$ rays that cannot even penetrate the skin if they are external to the body, if they are deposited within an organ including the skin, large doses can result. As an example, microcurie quantities of $^{226}$Ra ingested by radium watch dial painters have caused bone cancer. Once inside the body many radioisotopes are difficult to remove. Removal may be impractical, difficult, or even dangerous to the individual. Thus the main effort must be to avoid intake of radioisotopes. Intakes may occur through ingestion, inhalation, absorption through skin, and cuts or puncture wounds. Procedures are given later in this chapter for minimizing the intake of radioisotopes and exposure to external sources.

## IV. PRINCIPLES OF EXPOSURE CONTROL

### A. As Low as Readily Achievable Philosophy

Not all interactions of radiation with biological systems are well understood. There may be a threshold dose rate below which radiation damage is repaired so that permanent injury does not result. It is, nevertheless, in the interest of workers and the general public to maintain radiation exposures as low as possible. In radiation protection it is generally assumed that there is some damage associated with any level of exposure. The as low as readily achievable philosophy (ALARA) differs, however, from the strict concept of as low as possible, since it considers the economic or other benefits of radiation exposure. It should be a standard method of operation to prevent all unnecessary exposure and limit necessary exposure to the greatest extent. ALARA requires that occupational exposures be reduced to as low a level as is possible while giving consideration to economic and societal benefits to the low-level exposures; high-level exposures cannot be allowed except in dire emergencies.

### B. Time

The magnitude of a dose is directly related to the duration of the exposure. This is true for all forms of radiation, although for practical applications only doses from penetrating radiation or skin contamination are easily controlled by reducing the duration of exposure. Much of the radioisotope work in biological research involves the $\beta$ emitters, $^3$H, $^{14}$C, and $^{35}$S. If sources of these radioisotopes are more than a few centimeters away from the skin, there will be no dose to the body regardless of the time spent working with the material.

Controlling the duration of exposure for laboratory workers is often impractical. It is probably better to be careful and take time in order to perform well rather than to rush through an experiment and create a hazard.

The following methods for dose control tend to be more effective and easier to implement than the control of length of exposure.

## C. Distance

Radiation intensity decreases as one removes himself from the source. For a very small point source of $\gamma$ or X rays the radiation intensity decreases as the inverse square of the distance, $I = I_o/d^2$. For plane or line sources the intensity decreases linearly with increasing distance from the source. Alpha and $\beta$ radiations have fairly definite ranges in air; for short distances the radiation intensity decreases rapidly until at the maximum range the intensity suddenly drops to near zero.

It is extremely easy to reduce exposures by controlling the distance from the body to the radiation source. Consider an experimenter working with 5 mCi of $^{32}$P. At 1 mm from a typcial 1 ml volume the dose rate is about 15 rad/hr. At 1 cm the dose rate is about 1.5 rad/hr, and at 10 cm the dose rate drops dramatically to about 0.015 rad/hr. The use of tongs or forceps in handling a 5 mCi vial of $^{32}$P could reduce the dose by a factor of 1000 compared with the dose received by touching the vial directly with the fingers. The case of $^{32}$P is very dramatic; similar reductions can be obtained for $\gamma$ emitters. For low-energy $\beta$ emitters such as $^3$H and $^{14}$C distances of 10 cm reduce the dose rate to zero. In addition to using tongs and other mechanical devices to keep fingers away from radiation sources, sources should be positioned away from personnel. Radioactive solutions should be stored at the rear of hoods or in areas not frequented by personnel. When several experiments are set up near each other, care must be taken to plan work areas to maximize distance from the areas of highest levels of radiation. Desks and study areas should be located away from radioisotope areas. The use of distance as a dose reduction technique can be a powerful tool that need not impinge on the ease of conducting an experiment.

## D. Shielding

The absorption of radiation by various materials is an often used method for controlling personnel exposures. Alpha and $\beta$ radiation have, for the most part, very short ranges in nearly all materials—paper, cardboard, plastic, etc. $^{32}$P is one of a few $\beta$ emitters that is rather penetrating, having ranges in various materials given in Table V.

Gamma radiation is absorbed exponentially in absorbing material. The ability of material to stop $\gamma$ photons is related to the density and electron density of the material.

Storage of $\alpha$- and most $\beta$-emitting isotopes in plastic vials prevents ex-

TABLE V

Range of $^{32}$P $\beta$ Particle in Various Materials

| Material | $\beta$ range (cm) |
|---|---|
| Air | $8 \times 10^3$ |
| Water | 0.76 |
| Lucite | 0.61 |
| Aluminum | 0.28 |
| Cardboard | 0.46 |

ternal radiation levels and these vials have the advantage of being break resistant. When high-energy $\beta$ or $\gamma$ radiation is used, experimenters should employ shielded containers and work behind shielding barriers. The shielding material and the shield thickness will be determined by the type of radiation, amount in use, and times spent in the work area. Figure 1 provides the recommended thickness of readily available material needed to reduce the radiation intensity from various $\gamma$ emitters by a factor of 2. For workers using $\alpha$ and low-energy $\beta$ emitters shielding is not needed and is really "overkill." Where shielding is used, the design should minimize the material needed. The optimum configuration is that in which the radiation source is close to the shield. The shield area needed increases as the square of the distance from the source for small sources. In fume hoods the floors and walls as well as the front may require shielding. Similar considerations must be made for lab bench shielding. Radiation passing through the hood floor or lab bench can expose the lower portions of the body. Shielding should be high enough to protect the eyes and head. If the shield restricts vision, a mirror can be placed at the rear of the work area to allow observation.

Although shielding works very well to reduce dose rates, it has disadvantages too. These include expense, weight, reduction of experiment flexibility, and potential for becoming contaminated. Great care should be taken not to overshield and to use readily available materials. Porous materials such as concrete should be sealed to make decontamination easy.

## E. Containment and Ventilation

Of the techniques for exposure control, containment is the most widely employed, the least expensive, and the most effective. Containment refers simply to maintaining loose radioactive material in a controlled situation or maintaining the worker in a controlled environment when in the midst of radioactive material. Containment methods described as follows are the

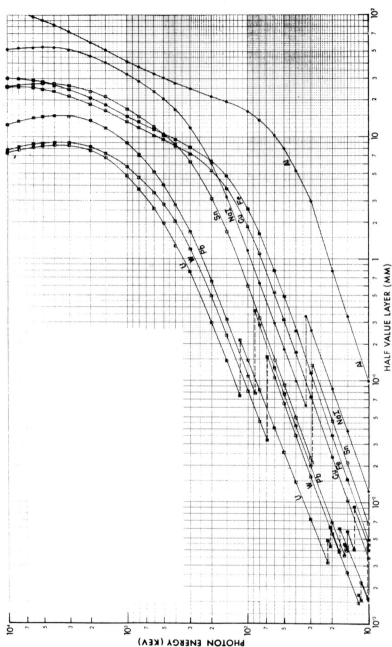

**Fig. 1** This graph gives the thickness of various materials to reduce the radiation intensity by one-half (half value layer) as a function of $x$ or $\gamma$-ray (photon) energy. To calculate the thickness needed to obtain a given intensity reduction, calculate $x$ in $I/I_0 = 1/2^x$ where $I/I_0$ is the reduction in intensity, $x$ is the number of half value layers needed. Then use this graph to find one half value layer and multiply by the value of $x$ to find the total thickness required. (Data from "Radiological Health Handbook," U.S. Govt. Printing Office, 1970).

most common; ingenious workers may develop techniques that apply specifically to their experiments.

## 1. Drip Trays

Activities involving liquid and powder radioisotope sources should be performed on deep trays lined with absorbent paper. When necessary, suitable shielding should also be employed. Trays are available in many sizes appropriate for bench top and fume hood use. Standard cafeteria trays are acceptable and are available at low cost. The drip trays serve also to denote radioisotope work areas. Entire experiments may be picked up and placed in secure storage areas during times when bench or hood space is needed by other workers. At the conclusion of work the absorbent paper lining may be carefully folded in on itself and transferred to a radioactive waste container. If spills have taken place on the trays, their nonporous surfaces can easily be cleaned. It is wise to have several such trays for various levels of radioisotope usage. Deep trays may be used in refrigerator or freezer for storage of radioactive materials. Spills or leaks do not then contaminate the entire contents in storage.

## 2. Fume Hoods

Almost all modern biology and chemistry laboratories are equipped with at least one fume hood. Hoods should be used for work with volatiles, aerosols, and gases. Work with powders should be restricted to draftfree areas. The purpose of the fume hood is to remove hazardous materials from the work environment and dump them in very dilute concentrations to the air. Many hoods are designed to accept filters to trap dusts, aerosols, and in some cases reactive gases and vapors. If such filters are used, it is imperative that the fans and motors be large enough to accommodate the resistance to air flow that the filters invariably cause. The face velocity of the hood should be in the range of 120–150 linear ft/min. Lower velocities will not remove the material to the stack and will allow it to flow into the occupied room. Higher velocities may cause turbulence, eddy currents, and downdrafts that tend to pump the hood contents into the work area. Face velocities are adjusted by the hood sash position, and it is an excellent idea to mark the sash position where acceptable flow rates exist. The interior of the hood should contain as few items as possible. Great care should be taken not to obstruct baffles at the hood lip and at the rear of the hood floor, since the open baffles are designed to control air flow patterns. If hoods are lined with aluminum foil or absorbent paper, intake or exhaust ports should not be blocked. Air flow in modern hoods is adjusted and controlled by various methods, including front and rear baffles, upper front positive air feeds, and dampers. All of these must operate as intended if the hood is to contain and remove the radioactive material. Hoods equipped with filtration devices should have differential pressure gauges to indicate when filters are filled and need to be changed. Hoods should also have a light to indicate when the hood fan is on. Many hoods will have a small natural draft due to the wind passing over the exit duct. Hoods may be ideal places to work with moderate-sized sources of gamma emitters in loose form. If shielding is used, its placement must not obstruct necessary flow routes. Remember that the hood floors may require some shielding to protect the lower body. There are three common measurements for evaluations necessary for proper hood use. They are face velocity, interior flow patterns and release of radioisotopes to unrestricted areas. Vane velometers are inexpensive and provide adequate indication of linear air flows in the range of 75–400 ft/min. Institutions with numerous hoods and ducts that need to be tested periodically might consider purchase of a thermal anemometer. Such instruments are expensive but very accurate and flexible. Flow patterns within hoods can be checked by commercially available smoke tubes. The methods and instrumentation needed for evaluating releases of radioactive material to unrestricted areas are detailed in Section IV,F.

## 3. Gloves and Protective Clothing

It is impossible to contain completely the radioactive material being used. Open vials, beakers, test tubes, syringes, TLC plates, and the like, allow the radioactive material to be exposed to air and skin. Even though remote-handling devices should be used where practical, there will arise situations when contaminated or potentially contaminated items must be handled. Containment is still possible; however, now it is the hands and bodies of the experimenters that are contained. Containment is achieved by wearing gloves, lab coats, and protective eye wear. For nonpenetrating and weakly penetrating radiation these containment devices also provide some shielding.

Many types of lab gloves are available. Gloves may be counterproductive if they easily fall off or if fingers holes are so large that they are caught in the experimental apparatus. The use of close-fitting gloves is strongly recommended rather than the less expensive one-size-fits-all type. Physician's examining gloves are close fitting, inexpensive, and resistant to tearing. Surgical gloves are not necessary. If high-energy $\beta$ rays will be encountered, as with $^{32}P$, it may be desirable to wear somewhat heavier gloves where possible. If heavier gloves are too restrictive and cumbersome, they may prolong exposure times and be counterproductive. It is desirable to use talcum powder to absorb perspiration and to make wearing more comfortable and removal easy. Glove removal can be a source of personnel contamination if not done properly. The best method is to grab the loose portion of one glove near the wrist with the other gloved hand. Slowly pull the glove off inside out. The contaminated portion of the removed glove is now on the inner side. Hold the removed glove in the still-gloved hand. With the clean hand slip one or two fingers under the wrist area of the other glove and remove it inside out as before. What you now have is a neat package with any contamination on the inside that can be placed in the special radioactive waste disposal container.

Lab coats are also an inexpensive yet worthwhile containment measure. Protective eye wear is desirable and is required in many laboratories. For high-energy $\beta$ emitters lab goggles and glasses effectively shield the eyes—radiosensitive organs.

There are other containment devices in common use, although not often in the biological laboratory. Glove boxes are used where high levels of radioisotopes are used or where dusts and powders are present. Glove boxes are now coming into use in situations where viral and recombinant DNA research is performed. Hot cells—heavily shielded rooms with lead glass viewing ports and mechanical manipulator hands—are used where extremely high radiation fields are present. Recently, with the increasing use of radioiodine, Lucite glove boxes for in-hood use have become popular. Their application is discussed in greater detail in the last section of this chapter.

## 4. Traps and Filters

The use of traps and filters has been briefly mentioned in connection with hoods. Traps are common where relatively large quantities of radioisotopes are in use. Several examples of workable traps may best indicate their value. Dry ice–acetone baths surrounding a vacuum trap are common for trapping acetic anhydride labeled with $^3H$ or $^{14}C$. Reaction flasks are connected in series to the trap. The frozen trapped product may be carefully liquefied for transfer to a storage vial or to an absorbent material in a sealed vial for waste disposal. Liquid nitrogen traps are commonly used to remove $^3H$ as tritiated water or other compounds, $^{14}CO_2$, $^{35}SO_2$, and the like. Activated charcoal is advised for radioiodine, other halogens, and gases such as xenon. If, for example, $^{18}F$ (108 min half-life) were in use, a trap of activated charcoal might be used to hold the fluorine until it decayed to acceptable levels. Molecular sieves may be used at various temperatures to remove $^3H$ and nonreactive gases such as $^{133}Xe$ and $^{41}Ar$. This is an expensive technique but effective. The sieves may be heated to release the trapped product and regenerate the sieves.

The use of traps and filters is a direct consequence of attempting to restrict releases of radioactive material to the environment consistent with ALARA philosophy. Filters are generally of two types—particulate and reactive gas. In the former multiple barriers of fine strands absorb particles from the air. Filters are specified by their ability to remove a given fraction of particles of given aerodynamic median diameter. Where very high-efficiency filters are in use, it is necessary first to use "roughing" filters to remove relatively large particles that would otherwise prematurely clog the better filter. Reactive gas filters use activated charcoal to absorb and hold the gas in a fixed state. Filter selection and installation is best done in cooperation with an expert. The variables entering into the selection process are many and often complex.

## F. Surveys

Even when an experiment is well designed and carried out by trained personnel, physical measurements must be made to ensure that exposure of workers and the public is kept as low as reasonably achievable and maintained within regulatory limits. The complete survey is more than a simple measurement with a Geiger counter or wipe; it is an evaluation of the hazards of a particular operation based on knowledge of the system, calculations, and confirmatory physical measurements. The measurements themselves serve only as data. A satisfactory survey is one that continues to the next step—analysis to determine the meaning of the measurements. Often this last step is automatic and is performed with a minimum of effort. In other cases, however, once the data have been obtained, the surveyor must consider whether the data are representative, determine detector efficiencies, release rates, occupancy by potentially exposed individuals, and the like. It is not as complicated as it appears when in print, still it does take careful thought to complete a good survey. Table VI provides a basis for setting survey frequencies. Consider a survey for surface contamination in a laboratory performing radioimmunoassays with millicurie quantities of radioiodine. Following the wiping of lab benches and other potentially contaminated surfaces, counting of the wipes, and calculation of surface activity, the surveyor must consider whether airborne activity or a spill was the cause of detected contamination. Surface contamination may be indicative of potential ingestion or inhalation of the radioisotopes by the workers. It may indicate sloppy technique and point to a cross-contamination problem. The last steps of the survey are to make the evaluations that establish the levels of personnel exposure if any and take corrective action if needed. Of course, such surveys should be documented. Obviously, it is not possible to delineate every type of survey that may be needed in the laboratory. The specifics depend on the isotopes or radiation sources and the levels of radioactivity or radiation present. The general guidelines that follow should allow experimenters to select the survey best suited to their needs.

**TABLE VI**

**Recommended Frequencies for Surveys**

| Radionuclide | External radiation surveys[a] (nuclides with asterisks only) | | Air Sampling[b] (in mCi) | | | Contamination (in µCi) Surface | | |
|---|---|---|---|---|---|---|---|---|
| | Weekly | Monthly | Weekly | Monthly | Quarterly | Weekly | Monthly | Quarterly |
| H-3, C-14, F-18,* K-42,*, Cu-64,* Tc-99m,* In-133m* | If point source of activity could exceed 5 mrad/hr at 1 m | If point source of activity could exceed 0.5 mrad/hr at 1 m | ≥10 | ≥1 | <1 | ≥100 | ≥10 | <10 |
| | | | | <10 | | | <100 | |
| Br-82, Cr-51,* Fe-55, I-123,* Hg-197* | " | " | ≥1 | ≥0.1 | <0.1 | ≥100 | ≥10 | <10 |
| | | | | <1 | | | <100 | |
| S-35, Au-198, Ca-47, I-132, Ce-141, Mixed fission products,* Sr-85, La-140, Nb-95, Zn-65,* Co-57,* Fe-59,* Na-24,* Co-60,* Se-75,* Mo-99* Tc-99m,* | " | " | ≥0.1 | ≥0.01 | <0.01 | ≥100 | ≥1 | <1 |
| | | | | <0.1 | | | <100 | |
| Hf-181, Pm-147, P-32,* Ba-140,* Th-234, Kr-85, Ir-192,* Ci-36, Y-91, Ta-182, Ca-45, Sr-89, Cs-137, Co-60.* Ce-144,* I-126, Eu-154, I-131,* I-125,* Tm-170, Na-22,* Mn-54,* Ag-110m,* Hg-203* Rn-222,* Sn-113* | " | " | ≥0.1 | ≥0.01 | <0.01 | ≥100 | ≥10 | <10 |
| | | | | <0.1 | | | <100 | |

Column span heading: Amounts in process at any one time within any room requiring surveys

[a] Nuclides with asterisks only.
[b] Air sampling methods include use of particulate filters, water or chemical bubblers, adsorbers such as activated charcoal and silver zeolite, and cold traps.

*1. Direct Radiation:* Radiation from sources external to the body, both penetrating and nonpenetrating, is classified as direct. In most cases $\alpha$ and low-energy $\beta$ sources have little or no direct radiation of significance. For a source external to the body the radiation will be hazardous only if it can penetrate the protective layer of dead skin. Thus, direct radiation sources worthy of note include $\gamma$ and X-ray emitters, X-ray machines, higher-energy $\beta$ emitters, and neutron sources. Radiation dose rates should be determined by calculation and measurement with a suitable instrument at distances from the radiation source where personnel exposures are likely. The areas to measure include locations of hands, whole body, and eyes relative to the position of the sources of radiation. Estimates of the duration of exposure are needed so as to establish a reasonable estimate of total dose to each affected part of the body. Instrumentation best suited to this task includes ion chambers and Geiger-Müller (GM) counters. These are available in many models. Where $\beta$ radiation or low-energy X and $\gamma$ rays are being measured the detector should have a thin wall or window. Mylar or mica windows are commonly used. The instrument should be calibrated for the type of radiation being detected. Most ion chambers and GM counters are originally calibrated for $\gamma$ radiation from $^{137}$Cs (0.662 MeV) and $^{60}$Co (1.25 MeV). They tend to overrespond to photons below 0.3 MeV. Response to $^{32}$P when within 1 m approximates the original calibration. Such instruments have scales that usually read in millirem per hour. The user should be familiar with instrument characteristics. Most of these instruments have multiple scales. An instrument switch will be present that may be set at $\times$ 0.1, $\times$ 10, $\times$ 100, etc. In nearly all cases these settings indicate that the scale is multiplied by the factor given and not that instrument sensitivity is greater at $\times$ 100 than $\times$ 1. The key to correct interpretation of the meter reading is familiarity with the instrument itself. Most instruments operate on batteries, and if they are not in good condition the instrument is useless. It is also an excellent idea to have a check source to test instrument function. Use of the check source *is not* a calibration; it is only a test confirming that the instrument is responding reproducibly to radiation. It serves also as a check of constancy and can alert the user to a failing instrument. When making measurements of direct radiation, one should sweep the area with the detector probe slowly. Most instruments have electrical time constants of from several to ten seconds. Thus, if the area to be measured is traversed too quickly, it is possible not to detect radiation present. If a constancy check source is used routinely, one can estimate the time necessary for a measurement by observing how long it takes for the meter to reach its final value when tested.

Now that the method of measuring has been discussed, we will examine acceptable radiation levels. There are two cases for acceptable levels—areas restricted for the purpose of radiation protection and unrestricted areas. Ideally no radiation levels in any areas are best. In practice some levels may have to be present in order to work with the materials. The Nuclear Regulatory Commission in Title 10 of the Code of Federal Regulations Part 20 (10CFR 20) sets forth the maximum radiation levels in unrestricted areas and the maximum total dose that may be received by a worker in a restricted area (work area) in any calender quarter. These limits are based on recommendations of the National and International Commissions on Radiation Protection (NCRP and ICRP). The commissions have determined that from the available data no immediate and very few latent effects will be observed if annual doses are limited to 5 rem for the whole body, blood-forming organs, and eyes. In practice 5 rem/yr is well in excess of the doses typically found in most laboratories. It is relatively easy to maintain whole-body doses to 0.5 rem/yr or less in most biologic laboratories. Where $^{32}$P is used in 10–50 mCi quantities routinely, such as in studies of ATP, annual whole-body doses may reach 1–2 rem and it is possible to find areas in the laboratory with radiation levels of 20–50 mrem/hr. The methods of exposure control discussed in previous sections can be applied after surveys are made to keep doses ALARA.

*2. Surface Contamination:* Such contamination may result from spills or the settling of aerosols and dusts containing radioactive material. Surface contamination may be resuspended in the air by air movement or personnel rubbing against the surfaces. Airborne radioactivity may be inhaled or ingested. Surface contamination may be transferred to the skin and absorbed or ingested. Thus, surface contamination is indicative of potential internal uptakes of radioisotopes. It may be indicative of unknown spills or poor experimental techniques. Surface contamination may be fixed or transferable. The former may be measured by techniques used to measure direct radiation; the latter is that which may ultimately enter the body and is measured by wiping the surface in order to remove the material. The wipe (a filter disk, cotton swab, etc.) is then counted with a suitable instrument to assay the material. Wipe materials are variable. Hard filter paper disks or glass fiber filter paper commonly found in the laboratory are best choices. Using the filter with moderate pressure, about 100 cm$^2$ of the surface is wiped. Wipes may also be used to test tools such as tongs, test tube racks, and machinery such as centrifuges. The wipe may be counted by holding it up to a portable GM survey meter, but the limit of detectability is poor. For laboratories having a liquid scintillation or $\gamma$ counter the wipe may be counted as would the radioisotope in use. Low background counters are also used, as are shielded GM tubes coupled to scalers. There are no federal regulations at the present time that address transferable contamination. Recommended acceptable contamination levels are 10 disintegrations per minute (dpm) per 100 cm$^2$ wiped for $\alpha$ emitters and 100 dpm/ 100 cm$^2$ wiped for $\beta$-$\gamma$ emitters. The Nuclear Regulatory Commission's guidance allows contamination levels 10-fold higher. Again ALARA should be observed. When unacceptable contamination levels are found, the lab worker should clean the area. Cleanup may be done with soap and water, foam cleanser, or solvents in which the labeled compound will dissolve. Care should be taken to avoid spreading contamination while cleaning. The worker should not forget to wear gloves and lab coat. The question of how often wipes should be taken arises. This depends on the scope of radioisotope use. The individual experimenter should make one or two wipes each day that radioisotopes are used. Once a week or month, depending on the volume of radioisotope work, a more extensive wipe survey should be conducted. For this purpose it is useful to draw a map of the laboratory noting where wipes have been taken and writing down the contamination levels above background. The map should be kept in a record notebook.

*c. Airborne Radioactivity:* Wiping surfaces to check for contamination helps in checking for airborne radioactivity. In some situations airborne activity does not result in subsequent surface contamination. Such is the case with radioiodine, $^{14}CO_2$, $^{35}SO_2$, and airborne activity in hood exhausts to unrestricted areas. When airborne radioactivity measurements are appropriate, it is common to draw air samples through filters or traps. The air flow rate in the sampler and collection efficiency must be known. Filters or traps are counted and the total collected activity that passed through the collection medium determined. The average air concentration is calculated from the following equation:

$$A = \frac{C - B}{\lambda E_s F T E_c \,(2.22 \times 10^6 \text{ dpm}/\mu\text{Ci})} = \mu\text{Ci/ml} \qquad (1)$$

where $A$ = activity concentration in $\mu$Ci/ml, $C$ = sample count rate, $B$ = background count rate, $\lambda$ = the decay constant of the radioisotope in use, $E_s$ = counter efficiency (obtained by using a suitable counting standard), $F$ = air sampler flow rate in ml/min, $T$ = air sampling time in minutes, $E_c$ = air sampler collection efficiency (data from manufacturer or literature), and $2.22 \times 10^6$ = conversion from dpm to $\mu$Ci.

The value obtained from this calculation is compared with acceptable limits found in regulations or other guides. The limits for airborne activity are published in the Nuclear Regulatory Commission's 10 CFR 20 Appendix B and in equivalent state statutes. Limits for occupational exposures are set, so that an individual present at the given concentration for 40 hours per week, 50 weeks per year will receive an annual dose to the critical organ less than or equal to that set by ICRP and NCRP. The limits for airborne activity concentrations in unrestricted areas are at least 10 times lower than for restricted areas. For most laboratory uses of radioisotopes, the regulatory limits are considerably higher than readily achievable by using well-designed apparatus and techniques. Not all uses of radioisotopes in unsealed form require airborne radioactivity measurements. Often it is possible by calculation to demonstrate that there is no credible way in which airborne activity limits will be exceeded. This may be due to the non-volatility of the compound in use, the very low radioactivity present, or the nature of the experimental setup. When such evaluations are made they constitute an acceptable survey in lieu of a physical measurement.

*4. Effluents and Releases to Unrestricted Areas:* The techniques for measuring airborne radioactivity in releases to unrestricted areas are identical to those mentioned earlier. Samples may have to be larger but methods of collection, counting, and calculation do not differ. Liquid effluents arise from wash water, excreta from animal experiments, and liquid wastes. Samples may be collected continuously by tapping a main release point or grab samples may be taken as needed. Analysis for the most frequently used radioisotopes in biological laboratories is usually by liquid scintillation counting methods or planchet counting with GM or proportional counters. In the latter case samples are usually evaporated to dryness in the planchet if they do not contain volatile radioisotope-tagged components. Calculations may suffice and physical measurements may not be necessary where release rates, dilution, and occupancy factors are well known or can be easily established.

*5. Bioassays:* The analysis of radioisotopes in the body by *in vitro* counting of body fluids, breath, or blood, or by *in vivo* counting of portions of the body constitutes bioassays. Bioassays have a threefold use: to determine the deposition of specific radioisotopes within the body, to calculate the resulting dose to organs, and to evaluate the presence of radioisotopes in the environment that resulted in the uptake in the body. To perform a satisfactory bioassay one must know the radioisotopes in use, chemical and physical form, and metabolism within the body. Metabolic data must include concentration factors in specific organs, and release, exchange, and excretion functions. These latter factors are summarized by the biological half-life for a particular compound. The ICRP Publications 2, 10, and 10A provide excellent sources of information on metabolism and distribution functions that allow back calculation to air and water concentrations from organ burden and excretion data. The Medical Internal

Radiation Dose (MIRD) Committee of the Society for Nuclear Medicine has published two fine companion pamphlets (Nos. 10 and 11) that give radioisotope decay schemes and a procedure for calculations of internal organ doses from deposited radioisotopes.

The bioassays most commonly performed in biological laboratories are for $^3$H, $^{14}$C, and $^{32}$P in urine; $^{14}$C as $Co_2$ in exhaled breath; and thyroid counting for $^{125}$I and $^{131}$I. The techniques for urinanalysis for $^3$H, $^{14}$C, or $^{32}$P require that one know or assume a fractional excretion rate via urine for the particular tagged compound. Raw urine may be counted using liquid scintillation techniques. Quenching (the reduction in light output due to color centers introduced into the sample) may be severe for samples containing greater than 20% urine by volume. In such cases clarification by passing urine over activated charcoal is acceptable if none of the tagged compound is removed. This technique or distillation works well where tritiated water is the compound being analyzed. Samples of urine may be evaporated in planchets and counted by GM or proportional counters for $^{32}$P. Whenever one performs liquid scintillation or planchet counting, it is best to count replicate samples and blanks and to count spiked samples (samples with known radioactivity added) to obtain counting efficiencies. $^{14}CO_2$ may be analyzed by exhaling through Hiamine hydroxide solution. The volume of exhaled air must be known as well as absorption efficiency for the Hyamine hydroxide [p-(diisobutylcresoxyethoxyethyl)dimethylbenzylammonium hydroxide]. This technique is useful when $^{14}CO_2$ is formed as the result of the uptake of some other $^{14}$C precursor of $CO_2$. In most biological laboratories $^{14}$C is not used in sufficient quantities to warrant such elaborate bioassays. Radioiodine bioassays are discussed specifically in the last section of this chapter. The need for and frequency of bioassays is determined by the following factors: nature of tagged compound (volatility, skin absorption, etc.), level of radioactivity in use, nature of the experiment, availability (or unavailability) of other types of surveys, including calculations to demonstrate that exposures are within acceptable limits. When one is not sure whether or not to perform a bioassay, it is best to do one instead of being unable to obtain a proper sample in the future.

*6. Inventories and Accountability:* One last method regarding principles of exposure control is worthy of note. The making of routine physical inventories of radioactive materials on hand is one mechanism by which an experimenter can help to keep track of the locations and uses of radioisotopes. It is wise to maintain a log book to record receipts, use, and disposal of radioisotopes. Figure 2 is a sample inventory form that may be used to maintain receipt, check-in, use, release, and disposal records in a loose leaf binder. One page is used for each separate compound. It is also

Supervisor _____  Identification # _____

Location of Isotope _____  Isotope _____

Dept. & Order No. _____  Activity _____

Comments _____  Date Received _____

Chemical Form _____  Supplier _____

## RECEIVING CONTAMINATION SURVEY

Surveyed By _____ Date _____

Counted By _____

Transferrable Contamination Per Smear

| No. | d/m | Location |
|-----|-----|----------|
|     |     |          |
|     |     |          |
|     |     |          |
|     |     |          |
|     |     |          |
|     |     |          |
|     |     |          |
|     |     |          |
|     |     |          |
|     |     |          |
|     |     |          |
|     |     |          |
|     |     |          |
|     |     |          |

**SPECIAL INSTRUCTIONS**

ENTER BELOW WHEN ANY MATERIAL IS USED, PLACED IN RADIOACTIVE WASTE CONTAINERS, OR RELEASED TO SINK DRAIN OR ATMOSPHERE. RETURN COMPLETED FORM TO THE HEALTH PHYSICS OFFICE WHEN ALL RADIOACTIVE MATERIAL HAS BEEN USED, DISPOSED OF AS RADIOACTIVE WASTE, TRANSFERRED TO ANOTHER ISOTOPE USER OR HAS DECAYED TO INSIGNIFICANT ACTIVITY. (ALL TRANSFERS MUST BE APPROVED BY THE UNIVERSITY HEALTH PHYSICIST).

| Date | Amount | Comments | Entry By |
|------|--------|----------|----------|
|      |        |          |          |
|      |        |          |          |
|      |        |          |          |
|      |        |          |          |
|      |        |          |          |
|      |        |          |          |
|      |        |          |          |
|      |        |          |          |
|      |        |          |          |
|      |        |          |          |

## SUMMARY

Date _____

Activity released to hood _____

Activity released to sewer _____

Activity to radioactive waste containers _____

**Fig. 2.** "Radioisotope Inventory Accountability Form" courtesy The Pennsylvania State University Health Physics Office.

recommended that stock radioisotopes be given identification numbers written on the vial labels that can be matched to numbers on the inventory/accountability form. When radioactive material is accounted for, users can be one step closer to being sure that unintentional exposures or releases did not occur.

## V. REGULATIONS

### A. Sources of Regulations for Ionizing and Nonionizing Radiation

#### 1. Nuclear Regulatory Commission (NRC)

The NRC has been given regulatory authority over by-product, source, and special nuclear materials. The authority for by-product material regulation may be delegated to state governments by agreement. By-product material refers to any radioactive material (except special nuclear material) yielded or made radioactive as of the result of the fissioning or fabricating of plutonium and $^{233}U$ and $^{235}U$. Thus radioisotopes produced by accelerators such as $^{57}Co$ and those found in nature and not also produced by using plutonium or uranium are not under the NRC's authority. Source material means uranium or thorium as metals, as certain ores, or in combination. Special nuclear material refers to plutonium, or the fissionable isotopes of uranium ($^{233}U$ and $^{235}U$). For individuals involved in typical biological research most radioisotopes used will be by-product material (NRC regulated) or accelerator produced (not regulated by NRC). Some states, but not all, have elected to regulate use of certain naturally occurring (mainly $^{226}Ra$) and accelerator produced radioisotopes.

#### 2. The Food and Drug Administration (FDA)

The FDA, through the Bureau of Radiological Health (BRH), has been given regulatory authority over machine-produced radiation, including diagnostic and theraputic X-ray machines, lasers, accelerators, and the like. BRH sets performance standards for equipment makers; individual states regulate individual users.

#### 3. The Environmental Protection Agency (EPA)

The EPA has a twofold responsibility regarding radiation. First, it sets standards for radioactivity and radiation in the environment; second, it establishes guidelines for emissions and concentrations that are submitted to the president for approval and then are to be used by all federal agencies.

## 4. The Department of Transportation (DOT)

The DOT regulates the interstate transport of radioactive materials. DOT regulations may be found in Title 49 of the Code of Federal Regulations (49CFR); NRC regulations are found in 10CFR or agreement state statutes. Figure 3 indicates the organization of the United States according to the five regions of the U.S. Nuclear Regulatory Commission.

## B. Scope of Regulations for Radioactive Materials

The NRC regulations contained in 10CFR and, where applicable, agreement state regulations are the most often encountered by radioisotope users. Table VII lists the particular aspect of control involved in using byproduct material and references the applicable sections in 10CFR to consult for details. The individual in charge of a laboratory using radioisotopes or the facility or institution radiation safety officer should be familiar with the applicable regulations and enforce them.

This text does not provide complete regulatory requirements for users of radioactive materials. The text does provide guidance that if followed will lead to compliance with the radiation protection aspects of NRC and agreement states.

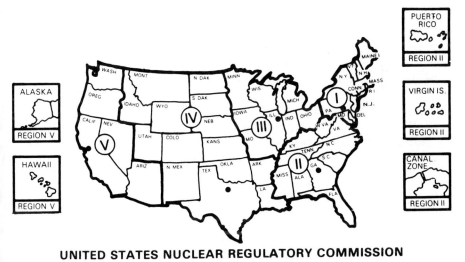

## UNITED STATES NUCLEAR REGULATORY COMMISSION

**Fig. 3.**   Organization of the United States for the Regulation of By-product Materials.

TABLE VII

**Major Regulations Affecting Use of By-product Material Use**

| Activity | Appropriate regulation |
|---|---|
| Licenses for use of by-product material | 10CFR 30 |
| Radiation protection | 10CFR 20 |
| Dose Limits | 10CFR 20.101 |
| Concentrations of by-product material allowed in work areas | 10CFR 20.105 |
| Concentrations of by-product material allowed in public areas | 10CFR 20.106 |
| Conduct of surveys | 10CFR 20.201 |
| Signs, posting, labels | 10CFR 20.203 |
| Receipt of by-product material | 10CFR 20.205 |
| Disposal of by-product material | 10CFR 20.301-.305 |
| Human use of by-product material | 10CFR 35 |
| Transportation | 49CFR 173-199 |

## VI. INSTRUMENTATION

### A. Instrument Functions

Instrumentation used in radiation safety and research with radioisotopes may be classified into two broad categories: (1) collection and detection and (2) identification and quantification. The first category encompasses qualitative analysis equipment; the second, quantitative. Table VIII provides a listing of common radiation research instrumentation found in each of the two categories, their applications and limitations, and typical costs. Note that much of the qualitative instrumentation tends to be portable, whereas most quantitative instrumentation is fixed equipment.

There are several sources of information on instrument characteristics and performance. These include the Lawrence Berkeley Laboratory report "Instrumentation for Environmental Monitoring" (1973), instrument manufacturer's catalogues and data sheets, and the annual buying guide issues of the journals *Science* and *Nuclear News*.

### B. Collection and Detection

The most common laboratory radiation detector is the Geiger-Müller counter (GM). The GM is able to detect medium energy $\beta$ radiation, X rays, and $\gamma$ radiation. There are numerous designs available, including detectors able to discriminate $\gamma$ radiation in the presence of $\beta$ radiation. GM counters are rugged and inexpensive as well as being very sensitive. They

**TABLE VIII**

**Comparison of Common Laboratory Radiation Counting an Detection Instruments**

| Instrument | Radiation detected, use | Detection efficiency (%) | Qualitative, quantitative | Price range ($) | Advantages, limitations |
|---|---|---|---|---|---|
| Geiger | | | | 300–600 | Low cost, rugged, reliable, retains calibration over long periods, may be saturated at high dose rates. May be directionally dependent |
| End window | $\alpha$, $\beta$, $\gamma$, X ray | 1–7 | Qualitative | | |
| No window | High-energy $\beta$, $\gamma$, X ray | 0.5–7 | | | |
| | $\gamma$, X ray | 0.5–7 | | | |
| Sodium iodide | | | | 400–3000 | High efficiency, $\alpha$ and $\beta$ discrimination. Allows $\gamma$ spectroscopy when coupled to single or multichannel analyzer, Fragile, Often needs expensive auxiliary equipment |
| Thin crystal | Low-energy X and $\gamma$ ray (<−1 MeV) | 3–15 | Qualitative | | |
| Large crystal | High-energy X and $\gamma$ ray (> 0.1 MeV) | 3–10 | Quantitative | | |
| Well crystal | All energy | 15–35 | | | |
| GeLi | All energy X and $\gamma$ rays | 0.1–3 | Quantitative | 4000–100,000 | Very good resolution. Low efficiency, very costly, much attention to routine maintenance |
| Gas Flow Prop | | | | | Very flexible. Wide range of costs. Detector $\alpha$, $\beta$, $\gamma$, X ray. May be used with automatic counter. Requires continuous supply of counting gas |
| Manual | $\alpha$, $\beta$, $\gamma$, X ray | 5–60 | Quantitative | 1000–10,000 | |
| Auto | $\alpha$, $\beta$, $\gamma$, X ray | 5–60 | Quantitative | 1000–30,000 | |
| Portable | $\alpha$, $\beta$, $\gamma$, X ray | 1–10 | Qualitative | 1000–2000 | |
| Ion chamber | All $\beta$, $\gamma$, X ray | 3–15 | Quantitative | 600–4000 | Simple, accurate. Often heavy and bulky. Small beams may not be detected well |

may be used to detect contamination as well as direct radiation and, if properly calibrated, are suitable for measurement of radiation fields. For laboratories using radiation that is detectable by a GM counter, the thin, end window type is most flexible. With an end window detector low-energy $\beta$ rays such as those from $^{14}C$ and $^{35}S$ may be detected. The GM is useful for checking for contamination on surfaces (floors, benches), equipment, and personnel (hands, feet, clothing, body). Many GM counters are equipped with audible speakers that give an indication of the presence of radiation and its relative intensity. This feature frees the user from having to watch the meter constantly while making a survey. There are other instruments for detection of radiation, including proportional counters and small sodium iodide detectors. These are more costly than GM detectors generally but have advantages as very high probe area for surface contamination measurements and greater sensitivity for very small levels of radiation. Certain radioisotope users will need unusual or custom instruments for radiation detection; in such cases instrument manufacturers are excellent sources of information.

When airborne radioactivity is present, two types of collection apparatus are generally used: body-worn breathing zone samples and high-volume air samples. For both types of instruments air is pulled through a filter, activated charcoal, silica gel, water, or other material. The collection medium is then analyzed for radioactivity so that the concentration of the radioactive material can be calculated. Air sampling instrumentation should be selected with due consideration as to the total volume that must be collected to achieve desired sensitivity, portability, use with multiple collection media, and ability to measure the air flow rate or total flow through the collection medium.

Wipes, or smears as they are also identified, are used for collecting transferable surface contamination. Smears may be paper, cloth, cotton balls, and the like. Smears are generally counted in fixed, low-background instruments described below.

## C. Identification and Quantification

Fixed instrumentation, such as liquid scintillation counters, $\gamma$ counters, low-background counters, large ion chambers, and multichannel analyzers, is used for accurate analysis of the activity present as well as to identify the type of radiation present. Liquid scintillation counters are most useful for $\beta$ detection and are a virtual necessity where tritium is in use. Care must be taken when using the flammable and toxic solvents used in liquid scintillation counting. Samples should be prepared in well-ventilated areas. Some counting vials, such as high-density polyethylene, are easily permeated by the toluene, xylene, and benzene solvents in use. Glass and polycarbonate vials are much less permeable. Used scintillation vials should be disposed of in containers stored in well-ventilated areas. Obviously, the highly flammable solvents should be stored in approved fireproof cabinets. Sodium iodide detectors are found in $\gamma$ counters; they are common in those laboratories where $^{125}I$ is used in radioimmunoassay procedures. Both sodium iodide (NaI) and lithium-drifted germanium (GeLi) detectors are used in combination with multichannel analyzers for identification and quantification of $\gamma$-emitting radioisotopes. When only one $\gamma$ emitter is in used, a NaI detector coupled to a single-channel analyzer is a reasonable quantitative instrument. NaI and GeLi instruments are expensive; GeLi detectors must be kept at liquid nitrogen temperatures. When routine samples of filters, chromatograph sections, and the like, must be counted, automatic multisample, low-background, gas flow proportional counters are very useful. Refer to Table VIII for a comparison of the advantages and limitations of the instruments discussed here.

## VII. PERSONNEL MONITORING: PURPOSE AND NEED

Body-worn devices are designed to provide data from which one may calculate rather accurately the dose an individual may have received. Such devices may also be placed on objects in radiation areas to assist in evaluating the 24-hr, "continuous-presence" dose. The data obtained from such devices may constitute a legal record of doses received by the individual wearing the device. The use of such dosimeters is required by law for certain radiation workers. NRC regulations require that any individual working with licensed radioactive material must be provided with personnel monitoring devices if that individual is likely to receive more than 25% of any dose limit specified in the regulations. The regulations of the NRC and individual states are minimum requirements; there is nothing to prevent radiation workers from using personnel monitoring devices when dose levels below those specified in regulations will be encountered. There are situations, however, when wearing currently available personnel dosimeters will be of no value. Such is the case with users of $^3H$, $^{14}C$, $^{35}S$, $^{63}Ni$ and other low-energy $\beta$ emitters, which are not detected. Most $\alpha$ sources are also sources of X rays and $\gamma$ rays that are detected.

An ideal personnel dosimeter should be small; absorb radiation in a manner similar to body tissue; respond linearly with increasing dose; be inexpensive and unaffected by heat, cold, or other environmental stresses; respond to the types of radiation being used; and be free of energy-dependent characteristics. It is not surprising that finding a suitable dosimeter for a particular application takes some care. There are at the present three major types of personnel dosimeters, each with its good and bad points. At the present time there is no universally applicable commercially feasible personnel dosimeter. The three types of dosimeters are photographic film, thermoluminescent solid-state devices, and pocket ionization chambers. For most situations researchers and laboratory workers will find that commerical dosimetry services provide the best compromise between cost, reliability, manpower to operate, and convenience.

## VIII. GENERAL LABORATORY PROCEDURES FOR RADIOISOTOPE WORK

### A. Basic Guidelines

The following guidelines are essential for the health and safety of radiation workers in biological laboratories:

1. Mouth pipetting should never be allowed. Mouth pipetting of radioactive solutions may cause internal uptakes because of accidental swallowing or carrying of radioactive vapors into the lung. Contaminated pipettes may transfer radioactive material to the body.
2. Eating, drinking, or storage of food containers in radioisotope laboratories should not be allowed. Reasons as above.
3. Untrained individuals should not work with radioisotopes unless under direct supervision and in physical presence of an experienced individual.
4. Aid should be available when large quantities of radioisotopes are being used.
5. Protective clothing, including gloves and eye protection, should always be worn.
6. Where volatiles are in use, work should be done in a hood.
7. A "cold" run (no radioisotopes) of a new experiment should be conducted to detect possible difficulties.
8. The most effective dose reduction devices—time, distance, shielding, containment—should be used first.
9. Radiation and/or contamination surveys should be conducted as appropriate to confirm safe situations and to demonstrate compliance with regulations.
10. Survey instruments should be used to check hands and clothes for contamination at end of work (if radiation is suitable for detection).

## B. Emergency Spill Procedures

Whenever a spill occurs, utmost effort should be directed at minimizing personnel exposure and contamination and containing the spread of contamination. When spills are associated with other catastrophes, such as fire or explosion, radiation hazards are usually exceeded by other hazards or the need to care immediately for the injured. For facilities where radiation hazards must take prime consideration over other injuries or hazards, professional radiation safety experts will be available and their qualifications and duties will be specified in the licenses for use of radioisotopes.

The following procedures will be useful in dealing with spills or accident situations.

1. Emergency Priorities
    a. Save lives.
    b. Control severe injuries
    c. Prevent spread of major fire.
    d. Ensure safety of personnel from radiation.
2. Emergency Spill Procedures
    a. Stop all work.
    b. Remove all personnel from immediate spill area to one safe meeting location in or near lab.
    c. Shut off ventilation, close windows and doors, turn off hoods if possible. Do not do this if a radioactive gas is involved, as release to the environment is preferable in that case.
    d. Check all personnel for skin and clothing contamination.
        (1) For $^3H$ wipe skin and clothes count wipes in liquid scintillation counter.
        (2) For other isotopes detectable by portable survey meters check skin, clothing, shoes for contamination.
        (3) Remove contaminated clothing.
    e. Decontaminate personnel and resurvey; see decontamination procedures.
    f. Survey spill area by wipes for $^3H$ or portable survey instrument, mark contaminated areas with grease pencil or magic marker.
    g. Begin facility decontamination efforts. Start with low-level contamination, work toward center or high contamination areas. Wear gloves and protective clothing. For surface decontamination use soap and water and cleansers appropriate to the compound spilled. Do not use scrub brushes, which will raise droplets and spread contamination. Work slowly, deliberately. Survey continuously. Avoid volatile cleaners if they cause airborne activity.

h. Call a radiation safety professional when one is available; always call whenever there is a spill. He can evaluate the spill and decide whether the lab workers can conduct clean-up efforts. When the spread of contamination is extensive and would result in exposure of the general public, aid should be sought. It may be necessary to seek assistance from state radiation control officials. For facilities licensed and regulated by the NRC, the appropriate regional office should be contacted for advice as to sources of health physics assistance.

i. Follow instructions for removal of radioactive contamination from skin. Adhere to the following procedures in the order given, being careful not to scratch, abrade, or erode skin. Apply lanolin or skin lotion to prevent chapping. Clean areas with the most contamination first, protecting uncontaminated areas with tape or plastic sheet. Skin breaks that are uncontaminated should be covered with a waterproof dressing.

(1) Remove all loose particulate material by gentle brushing, wiping, or use of adhesive tape. Perform over sink, catch basin, or plastic bag. Survey.

(2) Wash with water only. Survey.

(3) Wash with mild soap and water, using a thick lather. A *soft* scrub brush may be used if needed. Use a stiff brush for fingernails and callous areas. Rinse and survey.

(4) Scrub with a paste of 50% commerical laundry soap and 50% cornmeal or use a commercially available skin decontamination foam. Survey.

(5) Discontinue any step when less than 5% of the contamination is removed per washing. Place all contaminated wipes, towels into a radioactive waste container. If wounds are caused by contaminated items and bleeding is not profuse, allow bleeding while flushing with clear water.

(6) Use a decontamination kit including
1 bar soap (Ivory or other mild type)
1 bottle germicidal soap (Phisohex or equivalent)
1 soft hand brush
1 stiff hand brush
1 stiff fingernail brush
1 jar 50% laundry soap (Tide) and 50% cornmeal
5 paper towels
1 bag cotton balls
25 gauze pads $2\frac{1}{2} \times 2\frac{1}{2}$ in.
1 roll 1-in. adhesive tape

1 wash basin
1 bottle skin lotion
1 box facial tissue
3 five-gallon plastic bags
1 copy of instructions

Place the items in a clearly marked box. Make sure that all lab workers know the kit location. The instruction sheet should have any appropriate emergency telephone numbers. This kit should not be used for routine cleanup. Make a note on a reminder calendar to check the box twice yearly.

Table IX can be seen as a general laboratory procedure guide for radio-isotope work. Table X provides general guidelines on the acceptable limits for surface contamination on various items including skin.

## IX. SPECIFIC PROCEDURES FOR USERS OF LARGE QUANTITIES OF $^3$H, $^{32}$P, and $^{125}$I

### A. $^3$H (Tritium)

#### 1. Levels Requiring Special Care

Most research with tritium ($^3$H) involves amounts of several microcuries to 10 mCi at any one time. Very often tritiated precursors such as acetic anhydride and sodium borohydride are sold commercially in lots of 25, 50, and 100 mCi. Where no volatile compounds or processes are involved, several mCi of $^3$H may be handled with relative ease. As the specific activity and total activity increases, however, physical properties, such as vapor pressure at room temperature, become sources of measurable and trouble-some levels of airborne $^3$H. When such levels are present, the additional precautions that follow are in order.

To give an example of a problem in handling $^3$H, consider tritiated water (HTO) in a 1-ml solution containing 1 Ci. At 22°C the vapor pressure of water is 19.8 mm Hg. If 1 Ci of HTO is in a bottle with 1-ml air space, then at 22°C, nearly 20 $\mu$Ci is present as vapor. Simply opening the vial in the open laboratory will result in an uptake of $^3$H by inhalation. Tritiated water is also absorbed through the skin readily, so uptakes are not unusual. Continued use of the open vial will allow more $^3$H into the lab air. Boiling or vigorous reactions, especially those intended to release $^3$H for binding to other molecules, will result in considerable airborne $^3$H. It has been observed that great care must be taken to avoid internal uptakes when using

## TABLE IX

### Recommended Laboratory Procedures at a Glance

| Protective Techniques[a] | Tritium | Low-energy $\beta$ radiation |
|---|---|---|
| **Containment** | | |
| Gloves | For greater than 10 $\mu$Ci amounts | For greater than 10 $\mu$Ci amounts |
| Lab coats, drip trays, covered benches | For 50 $\mu$Ci quantities and greater | For 50 $\mu$Ci quantities and greater |
| Glove boxes | Millicurie quantities of volatile compounds. For 25 mCi or more of tritiated water | For volatile compounds as dictated by the nature of the experiment Powered materials or dusts |
| **Ventilation** | | |
| Fume hoods or dry boxes | For volatile compounds. When 100 mCi or more is used at one time in volatile form | Volatile materials |
| **Shielding** | None needed | 1-5 mm lucite or aluminum |
| **Eye protection** | Normal chemistry laboratory eye protection | Normal chemistry laboratory eye protection |
| **Personnel Monitoring Devices** | | |
| Whole body | None required | None required |
| Extremity | None required | None required |
| **Survey Techniques** | Wipe work areas, floors; count in liquid scintillation counter. Benches should not exceed 100–200 dpm/100 cm². Airborne H-3 can be collected using water bubblers with 50% efficiency dry ice/acetone cold traps or liquid nitrogen cold traps with 100% efficiency unless ³H₂ is in use | Wipe work areas, tools, floor. Wipes can be counted in liquid scintillation counter or gas flow proportional counter. Same contamination limits - 100–200 dpm/100 cm-2 - as for tritium. Airborne activity can be collected in water bubbler, impinger, glass fiber filter or by chemical bubbler specific for the labeled compound in use |
| **Survey frequency** | Weekly for less than 1 mCi. Daily for 1–5 mCi. After each use for more than 5 mCi | Weekly for up to 1 mCi. Daily for 1–10 mCi, after each use for more than 10 mCi |
| **Waste disposal** | Plastic bags for nonliquids. Absorb liquids onto kitty litter; then transfer to metal cans | Heavy cardboard box or barrel for nonliquids, line the box with a heavy plastic bag. Liquids should be absorbed onto kitty litter which is placed in metal cans |
| **Bioassays** | Weekly urine counts for users of 10 mCi or more at any one time. Use liquid scintillation counting | Need and frequency must be based on potential for uptake on case by case basis |

[a] The protective techniques listed in this column correspond to the quantities of radioisotopes in the columns to the right.

| P-32 | Radioiodines | γ Emitters |
|---|---|---|
| For greater than 1 μCi amounts<br>For 10 μCi quantities and greater | Gloves should be used for all work with radioiodine except manufactured kits containing less than 1 μCi total Lab coats are essential | For quantities greater than 1 μCi in unsealed form<br>For quantities greater than 1 μCi in unsealed form |
| Only when volatile compounds are used | Dry boxes with activated charcoal trap ventilation for use in protein iodination | When dusts, aerosols, chips may be encountered |
| When aerosols, volatile materials, or strong boiling is being used | For all protein iodinations. Activated charcoal traps required. Filtered dry boxes within fume hoods are recommended | Need will depend on volatility or likelihood of creating airborne activity |
| 1-2 cm of lucite on front and sides of experiments using more than 1 mCi | A 1/16 in. lead sheet reduces dose rate by a fractor of 1000 for ¹²⁵I, ½ in. lead reduces radiation intensity from ¹³¹I by a factor of 30 | Lead sheet or bricks are most effective; enough shielding is required to reduce the maximum total dose to the whole body of the worker to 100 mrem per week. It should be possible to reduce weekly doses to 10-20 mrem with great ease |
| Plastic goggles or safety glasses when using 100 μCi or more in the open | As needed, based on chemical hazard | As needed, based on chemical hazard |
| Yes, for 10 μCi or more<br>Yes, for 1 mCi or more | Yes, for all uses<br>When performing hand operations near the source | Yes, for all uses<br>When performing hand operations near the source |
| Survey benches floors with GM survey meter at least once per day when material is in use and prior to leaving area. Take wipes for counting on proportional counter or liquid scintillation counter. Use the GM survey meter with an end window probe and with it check hands and clothing. Most uses of this isotope do not result in airborne hazards | GM or portable sodium iodide crystal surveys of benches, floors, tools, hands, clothing. Wipes counted on liquid scinitllation counter or γ counter daily when using 5 mCi or more per day, or weekly if using less; also after each use. Contamination limits, 500 dpm/100 cm². Airborne activity surveys use 1-2 ft³/min air sample through charcoal trap in work areas and 20 ft³/min through charcoal in unrestricted areas. These are analyzed on the autogamma counter or on a multichannel analyzer | GM surveys of benches, floors, tools hands, and clothing daily. Weekly wipe surveys of lab area counted on gas flow proportional counter or autogamma |
| GM survey daily or after each use. Wipes weekly for less than 1 mCi; daily for 1-10 mCi. After each use check hands and clothing with survey instrument | As indicated above | As indicated above |
| Heavy cardboard boxes lined with plastic bags or metal drums for nonliquids. Absorb liquids onto kitty litter. ³²P waste may be held and allowed to decay to background levels, then surveyed to confirm background, and then placed in normal waste | Boxes, drums, or barrels lined with plastic Sealed containers in ventilated areas. Store liquids at basic conditions in fume hoods | Boxes, drums, or barrels Shielding may be required |
| Monthly for users of more than 25 mCi per week. Urine sample counted by liquid scintillation | Monthly thyroid counts using thin sodium iodide crystal with single or multichannel analyzer. Urine counting on autogamma is possible but less sensitive | When internal uptake is suspected, urine sample counted on sodium iodide or GeLi system. Whole body counting in unusual cases |

**TABLE X**

**Recommended Limits for Removable Surface Contamination**

| Type of surface | $\alpha$ Emitters[a] d/m per 100 cm² | High-risk[b] $\beta$-$\gamma$ emitters d/m per 100 cm² | Low-risk[c] $\beta$ emitters d/m per 100 cm² |
|---|---|---|---|
| 1. Unrestricted areas (areas open to public | 10 | 100 | 250 |
| 2. Restricted areas (laboratory rooms, floors, unsecured benches) | 25 | 250 | 500 |
| 3. Protective clothing and skin | 10 | 100 | 100 |

[a] d/m = disintegration per minute.
[b] High-risk $\beta$-$\gamma$ emitters include $^{32}P$, $^{33}P$, $^{51}Cr$, $^{59}Fe$, $^{65}Zn$, $^{99m}Tc$.
[c] Low-risk $\beta$ emitters are limited to $^3H$, $^{14}C$, $^{35}S$.

large quantities of $^3H$. Curie quantities of $^3H$ are encountered in many primary labeling processes in which high specific activity products are desired or in which labeling efficiency is low. In such instances $^3H$ is often used as a gas. Gaseous $^3H$ is readily converted to HTO when inhaled; therefore, its control is essential.

## 2. Containment

Work with curie levels of $^3H$ should be done in a glove bag or dry box. The bag or box should be used inside a fume hood or connected to it. Inlet air to the containment need not be treated; however, outlet air should pass through a prefilter, then a suitable $^3H$ trap, such as a water bubbler. Double gloves should be used. Arm shields are suggested where arms will be exposed to $^3H$-filled air. As with other uses of radioisotopes, drip trays are a must. Absorbent paper should be used to line trays and hoods. Lucite dry boxes are relatively inexpensive and are readily available. Suppliers of these new items may be found by consulting the buying guides of *Science, Nuclear News,* and the *Health Physics Journal.*

## 3. Handling Precautions

Large quantities of $^3H$ should be stored in double containers. The inner container holding the labeled material should be in a larger, tightly covered outer container that contains absorbent material, such as kitty litter or cotton balls. This will help prevent loss of material resulting from spills. Long-term storage of aqueous material may be done in freezers if no damage to the labeled material will ensue. Storage of double containers in fume hoods is also acceptable. When such materials are stored in refrigerators or freezers, it is normal for condensation and frost to become contaminated. These should be checked by smears of condensation or collection of frost with counting in a liquid scintillation counter.

Care should be taken to open stock solutions only in dry boxes or fume hoods. This will reduce the potential for inhalation. Tritiated water is readily absorbed through the skin, which as a pathway to body fluids is equal to inhalation. Tritium absorbed through skin or inhaled readily equilibrates in body water within 3–5 hr. To avoid contamination of glassware, disposable labware is recommended. Disposable beakers, pipettes, and tools should be placed in plastic bags that are tied or taped closed for disposal into normal radioactive waste. Ensuring a complete bagging of radioactive waste will minimize $^3$H vapors in the waste storage containers.

## 4. Ventilation

The rules for ventilation do not differ greatly from normal. Hoods should operate at 125–150 ft/min face velocity. Wherever possible, dry boxes should have separate exhaust blowers that exit directly into the fume hood duct. The flow rate of the hood should be adequate to ensure that the $^3$H concentration exiting the hood stack is below $2 \times 10^{-7}$ $\mu$Ci/ml.

## 5. Survey Techniques

Surface contamination is difficult to prevent when large quantities of even slightly volatile tritium-labeled compounds are used in hoods and dry boxes. Smears should be taken using glass fiber filter paper or styrofoam sheet squares. Gloves must be worn and the handling of smears with forceps is recommended. Counting by liquid scintillation techniques is advisable. Care should be taken to place smears into counting vials directly without touching the outsides of counting vials or gloves. Counting efficiencies of 25–40% are possible.

Airborne $^3$H is a distinct possibility. Air samples should be taken in the individual worker's breathing zone by one of two general methods: fixed air samples of 0.2–2 liter/min placed so the sample inlet is near the nose (within 1 m) and preferably between the source and nose; body-worn air sample with lapel-mounted inlet sampling at 250 ml to 2 liter/min. The exit of the sampler pump should be directed to a hood where possible. This is often not possible for bodyworn portable samplers. For most common $^3$H compounds a bubbler containing 100 ml deionized water with a sample flow of 0.2–1 liter/min is reasonable. Collection efficiencies are about 90%. This may be checked by using two bubblers in series. Care is needed to avoid foaming in the bubblers. Samples should be taken for the duration of the work or for at least 30 min. If one particular phase of an experiment is more likely than others to cause release of volatile compounds containing $^3$H, that phase should be monitored. Samples may be analyzed by taking an aliquot of the bubbler (1–5 ml) and counting in 10–15 ml of a liquid scintillation cocktail designed for use with water (such as NEN Aquasol™). A duplicate sample made with deionized water and "spiked" with about $10^{-2}$ ml of standard tritiated water (about 40,000 d/m is needed) can be used to establish counting efficiency. The airborne tritium concentration is obtained by applying Eq. (2)

$$[^3H] = \frac{N_s - N_b}{EfV(2.22 \times 10^6 \text{ dpm}/\mu\text{Ci})} \quad \mu\text{Ci} \quad (2)$$

where $N_s$ = count rate in counts/min of the bubbler sample, $N_b$ = count rate in counts/min of a blank sample containing the same volume of deionized water as $N_s$, $E$ = counting efficiency determined from the "spiked" sample, $f$ = fraction of the total bubbler contents counted, $V$ = integrated flow through the sampler in ml, $2.22 \times 10^6$ = conversion factor from disintegrations/min to microcurie, and $[^3H]$ = airborne concentration of $^3H$.

Surveys of hood exhausts to unrestricted areas may be necessary to demonstrate compliance with NRC or state regulations. Since concentrations of $^3H$ leaving the hood stack will be low, larger bubblers are needed. Bubblers of 1 liter with air flow rates of 2–5 liters/min are recommended. Sample counting and calculation of concentration of $^3H$ in the unrestricted air are the same as those presented above. Based on the results of contamination and airborne $^3H$ survey measurements, the user can take remedial action, including increasing total air flow in hoods, using more efficient traps on exhaust, or limiting work time with high-level sources.

## 6. Bioassays and Their Meaning*

Most users of large quantities of $^3H$ use the material as tritiated water (HTO) or other organic compounds that tend to deposit uniformly through the body or readily exchange with body water to result in uniform deposition throughout body water. Inorganic compounds such as sodium borohydride release tritium as a gas, which is rapidly converted to HTO. As mentioned earlier, HTO entering the body via inhalation or skin absorption equilibrates in extracellular body water in 3–5 hr. The analysis of HTO by liquid scintillation counting of urine is an easy, rapid, and reproducible method of evaluating exposure to environmental tritium. Calculations may be made to establish concentration of $^3H$ in air using assumption of

*References for *Tritium Bioassay Procedures:* Osborne, R. V. (1966). *Health Phys.* **12**, 1527; Recommendations of the International Commission on Radiological Protection (1968). Report of Committee IV, ICRP Publication No. 10. Pergamon, Oxford; The intake rate in liter/minute refers to the uptake by the body of all the tritiated water vapor contained in that volume of air in one minute; Recommendations of the International Commission on Radiological Protection (1960). Report of Committee II, ICRP Publication No. 2. Pergamon, Oxford.

breathing rate and skin absorption rate. The dose to the body if tritiated material is uniformly deposited in body water is approximately 190 mrem/mCi. Tritium has a biological half-life of 10 days in the "average" individual. Using the information as to date and time of exposure and date and time of urine sampling, assuming equal uptakes by skin absorption and inhalation, and assuming a 10-day biological half-life, one may use evaluation of $^3H$ in urine to "back-calculate" airborne concentration of $^3H$ to which an individual was exposed. The procedure and calculations are given below.

A sample of raw urine should be added to a liquid scintillation medium designed to count water. The urine may be clarified by distillation or by using charcoal, if this is necessary, but raw urine is preferred to eliminate the possibility of the loss of tritium incorporated in molecules other than water. The volume of sample that can be efficiently counted depends on the volume and type of scintillator used and ranges from 0.5 ml to 5 ml in a 15-ml scintillation cocktail. The count rate for this sample will be called $U_x$ in subsequent calculations.

Another sample identical to the first with the addition of 0.01 ml of calibrated $^3H_2O$ solution with activity $A$ d/m as an internal standard is used to determine the counting efficiency. The count rate of this sample is $U_s$ and the counting efficiency for urine $E_u$ is calculated as follows:

$$E_u = \frac{U_s - U_x}{A} \qquad (3)$$

The background is determined using a sample $W_B$ prepared with a volume of deionized water equal to the volume of urine used. A sample with an internal standard $W_s$ is likewise prepared. The efficiency for water $E_w$ is determined in a manner analogous to that for urine.

$$E_w = \frac{W_s - W_B}{A} \qquad (4)$$

The background for water will be different from that for urine because of the different amount of quenching. The urine background count rate $U_B$ is detemined using the water background and the efficiencies for water and urine.

$$U_B = W_B \ \frac{E_u}{E_w} \qquad (5)$$

If the scintillation counter used has a source for external quench correction, this method may be used to determine the quench of the water and urine samples to calculate $U_B$, but the method described is preferred.

The tritium concentration $C_t$ in $\mu$Ci/liter in the urine at time $t$ days after exposure is then determined as follows for a sample volume of $V$ ml.

$$C_t = \left(\frac{U_x - U_B}{E_u}\right)\left(\frac{1}{V}\right)\left(\frac{10^3 \text{ml}}{\text{liter}}\right)\left(\frac{\mu\text{Ci}}{2.22 \times 10^6 \text{ dpm}}\right) \mu\text{Ci/liter} \qquad (6)$$

The 95% confidence level ($2^\sigma$) is estimated as follows:

$$2^\sigma = 2E_u^{-1}\left(\frac{U_x}{t_x} + \frac{U_B}{t_b}\right)^{1/2}\left(\frac{1}{V}\right)\left(\frac{10^3 \text{ ml}}{\text{liter}}\right)\left(\frac{\mu\text{Ci}}{2.22 \times 10^6 \text{ dpm}}\right) \qquad (7)$$

$$2^\sigma = \left(\frac{9 \times 10^{-4}}{E_u V}\right)\left(\frac{U_x}{t_x} + \frac{U_B}{t_B}\right)^{1/2} \mu\text{Ci/liter} \qquad (8)$$

where $t_x$ and $t_B$ are the counting times for $U_x$ and $U_B$, respectively.

## 7. Dose Calculation

The tritium concentration in the urine at the time of uptake (assuming a single uptake $C_0$ is calculated using a biological half-life of 10 days.

$$C_0 = C_t e^{0.0693t} \mu\text{Ci/liter} \qquad (9)$$

Using the data of ICRP-10 the integral whole body dose for the standard man is calculated as follows:

$$D = \left(\frac{C_0\mu\text{Ci}}{\text{liter}}\right)\left(\frac{\text{liter}}{10^3 \text{ g}}\right)\left(\frac{14\mu\text{Ci day}}{\mu\text{Ci}}\right)\left(\frac{3.2 \times 10^9 d}{\mu\text{Ci day}}\right)\left(\frac{0.01 \text{ MeV}}{d}\right)$$

$$\left(\frac{1.6 \times 10^{-6} \text{ erg}}{\text{MeV}}\right)\left(\frac{\text{g} - \text{rem}}{100 \text{ erg}}\right) = 0.00717 \, C_0 \text{ rem} \qquad (10)$$

$$D = 7.17 \, C_0 \text{ mrem} \qquad (11)$$

Osborne measured a total intake rate of 15.8 liter/m for an unprotected person in an air volume contaminated with tritiated water vapor. Of this total, 9.6 liter/min was due to skin absorption. Therefore, in an atmosphere containing the maximum permissible air concentration ($MPC_a$) for tritiated water, the uptake for 1 hr would be

$$1 \text{ MPC}_a\text{-hr}\left(\frac{15.8 \text{ liter}}{\text{min}}\right)\left(\frac{10^3 \text{ cm}^3}{\text{liter}}\right)\left(\frac{5 \times 10^{-6} \mu\text{Ci}}{\text{cm}^3}\right)$$

$$\left(\frac{60 \text{ min}}{\text{hr}}\right) = 4.74 \, \mu\text{Ci} \qquad (12)$$

and 40 MPC-hr would give an uptake of 190 $\mu$Ci. The exposure to tritiated water vapor can be estimated by

$$\text{MPC-hr} = \left(\frac{C_0 \, \mu\text{Ci}}{\text{liter}}\right)\left(W \text{ liters}\right)\left(\frac{\text{MPC-hr}}{4.74 \, \mu\text{Ci}}\right) = 0.211 \, C_0 W \qquad (13)$$

The average airborne concentration can be estimated if the total exposure time is known

$$\mu Ci/cm^3 = \left(\frac{MPC\text{-}hr}{hr\ exposure}\right)\left(\frac{5 \times 10^{-6}\ \mu Ci/cm^3}{MPC}\right) = \frac{5 \times 10^{-6}\ MPC\text{-}hr}{hr\ exposure} \quad (14)$$

This bioassay procedure applies only to exposure to tritiated water vapor and may be in error for exposure to other tritiated compounds. Certain experimental conditions may also require a different estimate of the intake rate. However, for most instances, the preceding method is satisfactory for routine bioassay for tritium.

## B. Phosphorus-32

### 1. Hazards from External Sources

$^{32}P$ emits a $\beta$ particle with a maximum energy of 1.71 MeV. It has a range in air of about 6 m. This particular $\beta$ is sufficiently energetic to penetrate the epidermis and expose living tissue. The eye is a critical organ as well, since the $\beta$ can penetrate to the retina. In addition to the $\beta$ hazard itself, bremsstrahlung* is frequently found. Bremsstrahlung is created when the $\beta$ (a charged electron) is slowed down in absorbing medium. The more effectively (rapidly) this slowing down occurs, the more bremsstrahlung is produced. The characteristics of bremsstrahlung are similar to low-energy X rays. Shielding for $^{32}P$ is best designed by using a relatively low-density primary barrier such as Lucite or aluminum followed by iron or lead to absorb the small amount of bremsstrahlung produced. Sources of $^{32}P$ external to the body may be easily shielded and such is clearly advised. Table XI gives the dose rate per millicurie to skin from small volumes of $^{32}P$ aqueous solution.

**TABLE XI**

**Skin Dose per millicurie of $^{32}P$ as Function of Distance**

| Distance | Dose rate (mrem/hr) |
| --- | --- |
| 10 cm | 17 |
| 1 cm | 450 |
| 0.1 cm | 1000 |
| Contact with bare skin | $1-2 \times 10^4$ |

* Bremsstrahlung is electromagnetic radiation given off when a charged particle such as a beta experiences a change in velocity.

## 2. Hazards from Sources in the Body

$^{32}$P as phosphate or in similar groups tends to be absorbed in the bone. According to ICRP-10, nearly 70% of initially ingested $^{32}$P is excreted from the body within two days of uptake. A deposit of 10 $\mu$Ci in the bone will give a dose of 5 rem. Routes of intake are almost exclusively via ingestion, since most compounds of $^{32}$P are not volatile, nor are they absorbed through the skin. Ingestion usually comes about from contaminated skin to mouth. Although this may seem difficult to believe, careful observation of people reveals frequent hand-to-mouth contacts in a normal day. A finger contaminated with $^{32}$P may easily transfer material to the mouth while eating or even licking fingers while turning pages in a book.

## 3. Handling Procedures

Following these rules will minimize external doses and internal uptakes:

a. Use tongs, forceps, or other handling tools. Note the dramatic dose rate reduction as a function of distance.
b. Wear gloves as a form of containment.
c. Wear safety glasses or goggles to protect against chemical splashes and greatly reduce $\beta$ dose to eyes when using $^{32}$P in quantities of 1 mCi or more. Plastic goggles reduce the incident dose rate by a factor of approximately 2.
d. Conduct routine surveys for contamination using an end window GM survey meter. Check work areas, tools, floors, clothing, and skin.
e. Follow skin decontamination procedure if contamination is detected.
f. Lab bench shields of lucite 1-cm thick are very suitable. When shielding is used, place low-density material such as Lucite, Masonite, or aluminum nearest the source followed by thin iron or lead if needed.

## 4. Personnel Dosimetry

**a. Whole Body Badges.** For individuals working with 10 $\mu$Ci or more, whole-body film or TLD badges should be worn. Persons working with millicurie quantities should wear ring dosimeters as well. Wrist dosimeters are available but are of less value than finger dosimeters. Finger dosimeters should be worn with the sensitive portion facing into the palm when a fist is made. Certain situations may require that a small dosimeter be attached to safety glasses to evaluate eye dose. However, if the film or TLD badge is worn at chest or collar level eye dose may be calculated fairly easily.

**b. Bioassays.** Bioassays may be used as a form of internal dosimetry. Urine counting using liquid scintillation techniques is the preferred technique; however, evaporation of samples in deep planchets and proportional counting works well too. Users should consult ICRP-10 for equations and bioassay methods.

## 5. Waste Disposal

The collection and storage of $^{32}$P waste present serious problems. Many $^{32}$P experimental techniques in use today result in relatively large volumes of aqueous waste containing 50–100 mCi. The dose rates from bottles containing these solutions may be 100–500 mrem/hr near the surface. Spills of such bottles present a serious contamination hazard. It is recommended that liquid waste be stored in plastic bottles in deep plastic tubs behind suitable shielding. Waste containers should be kept in clearly marked areas away from personnel. Solid wastes (syringes, sample vials, ion exchange columns, pipettes) may contain large quantities of $^{32}$P also. Waste drums or boxes should be lined with heavy-gauge plastic bags. Waste containers may require shielding using building plywood and $\frac{1}{16}$-in. lead sheet barricades. The lead should face out to minimize bremsstrahlung. Liquid wastes may be stored for decay to background if convenient storage space is available. Prior to dumping into normal solvent waste or sanitary sewer, the contents of bottles must be checked by taking aliquots and counting. Release should not be made if concentrations of $^{32}$P are greater than twice background.

## C. Radioiodine

### 1. Radioactive Iodine Isotopes in Common Use

Two radioactive isotopes of iodine—125 and 131—are in use today, especially in routine and experimental radioimmunoassay (RIA) work. $^{125}$I is preferred because it has a relatively long 60-day half-life compared to 8 days for $^{131}$I and weak $\gamma$ and X-ray emission of 27–35 keV compared to a 806 keV $\beta$ and 364 keV $\gamma$ emission from $^{131}$I. From external radiation hazards $^{125}$I is clearly preferred over $^{131}$I.

### 2. Organ Specificity

Radioiodine is of particular concern in radiation safety because one small organ—the thyroid—concentrates approximately 30% of free iodine taken into the body. Stable, nonradioactive iodine has a lifetime in the thyroid of about 100 days. So radioiodine taken into the thyroid will cause a long-term dose. From an internal dosimetry standpoint, both $^{125}$I and $^{131}$I deliver approximately the same dose. This has been calculated at 5 rem/$\mu$Ci in the thyroid. Since such a large dose results per unit activity in the thyroid, great care must be taken to avoid exposure. To help keep exposures low, regulatory limits for radioactive iodines in the work area and environment are especially low. NRC regulations for $^{125}I$ allow $5 \times 10^{-9}$ $\mu Ci/ml$ and $8 \times 10^{-11}$ $\mu Ci/ml$ in air in restricted and unrestricted areas, respectively. With the advent of the very popular RIA techniques that require the liberation of free iodine in the labeling process, we have seen great efforts needed to minimize uptakes.

## 3. Sources of Internal Uptake

Iodine has a very high vapor pressure in solution. In very basic (pH 11) solutions measurable iodine is always present in the air above; in acidic solutions the iodine practically jumps out of solution. Elemental iodine sublimes, giving airborne $I_2$. RIA techniques utilize strong oxidizing agents, such as chloramine T, to liberate free iodine. Free iodine can be inhaled or readily absorbed through intact skin. It has been estimated that up to 15% of total thyroid uptakes of radioiodine is due to skin absorption. Of course, iodine can be ingested following contact with contaminated items.

## 4. Handling Techniques

Radioiodine, even when incorporated into stable molecules, tends to become airborne. Much radioiodine in use in RIA comes initially as NaI in high specific activity solutions of NaOH. The simple act of opening the sample vial in an open room can result in measurable airborne concentrations of iodine. It is strongly recommended that radioiodine be used in vials suitable for carrying out reaction or by withdrawing samples without opening the sample vial itself. Withdrawals may be performed by using a microsyringe and needle through the vial septum. Addition of reagents and substrate may be made through the septum into the vial. A vent needle may be placed through the septum. All radioiodine work should be done in a hood with a face velocity of 125–150 ft/min. Radioiodine involving more than 1 mCi should be used with activated charcoal filter traps to reduce release of iodine to the environment. There are several ways in which this may be accomplished. For infrequent radioiodine users, a syringe plugged with glass wool and filled with 10 cm³ of finely divided activated charcoal may be attached to the reaction vial using a needle through the septum. Reactants exiting the vial pass through the charcoal, trapping most iodine. The syringe containing the charcoal should be bagged tightly and placed in radioactive waste. For moderate-frequency users, in-hood lucite dry boxes equipped with activated charcoal filters and blowers are very useful. There are several suppliers of ready-made dry boxes with built-in charcoal filters for RIA users. For frequent users, hoods should be equipped with preplenum activated-charcoal-impregnated filters. Such hoods should be operated only when work is being performed so as to prevent premature depletion of charcoal effectiveness.

## 5. Survey for Airborne Activity

Surveys for airborne radioiodine in work areas and hood stacks or ducts are necessary when working with radioiodine. Work area samplers may be constructed of long, narrow tubes (15–30 cm long, 1–2 cm in diameter) plugged with glass wool at both ends and filled with activated charcoal. Sampling rates of 10–30 liters/min (0.3–1 ft$^3$ min) are recommended. Placing two traps in series allows one to evaluate the trapping efficiency by evaluating the ratios of first trap activity collected to that of the second trap. In building a collection trap it is not the total volume that is of primary importance but the path length that the sample must follow and therefore sample–charcoal contact time. For equal volumes a long, narrow tube is the preferred trap geometry. For stack sampling, commercially available charcoal cartridges (Scott, MSA, Gellman) may be used with high-volume (up to 500 liters/min) air samplers. Collection efficiencies for most samplers exceed 90% when used as directed. Table XII gives the ratio of first trap activity to second trap and the corresponding collection efficiency. This assumes equal collection efficiencies in the two traps.

Following collection the activated charcoal may be placed in a sample vial and counted with an automatic $\gamma$ counter or sodium iodine crystal and multichannel analyzer. Suitable calibration standards are needed also.

When working with radioiodine, gloves are essential, and double gloves are recommended. This will minimize skin absorption. Uptakes will be reduced by wearing long-sleeved lab coats. Self-administration of stable iodine as a thyroid blocking agent is to be avoided unless directed by a physician. Nonradioactive iodine in large amounts is toxic. The cure may be more harmful than its benefit; furthermore, if one follows recommended procedures, uptakes should be minimal and thyroid blocking agents unnecessary.

Control of solution pH is important, since iodine is much more stable at

**TABLE XII**
**Trap Ratio Data for Calculating Trap Efficiencies**

| Ratio of first trap activity[a] to second trap activity | Collection efficiency (%) |
|---|---|
| 1000 | 99.9 |
| 100 | 99 |
| 20 | 95 |
| 11 | 90 |
| 5 | 80 |
| 3.3 | 70 |
| 2.5 | 60 |
| 2 | 50 |

[a] Equal collection efficiencies in both traps assumed.

high (basic) pH. Solutions containing iodine wastes should not be mixed with acids.

## 6. Lab Design

For laboratories performing frequent experiments that utilize 1 mCi or more of radioiodine at one time, the following is a summary of lab features that will minimize exposures and releases of iodine to the environment.

1.  Use two hoods—one equipped with a preplenum charcoal trap for use with radioiodine, the second for other fume hood work. Separate ducts and blowers are needed. Or use an in-hood dry box equipped with its own charcoal trap in a hood. The hood may be used for other work.
2.  Store radioactive liquid wastes in the hood in deep tubs. Maintain pH in the range 8–11. Keep all bottles tightly capped. If shielding is necessary, use $\frac{1}{16}$-in. lead sheet stapled or tacked to epoxy-painted plywood. The $\frac{1}{16}$-in. lead will reduce the radiation intensity by a factor of about 1000.
3.  Store solid radioactive waste in steel or fiber drums with tight lids. Line the drums with heavy plastic bags. Place all ion exchange or chromatography columns and vials in small plastic bags that are tied closed, then place them in large waste drum. Keep the drum in a well-ventillated room away from normally occupied areas. If radiation levels are high, build a plywood-lead sheet enclosure for the drum.

## 7. Dosimetry

Users of radioiodine should wear body film or TLD badges. For individuals using 1 mCi or more, extremity finger dosimeters should be worn.

Bioassay by urinanalysis or thyroid counting is appropriate for individuals using 1 mCi or more of $^{125}$I or $^{131}$I at one time. The methods for urinanalysis are essentially the same as those for $^3$H. The necessary calculations may be found in ICRP publications 10 and 10a. Bioassay by thyroid counting is somewhat simpler and less time-consuming than urinanalysis. Since the thyroid is small and also concentrates iodine, it provides a ready-made "point" source for $\gamma$ photon counting. Where many individuals are involved a screening program may be used to reduce the numbers of people to be counted. Thyroid counting requires the following equipment: thin sodium iodide detector with beryllium window, single-channel or multichannel analyzer, thyroid-neck phantom, "simulated thyroid," set up for reproducible positioning of detector and subject. Thyroid screening needs, in addition a portable sodium iodide detector such as the Ludlum, Inc., Model 16 analyzer and 1 × 1-in. sodium iodide probe. A protocol for thyroid counting follows:

a. Preparation of thyroid-neck phantom: Commercially fabricated phantoms are available and usually consist of a solid Lucite right circular cylinder 15 cm in diameter with a hole in the top deep enough to accommodate a 30-ml bottle at a depth of 0.5 cm from the outer cylinder wall or a water-filled Lucite or plastic right circular cylinder 15 cm in diameter with similar provisions for a sample bottle. At least three 30-ml bottles should be prepared using one as a blank containing deionized water and two as standards containing known activities of $^{125}$I, $^{131}$I, or mock $^{125}$I or $^{131}$I. Mock standards are made with isotopes (such as $^{129}$I for $^{125}$I and $^{133}$Ba for $^{131}$I) that have relatively long half-lives and decay energies similar to those of the iodine isotopes they replace. Mock solutions or standards must be calibrated so that the activity expressed is corrected for the emission characteristics of the isotopes they are intended to replace.

b. Calibrate the counting system using the thyroid-neck phantom and standard solutions in the same geometry that will be used to count subjects. Suggested geometries are as follows: placing the detector in front of the neck area 15 cm away or, if a small-diameter crystal is used, in contact with the thyroid region of the neck. The important aspect of geometry is reproducibility. The detector is calibrated in terms of counts per time in the iodine channel per microcurie. A portable detector may be calibrated in a similar fashion, although in order to obtain sufficient counts it may be necessary to count on contact with the neck.

c. Subjects are counted by having them sit so that the detector can be placed near or against the neck using the calibration geometry. Count times of 1–10 min may be necessary, depending on background and counter efficiency. A minimum detectable activity of 0.01 $\mu$Ci should be sought. Preceding or following subject counting a long (20–100-min) background count is made. Calculation of thyroid activity is made by

$$A = \frac{C_S - C_B}{E}$$

where $A$ is thyroid acivity in microcuries, $C_s$ is subject count rate in counts/min, $C_B$ is the background count rate in counts/min, and $E$ is the calculate detector efficiency in counts/min/$\mu$Ci.

Subjects are screened using the portable detector held to the thyroid region of the neck. A positive reading (1.5–2 times background or more) is grounds for a thyroid count.

Such bioassays should be performed every 6–10 weeks for counts of routine users and within 48 hr of the use for occasional users. Screening should be done weekly.

One may calculate airborne iodine activity by knowing when the last iodine use occurred and by making assumptions of uptake route and breathing rate as found in ICRP 10 or 10a.

# BIBLIOGRAPHY

Arena, V. (1971). "Ionizing Radiation and Life." Mosby, St. Louis, Missouri.

Attix, F., and Roesch, W., eds. (1966). "Radiation Dosimetry," Vol. 2. Academic Press, New York.

Barnes, D. E., and Taylor, D. (1963). "Radiation Hazards and Protection." George Newnes, Ltd., London.

Casarett, A. (1968). "Radiation Biology." Prentice-Hall, Englewood Cliffs, New Jersey.

Cember, H. (1969). "Introduction to Health Physics." Pergamon, Oxford.

Chase, G., and Rabinowitz, J. (1964). "Principles of Radioisotopes Methodology." Burgess, Minneapolis, Minnesota.

Fitzgerald, J. J. (1969). "Applied Radiation Protection and Control." Gordon & Breach, New York.

International Commission on Radiological Protection (1959). "Permissible Dose for Internal Radiation." Pergamon, Oxford.

International Commission on Radiological Protection (1965). "The Evaluation of Risks from Radiation," ICRP Publ. 8. Pergamon, Oxford.

International Commission on Radiological Protection (1968). "Evaluation of Radiation Doses to body Tissues from Internal Contamination Due to Occupational Exposure," ICRP Publ. 10. Pergamon, Oxford.

International Commission on Radiological Protection (1970). "Protection Against Ionizing Radiation from External Sources," ICRP Publ. 15. Pergamon, Oxford.

Kobayashi, Y. and Maudsley, D. (1974). "Biological Applications of Liquid Scintillation Counting" Academic Press, New York.

Lawrence Berkeley Laboratory (1973). "Instrumentation for Environmental Monitoring Radiation." University of California, Berkeley.

Morgan, K. Z., and Turner, J. E., eds. (1968). "Principles of Radiation Protection." Wiley, New York.

National Academy of Sciences (1972). "The Effects on Populations of Exposure to Low Levels of Ionizing Radiation." BEIR (Advisory Committee on the Biological Effects of Ionizing Radiations.), National Academy of Sciences, Washington, D. C.

National Bureau of Standards (1954) "Permissible Dose from External Sources of Ionizing Radiation," Handbook 59. US Govt. Printing Office, Washington, D. C.

Price, W. (1958). "Nuclear Radiation Detection." McGraw-Hill, New York.

Shapiro, J. (1972). "Radiation Protection—A Guide for Scientists and Physicians." Harvard Univ. Press, Cambridge, Massachusetts.

Society for Nuclear Medicine (1975). Pamphlets 10 and 11. Medical Internal Radiation Dose Committee, SNM, New York.

United Nations Scientific Committee on the Effects of Atomic Radiation (1972). "Radiation Levels and Effects." United Nations, New York.

U.S. DHEW Public Health Service Bureau of Radiological Health (1970). "Radiological Health Handbook." US Govt. Printing Office, Washington, D.C.

Wang, Y. (1969). "Handbook of Radioactive Nuclides." Chem. Rubber Publ. Co., Cleveland, Ohio.

# PART TWO
# BIOLOGICAL
# LABORATORY
# SAFETY

# Chapter Four
# Viruses and Cancer

ROBERT GALLAGHER
RICCARDO DALLA FAVERA, AND
ROBERT GALLO

## I. INTRODUCTION

Oncogenic activity has been associated with viruses containing either DNA or RNA as their genetic information. Oncogenic potential has been associated with DNA viruses from at least three morphologically and biologically distinct families. RNA viruses with demonstrated oncogenic activity all fall into the retrovirus family.

For each virus family we consider basic morphological and biochemical properties. However, the emphasis is on the oncogenic potential of the viruses. Particularly stressed is available evidence for and against any association with human neoplasias. At the present time, there has been no definitive, etiologic relationship established between any DNA or RNA "tumor" virus and human neoplasia (with the exception of papilloma virus for some benign warts). However, there is mounting evidence that such viruses are at least a factor in some human neoplasias. The leading candidates are (1) Epstein-Barr virus in Burkitt's lymphoma and nasopharyngeal

LABORATORY SAFETY: THEORY AND PRACTICE

carcinoma; (2) herpes simplex virus-2 in cervical carcinoma; (3) a specific papilloma virus in an unusual genetic skin disorder, epidermodysplasia verruciformis; and (4) retroviruses in leukemia.

In evaluating the oncogenic potential of viruses, it is necessary to consider at least three elements: the virus itself, cellular factors affecting interactions with the virus, and host factors affecting the growth of transformed cells. Much information has come from *in vitro* studies on cell transformation. However, such studies can only identify the potential for producing cell changes defined as transformation in tissue culture; comprehension of viral oncogenesis in a biohazard context must also consider host factors that determine whether or not "transformed" cells can form a malignant tumor in the host. Recent studies indicate that immune surveillance is a particularly important factor in determining resistance to virus-induced neoplastic cells.

Viruses can be classified as oncogenic based on their tumor-inducing potential in animals. This classification covers a wide spectrum of "oncogenic" activity. Viruses may be tumorigenic in the species of origin under natural conditions. This has most consistently been demonstrated with retroviruses as the principal etiologic agents of certain naturally occurring leukemias in several mammals. Many viruses that seem nononcogenic in the native host are oncogenic in a heterologous species. Oncogenicity may or may not require treatment of the experimental host, for example, immunosuppression. Viruses are also classified as oncogenic based on *in vitro* assays. Virsues may have no demonstrable tumorigenic activity *in vivo* but may be able to transform cells *in vitro* which are tumorigenic on inoculation of an animal host. The complex interplay of factors determining whether a given virus has tumorigenic potential is only beginning to be understood.

## II. DNA ONCOGENIC VIRUSES

### A. General Comments

Three groups of viruses that contain DNA have been demonstrated to have the potential of producing neoplasia: papovaviruses, adenoviruses, and herpesviruses. However, only in a few instances have these viruses been conclusively related to the development of neoplasia in the native host, particularly under natural conditions. These include two herpesviruses, the Mareks disease agent that produces a lymphoma in chickens (Churchill and Biggs, 1967; Nazerian *et al.*, 1968), and the Lucke virus that produces adenocarcinomas in frogs (Mizell *et al.*, 1969), and at least two papilloma viruses that produce epithelial tumors in rabbits and cattle (Orth *et al.,* 1978a).

It has been suggested that an important factor in resistance to DNA on-cogenic viruses may be specific immune-response (Ir) genes directed against virus-specific antigens (Benacerraf and Katz, 1975; Klein, 1975). This hy-pothesis may account for the capacity of many of the DNA viruses to trans-form cells of the native host *in vitro* while lacking tumorigenic capability *in vivo.* It may also explain the capacity of DNA viruses to produce tumors with a highly variable incidence in different heterologous species.

The oncogenic DNA viruses have common properties in producing neo-plasia. Cell transformation involves a restriction in the viral replicative cycling, preventing lysis of the infected cell. A possible exception to this rule is the papilloma viruses, a subgroup of the papovaviruses; however, even in this instance restriction of virus replication in transformed cells ap-pears to occur (see below). A cell that supports the complete replicative cy-cle of the virus is termed *permissive.* The transforming potential of the DNA oncogenic viruses has been experimentally enhanced by infecting *nonpermissive* cells from certain heterologous species or by damaging the virus so that it is still infectious but unable to complete the lytic cycle.

The restriction in viral replication must permit the expression (transcrip-tion) of *early* viral genes while preventing the expression of *late* viral genes in order for transformation to occur. The products of the late viral genes are structural proteins required for assembly of complete virus particles. On the other hand, the products of the early viral genes are proteins that have regulatory functions. One of the most intensive efforts in molecular virology is currently being devoted to determining the nature and precise function of these molecules, since they may provide fundamental informa-tion about how these viruses transform mammalian cells. The model system for these studies is an early protein of the papovaviruses, the T (tumor) antigen. Expression of T antigen seems to be required for initiating and maintaining the transformed state. Recent evidence (Luborsky *et al.,* 1978) suggests that T antigen exists on the membrane and in the nucleus of transformed cells.

Another property of oncogenic DNA viruses possibly related to cell transformation is integration of viral DNA into cellular DNA. Integration is a universally associated feature of papovavirus and adenovirus-trans-formed cells, but there is controversy related to this feature in some herpes-virus-transformed cells (Roizman *et al.,* 1978). Studies using purified DNA fragments, representing less than the whole viral genome and defective for replication, indicate that such fragments can become stably integrated in the cellular genome. When such a cell is fused with a second cell containing a complementary but also defective portion of the virus genome, a complete infectious, lytic virus can be produced (Moyer *et al.*, 1978). This observation could be related to the property of *latency,* which is a common feature of potentially oncogenic DNA viruses.

## B. Papovaviruses

Morphologically, a papovavirus consists of a capsid with icosahedral symmetry enclosing a molecule of double-stranded DNA. The capsid contains 72 capsomeres and measures about 50 nm in diameter. The viral DNA consists primarily of a closed circle in supercoiled form, having a molecular weight of $3.6 \times 10^6$. Integration of the viral DNA into host cell DNA and production of T antigen have been universal accompaniments of transformation.

Two papovaviruses, polyoma, isolated from mouse cells, and the simian vacuolating agent 40 (SV40), isolated from rhesus monkey cells, have been the most extensively studied oncogenic viruses during the past decade. Even so, there is no substantiated evidence that these viruses produce tumors in their native hosts. Even in experimental hosts, tumors are produced only under artificial conditions (Eddy *et al.*, 1968). This limited tumorigenic capacity has been fortuitously important, since thousands of children were inadvertently inoculated with a polio vaccine contaminated with active SV40. Through more than 25 years' observation there have been many negative reports, with only one report suggesting a possible relationship to an increased incidence of neoplasia related to the contaminated vaccine (Heinonen *et al.*, 1973).

Serologic evidence indicates that SV40 itself rarely infects humans whereas papovaviruses related to SV40 naturally infect a high percentage (more than 80%) of humans during childhood (Shah *et al.*, 1973; Padgett and Walker, 1973). No clinical symptoms are associated with the initial infection. However, papovaviruses have been isolated from diseased humans. JC virus has been repeatedly isolated from the brains of persons who have died of progressive multifocal leucoencephalopathy (Padgett *et al.*, 1971; Dougherty and DeStefano, 1973). Another papovavirus, BK virus, has been repeatedly isolated from the urine of immunologically suppressed renal transplant patients (Gardner *et al.*, 1971; Takemoto and Mullarkey, 1973). These findings suggest that JCV and BKV may remain latent following childhood infection and may somehow be reactivated under these abnormal conditions.

Like SV40, BKV is capable of transforming some mammalian cells *in vitro* (Major and diMayorca, 1973). BKV and JCV can produce tumors in susceptible hosts (Shah *et al.*, 1975; Walker *et al.*, 1973). In view of this potential oncogenicity, there have been several studies for evidence of a possible association of these papovaviruses with human tumors. Tests for T antigen, which are invariably positive in known papovavirus-induced tumors in animals, have been performed by indirect immunofluorescence of cells derived from human tissues when grown in tissue culture or by complement fixation using fresh tumor extracts. Most tests gave negative results, though Weiss *et al.* (1975) initially reported three of nine meningiomas positive for the presence of T antigen. Corallini *et al.* (1976) reported finding antibodies to T antigen in a low percentage (about 1%) of sera from both normal and cancer patients. Others have reported failure to find such antibodies (Takemoto *et al.,* 1978). Fiori and diMayorca (1976) reported that DNA from 5 of 12 human tumors and from 3 of 4 human tumor cell lines contained BKV-related genes, but subsequent studies in other laboratories failed to corroborate the positive results (Takemoto *et al.,* 1978; Wold *et al.,* 1978).

In summary, there is no conclusive evidence that the papovaviruses, which are known to be broadly transmitted in the human population, are associated with human tumors in the manner in which they have been demonstrated to be oncogenic in experimental animals. Further studies are required to determine if papovaviruses cause sporadic tumors or if their role in natural disease might be related to mechanisms not demonstrable by the usual model systems.

Despite the low oncogenicity of papovaviruses *in vivo,* they have a consistent transforming effect on fibroblastic cells of certain species *in vitro.* The ready definition and quantitation of this *in vitro* effect and the relative genetic simplicity of the papovaviruses led to tremendous progress in deciphering how these viruses may relate to neoplastic transformation at the molecular level. This topic has been reviewed by Fareed and Davoli (1977).

## C. Papillomaviruses

The papillomaviruses can be distinguished from other members of the *papovaviridae* family by the somewhat larger size of the virion capsid (55–60 nm) and of the supercoiled, circular DNA ($5 \times 10^6$ daltons). Experimental work with the papillomaviruses is limited because no tissue culture system has been developed to permit their *in vitro* replication. Papilloma viruses are highly tissue specific, especially for keratinizing epithelium, which is difficult to maintain in tissue culture. Nevertheless, interest in these viruses has recently been revived, since these are the only papovaviruses that have been demonstrated to be tumorigenic in the species of origin, including man (a variety of benign warts). There has recently appeared an excellent detailed review of papillomaviruses as causative agents of tumors in animals and man (Orth *et al.,* 1978a). In rabbits and cattle, papillomaviruses cause malignancies. Shope papilloma virus can cause squamous carcinomas in rabbits, for example (Kidd and Rous, 1940; Syverton, 1952). The incidence of malignancy is enhanced by carcinogens: coal tar in the case of Shope papillomavirus (Rous and Friedewald, 1944) or a chemical in bracken fern in the case of papillomavirus-induced carcinomas of cows (Olson *et al.,* 1969; Jarrett, 1979). Additionally, host genetic factors seem to be important. Consequently, to identify a papilloma virus as an etiologic agent for any particular disease could be complicated. Tumors from domestic rabbits are much less likely to be permissive for the replication of complete virus, in keeping with the idea that oncogenic transformation is related to restriction of virus replication (Kidd and Rous, 1940; Syverton, 1952).

Most humans are infected with papillomavirus by adulthood (Pyrrhonen, 1976). In man several different types of benign warts have been associated with papillomaviruses, particularly warts of the hands and feet, of the anogenital area, and of the nasopharynx and larynx. There has been no demonstrated case of malignant conversion of papillomavirus-induced warts in normal individuals. Virtually all these warts spontaneously regress after a period of months to years, and this may somehow be related to the host immune response. Immunosuppressed patients have an increased incidence of warts but no case of malignant conversion has been documented (Spencer and Anderson, 1970; Morison, 1975). There is a rare human genetic disease, *epidermodysplasia verruciformis,* in which transformation of papilloma virus-associated warts to squamous carcinoma occurs in about 25% of the patients, particularly in sun-exposed areas. Further studies are required to determine the role of the virus in this process but it may provide a useful human model for virus-related malignant transformation (Orth *et al.,* 1978a). Serologic studies do not suggest any involvement of papillomavirus in other human malignancies (Pfister and zur Hausen, 1978), but detailed analyses of human tumors for papillomavirus DNA have not yet been reported.

## D. Adenoviruses

Adenoviruses are much larger and more complex than papovaviruses. The virion capsid has a cross-sectional diameter of 70–90 nm and consists of 252 capsomeres arranged on 20 symmetrical planes (icosahedron). The viral genome consists of linear double-stranded DNA and has a molecular weight of $23 \times 10^6$ daltons. This is sufficient to code for in excess of 30 viral proteins. During lytic infection, large quantities of virus-specific DNA, RNA, and proteins are produced, which have provided materials for extensive molecular biologic studies of viral replication (see Wold *et al.*, 1978, for review).

Adenoviruses have been isolated from a number of species, including 31 distinguishable serotypes from man. They are associated with a variety of acute, febrile illnesses, particularly of the upper respiratory and the gastrointestinal systems. They are not associated with the development of naturally occurring neoplasms in any species. However, in 1962 Tretin *et al.* demonstrated that human adenovirus type 12 could produce tumors on inoculation into newborn hamsters and since then all human serotypes have been examined for potential oncogenicity (Green and Mackey, 1978). Three classes of active viruses have been identified as follows: *class A* viruses (types 12, 18, and 31) are highly oncogenic in newborn rodents; *class B* viruses (types 3, 7, 14, 16, and 21) also produce tumors in newborn rodents but with less efficiency and with a long latent period; *class C* viruses (types 1, 2, 5, and 6) do not produce tumors on direct inoculation but can transform rodent cells *in vitro*. Under certain conditions these transformed cells can produce tumors *in vivo*.

As in the papovaviruses, cell transformation is linked to restriction of virus replication, integration of viral DNA into the host cell DNA, and the expression of viral-specific T antigen (McDougall *et al.*, 1978). A variety of human tumors, particularly respiratory and gastrointestinal tract tumors, have been tested for integrated adenovirus gene sequences, so far with entirely negative results (Green and Mackey, 1978; Mackey *et al.*, 1976).

## E. Herpesviruses

The herpesviruses are the most complex of the DNA oncogenic viruses. The viron measures 150–170 nm in diameter and consists of a lipoprotein coat enclosing a capsid made up of 162 capsomeres arranged with icosahedral symmetry. On cross-sectional electronmicroscopy additional intraviral membranes are visible, the innermost surrounding the nucleoprotein core with a diameter of 25 nm. Herpesvirus capsids are formed within the cell nucleus and mature by acquisition of the envelope as the virion passes through the nuclear membrane. The DNA of all herpesviruses has a molecular weight of approximately $10^8$ and consists of linear double-stranded DNA.

Five specific herpesviruses have been isolated from man: herpes simplex virus 1 (HSV-1), herpes simplex virus 2 (HSV-2), herpes zoster virus (HZV) or varicella, cytometalovirus (CMV), and Epstein-Barr virus (EBV). All but HZV have oncogenic potential.

Similar mechanisms may be involved in cell transformation by the herpesvirus to those described for papovaviruses and adenoviruses: restriction of viral replication, integration of viral DNA into host cell DNA, stimulation of host cell DNA synthesis, and expression of virus-specific T antigen. However, in many instances the data are not available for the herpes viruses and the possibility certainly cannot be excluded that alternative mechanisms could be involved (Roizman et al., 1978).

## 1. Herpes Simplex Virus (HSV)

HSV-1 is primarily an infectious agent of the oropharynx, whereas HSV-2 primarily infects the genital tract. Initial infection by HSV may be asymptomatic or produce a diffuse inflammatory reaction, characterized by the production of multiple blisterlike lesions of the epithelium. Congenital or neonatal infection by HSV may produce a severe systemic reaction and is particularly prone to produce central nervous system damage. After initial infection, HSV characteristically enters a period of latency in which infectious virus is not detectable in the target epithelial cells (Hayward et al., 1975). In the case of HSV-1, experimental studies indicate that the sensory nerve ganglia, particularly of the trigeminal nerve, are the site of persistent virus infection. The state of the virus during latency has not been defined, although it seems unlikely that integration of herpesvirus DNA occurs, since nerve cells are quiescent, whereas integration appears to require active cellular DNA synthesis. Recrudescences are characterized by localized herpes blister sores, "fever blisters." These recurrences have not been related to alterations in the immune response of the host, although this could be due to inadequate sensitivity of the tests utilized. Herpes sores from either primary or recurrent lesions actively shed virus that is infectious for uninfected contacts. HSV-2, for example, has been demonstrated to be venereally transmitted. The possibility that herpes simplex viruses may be associated with human neoplasia has been considered primarily because of an association between HSV-2 and carcinoma of the cervix (Rawls and Adam, 1978; Goldberg and Gravell, 1976).

Women with cervical cancer have a markedly increased incidence of infection by HSV-2 compared to age-matched controls. Antigens related to HSV-2 structural proteins or, possibly, nonstructural virus proteins have been detected fairly regularly in exfoliated cells from patients with cervical carcinoma. In one case of many that have been studied (Frenkel et al., 1972), HSV-2 DNA was detected in an integrated state in the DNA of cervical carcinoma cells. Recently, using the technique of in situ hybridization, RNA transcripts of HSV have been reported in several cervical tumors (McDougall et al., 1979); this potentially important finding requires extension and confirmation.

The preceding findings provide a strong link between HSV-2 and cervical carcinoma. However, the question remains whether the cancer and HSV-2 are etiologically related or covariables. Both HSV-1 and HSV-2 transform mammalian cells *in vitro* but not *in vivo*. Since virtually all mammalian cells are permissive for lytic infection by these viruses, transformation has been detected in systems utilizing inactivated (nonreplicative) virus (Duff and Rapp, 1971; Munyon *et al.*, 1971; Takahashi and Yamanishi, 1974). Like adenovirus-transformed cells and in contrast to papovavirus-transformed cells, complete HSV is not recoverable from transformed cells. Both molecular hybridization studies (Kraiselberg *et al.*, 1975; Davis and Kingsbury, 1976; Frenkel *et al.*, 1976; Minson *et al.*, 1976) and DNA transfection experiments with endonuclease restriction fragments (Maitland and McDougall, 1977; Wigler *et al.*, 1977) indicate that partial viral genetic information as few integrated copies (one to six) is related to the transforming event.

## 2. Cytomegalovirus

Cytomegalovirus (CMV) is a ubiquitous infectious agent in the human population, over 90% of persons having evidence of infection. In the vast majority of cases, the initial infection is completely asymptomatic. However, under certain circumstances, particularly in newborns, primary infection may produce disease. CMV was originally discovered as a rare infectious disease of newborns that produced a severe systemic illness characterized by jaundice, anemia, thrombocytopenia, hepatospleno-megaly, chorioretinitis, microcephaly, and mental retardation. Microscopic examination of placental and fetal cells disclosed characteristic inclusion bodies leading to the designation cytomegalic inclusion disease. After initial infection, cytomegalovirus, as is typical of herpesviruses, enters a sustained latent phase, probably in association with certain lymphoid cells (Pagano, 1975). It may be reactivated later in life, in which case it is most commonly associated with an infectious mononucleosis-like syndrome (Klemola *et al.*, 1970; Jordan *et al.*, 1973). Although this may occur spontaneously, it has most consistently been associated with the transfusion of large quantities of blood, as after cardiac bypass surgery, or in immunologically suppressed states, such as renal transplantation or in leukemia-lymphoma. Experimental studies suggest that the activation of latent CMV is associated with suppression of cellular immunity (Lang *et al.*, 1976; Jordan *et al.*, 1977) but may also be associated with mitogenic stimulation of lymphoid cells, as in homograft situations.

Human tumors have not been extensively analyzed for CMV information. However, some association has been made in at least three instances. CMV, apparently acquired *in vivo*, was recovered from the early passages of juvenile prostatic cultured cells (Rapp *et al.*, 1975). In later passages virus disappeared from these transformed cells, although CMV-specific antigens persisted. CMV was isolated from 2 of 10 cell cultures derived from cervical carcinoma biopsies (Melnick *et al.*, 1978). It is well documented that CMV can be transmitted venereally; however, in both positive cases the biopsies were derived from elderly, sexually inactive women so that these isolates presumably represent reactivated, latent virus. Serologic studies have related an increased titer of antibodies to CMV and a cutaneous tumor, a pigmented hemangiosarcoma, called Kaposi's sarcoma (Giraldo *et al.*, 1978). This disease occurs with a similar distribution to Burkitt's lymphoma in Africa, suggesting the possibility of an infectious vector.

## 3. Epstein-Barr Virus

The Epstein-Barr virus (EBV) has been closely associated with three human diseases: infectious mononucleosis, Burkitt's lymphoma, and nasopharyngeal carcinoma. EBV is the direct cause of infectious mononucleosis (Henle *et al.*, 1977). The disease almost invariably occurs in previously unexposed persons. After an incubation period of approximately 1 month a classical primary immune response to the virus develops. During the disease virus can be demonstrated in the throat by the ability of washings to transform cord blood lymphocytes. However, EBV has never been demonstrated directly in throat cells, and it is unclear whether a lymphoid cell or some epithelial cell is the principal repository of the virus. The majority of cells in the peripheral blood during the acute phase of the disease are T lymphoid cells and are not EBV-transformed B lymphocytes (Sheldon *et al.*, 1973). It has been postulated that this represents the host's cellular immune response to transformed B lymphocytes. EBV is recoverable from blood lymphocytes during the acute phase of the disease. However, viral antigen is present only in very rare cells, and currently there is controversy over the form and incidence of expression of EBV in infectious mononucleosis blood cells (Klein, 1975; Crawford *et al.*, 1978). Also, some evidence suggests that the genetic content of EBV DNA from different cases of infectious mononucleosis may differ from one another and from other sources of EBV (Pagano *et al.*, 1976).

Burkitt's lymphoma has a particular clinical and histopathological presentation and occurs primarily in a certain geoclime of Africa and New Guinea (Epstein, 1978). It is strongly suspected that EBV has an etiological function in this disease for four reasons: (1) EBV is capable of transforming B lymphocytes, which is the type of lymphocyte composing Burkitt's lymphoma; (2) EBV can produce B cell lymphomas on inoculation into certain subhuman primates; (3) lymphoid tumor cells from the vast majority

(>95%) of African Burkitt's lymphoma cases contain EBV DNA, whereas it is lacking in other types of lymphomas; and (4) most African patients with Burkitt's lymphoma have evidence of a heightened immune response to EBV compared to age-matched controls. An alternative explanation that has been proposed is that EBV could be an inactive passenger virus that selectively infects Burkitt's lymphoma and results in a concomitant rise in antibody titer.

Although there is apparently a strong association between EBV and Burkitt's lymphoma, we do not understand why this ubiquitous infectious agent would cause a tumor in such a restricted population. The high incidence of holoendemic malaria in the affected areas has been proposed as the necessary cofactor. This might act by producing immunosuppression and/or stimulating the proliferation of a specific subpopulation of B lymphocytes. It may also be that genetic factors may be important in susceptibility to disease. In summary, although the clinical and experimental evidence would appear to cover Koch's postulates, the classical criteria for accepting an infectious agent as the cause of an infectious disease, doubt remains regarding the essentiality of EBV for the development of Burkitt's lymphoma. It has been argued that the only test of causality is to prevent EBV infection by the development of an effective vaccine (Epstein, 1978). However, the ethical considerations of administering a vaccine with DNA of unknown tumorigenic potential has so far mitigated against such a program and at the moment efforts are being directed at reducing the incidence of arthropod-disseminated malaria (de-Thé et al., 1978).

EBV has also been associated with anaplastic carcinomas of the postnasal space. It has not been associated with more differentiated squamous carcinomas or other types of tumors of the postnasal space or with squamous carcinoma of any histologic grade from other head and neck sites (Andersson-Anvret et al., 1977). Unlike Burkitt's lymphoma, nasopharyngeal carcinoma is not confined to any geographic area but does have a predilection for certain ethnic groups, particularly Chinese. The association of these tumors with EBV has been consistent regardless of the geographical source of tumor. Initially, it was thought that EBV might be associated with the lymphoid cells that frequently heavily infiltrate the tumor. However, subsequent studies definitively demonstrated that EBV was associated with the malignant epithelial cells (Klein et al., 1974). As in Burkitt's lymphoma, most cases of nasopharyngeal carcinoma are associated with elevated titers of antibodies to EBV antigens (Henle et al.; 1977), and some studies suggest that the titer of certain antibodies may be a useful prognostic indicator of disease course (Henle et al., 1977; Pearson et al., 1978). Thus, as for Burkitt's lymphoma, the strong association of EBV with nasopharyngeal carcinoma tumor cells implies an etiologic role but does not exclude the possibilities that the virus is a commensal or active cofactor or promoter.

Studies of the molecular virology of EBV have been hampered by the lack of a tissue culture cell line fully permissive for viral replication. The effective *in vitro* host rate of EBV is restricted to a subpopulation of B lymphocytes from humans and some other primates (Steel *et al.*, 1977; Katsuki *et al.*, 1977). After exposure of fresh, mixed-blood lymphocytes from a previously uninfected person to a source of infectious EBV, a minor proportion of the cells manifest evidence of infection through the expression of an EBV-specific nuclear antigen called EBNA. Some of these infected cells are "converted" or "transformed," begin to proliferate, and, after a period of several weeks, develop into permanently established lymphoid cell lines. Such cell lines may also be derived spontaneously from the blood lymphocytes of persons who have previously been infected by EBV. They are derived with particular ease from the blood and lymphoid tissues of patients with infectious mononucleosis or African Burkitt's lymphoma, both of which have been associated with EBV infection *in vivo*. Lymphoblastic cell lines established *in vitro* from Burkitt's lymphoma cases have certain characteristics that distinguish them from newly established lymphoblastoid cell lines from normal or infectious mononucleosis donors; they are monoclonal (Nilsson, 1971), they readily form tumors in nude mice (Nilsson *et al.*, 1977), and they have cytogenetic abnormalities, most commonly on chromosome 14 (Kaiser-McCaw *et al.*, 1977). These human cell lines have been the basis for studying the events of EBV infection and transformation. Extensive reviews related to these findings may be found elsewhere (Roizman and Kieff, 1975; Frank *et al.*, 1976; Klein, 1975; Epstein, 1978).

In the vast majority of instances, EBV-transformed or converted lymphoid cell lines show no evidence of complete virus production. This has been termed *restringent* infection (Roizman and Kieff, 1975). Nucleic acid analyses indicate that, in contrast to cells transformed by other herpesviruses, the entire EBV genome is present in restringently infected lymphoid cells. Furthermore, the viral DNA is with few exceptions present in multiple copies (10 to 100 copies) in an extrachromosomal site, as supercoiled DNA (Roizman and Kieff, 1975). A small portion of the EBV DNA may also be integrated into cellular DNA, but this remains controversial because of the difficulty in studying integrated DNA in the presence of a large excess of the episomal form. In most instances, the only antigenic evidence of EBV expression in restringently infected cells is the presence of EBNA. Only about 10% of EBV DNA is transcribed into RNA which associates with polyribosomes (Hayward and Kieff, 1976), indicating that an extreme degree of expressional control is exercised on the EBV genome in restringently infected lymphoid cells.

In only a few instances do established lymphoblastoid cell lines spontaneously produce complete virus, most notably a cell line derived from a Burkitt's lymphoma patient (P3HR-1) and a cell line derived from infected marmoset lymphocytes (B95-8). In these cell lines only a minority (5–20%) of the cells are actively productive. Studies of these producer cells indicate that EBV has effects and properties similar to other herpesviruses following lytic infection. Only limited information is available about EBV proteins. EBNA, molecular weight of 48,000 (Luka *et al.*, 1978), exists in intact cells as a tetramer and binds to chromosomal DNA. In this respect EBNA resembles "T antigens" previously mentioned for other DNA oncogenic viruses. EBV produced by the cell lines P3HR-1 and B95-8 have important biologic differences; most important, B95-8 virus is able to transform other lymphoid cells whereas P3HR-1 virus cannot. Recent analyses of the DNA from these two sources of virus by endonuclease restriction mapping indicate that the transforming function of B95-8 virus is due to the presence of specific sequences lacking in P3HR-1 virus (Raab-Taub *et al.*, 1978). These same sequences are present in at least one other source of transforming EBV and related sequences appear to be constantly associated with restringently infected cells. Thus a current need is to isolate and characterize these gene sequences (e.g., by genetic recombination and cloning in bacteria) and to use them as molecular probes in human neoplasia.

In some instances, it has been possible to increase the level of EBV expression in restringently infected cells by treating with certain chemical inducing agents (Gerber, 1972; Hampar *et al.*, 1972) or by superinfecting with another source of EBV (Henle *et al.*, 1970). Following superinfection, complete virus has sometimes been produced (Traul *et al.*, 1977). The virus produced by superinfected cells may have different properties than the infecting virus, suggesting the possibility of complementation of the resident and infecting sources of EBV (Fresen *et al.*, 1978).

In addition to transforming lymphocytes *in vitro,* EBV has been demonstrated to be lymphomagenic in other primates, particularly marmosets (Frank *et al.*, 1976). Recently, similar B lymphotropic herpesviruses have been identified in other Old World primates, including baboons, orangutans, and chimpanzees (Rabin *et al.*, 1977). The baboon virus isolate has been associated with an outbreak of lymphoma in a baboon colony, although it is far from clear that this virus is etiologically related to the disease (Falk *et al.*, 1976). New World primates harbor similar herpesviruses which are trophic for T lymphocytes (Rangan and Gallagher, 1979). These viruses have not been demonstrated to be oncogenic in the native species but in some instances have been highly lymphomagenic in other New World primates.

## III. RNA TUMOR VIRUSES

From a biohazard safety viewpoint there is excellent reason for considering retroviruses as potentially oncogenic in humans. The term *retrovirus* refers to a group of RNA-containing viruses capable of synthesizing DNA copies of their RNA genome using a viral-coded enzyme, reverse transcriptase (RNA-dependent DNA polymerase). The DNA copy, the "provirus," can then interact with and become part of the chromosome of the infected cell. This RNA → DNA mode of replication distinguishes retroviruses from other RNA-containing viruses and may, in part, be responsible for their oncogenic nature.

It is certain that some retroviruses cause cancer in animals. In 1911 Peyton Rous isolated Rous sarcoma virus from web sarcomas in chickens. He later demonstrated that the purified virus was capable of causing the same disease in healthy chickens. Rous sarcoma virus has since become the prototype for studies into the mechanism whereby retroviruses cause sarcomas in animals and transform cells in culture. In general, transformation is not an inevitable consequence of viral replication and is restricted to particular target cells, whereas the host range for replication is less specific. The two phenomena (transformation and replication) have different viral genetic determinants, a viral gene called sarc (Src) being responsible for *in vitro* fibroblast transformation. It was not until the late 1960s that a retrovirus was identified as a causative agent in the etiology of a natural mammalian cancer. At that time feline leukemia virus was identified as the causative agent for cat leukemia. Subsequently, retroviruses have been identified as causes for leukemia in cows, mice, and subhuman primates.

RNA tumor viruses are capable of horizontal transmission using the common means of infection, e.g., via saliva, urine, blood. This has been demonstrated for cat leukemia virus (FeLV) (Hardy et al., 1973), gibbon ape leukemia virus (GaLV) (Kawakami et al., 1972), bovine leukemia virus (Olson et al., 1972), and avian leukemia virus (Rubin et al., 1961). A second mode of transmission is genetic transmission from parent to offspring. The spontaneous appearance of type-C viruses in "virus-free" animals of certain mammalian species and in virus-free cultured cells derived from these animals provided support for the hypothesis that information for the production of such viruses might be transmitted genetically (vertical transmission) from parent to progeny along with other cellular genes. These sequences, which are an integral part of the host species' chromosomal DNA, code for the production of type-C viruses, and are known as endogenous viruses or virogenes. Generally, the host is genetically resistant to infection with its own endogenous viruses. Type-C viruses derived from closely related species usually are more closely related than viruses released by distantly related species. The endogenous viruses are normally repressed but can be activated by a variety of intrinsic (genetic, hormonal) or extrinsic (other infecting viruses, chemical carcinogens, radiation) factors. In some cases, they have escaped from host control, presumably because of a genetic change in the virus, in the cell, or in both. As such, they become infectious for the species from which they arose or may even become infectious for other species. There are documented instances in which these viruses have been transmitted between species, have become incorporated into the germ line, and have subsequently been transmitted as cellular genes in the recipient species. Three known examples of this combined mode of transmission are (1) the endogenous group of feline viruses (RD114/CCC), transmitted from baboons to ancestors of the domestic cat (Benveniste and Todaro, 1974; Todaro *et al.,* 1974); (2) feline leukemia virus (FeLV), acquired by ancestors of domestic cats from ancestors of rats (Benveniste *et al.,* 1975); and (3) simian sarcoma-leukemia virus and gibbon ape leukemia viruses, transferred to gibbon apes and to a woolly monkey from virogenes that originated in mice. As a conclusion, the occurrence of interspecies transfer of type-C RNA viruses seems to be well supported and may be the basis for explaining some recent primate type-C virus footprints in humans, since there is no reason to exclude man as a potential recipient host.

Type-C viruses were isolated for the first time in primates during the past decade. The first isolate was from a New World woolly monkey, a household pet. The animal developed a spontaneous fibrosarcoma of the neck which yielded the virus (Theilen *et al.,* 1971). To this date there is only one isolate of the woolly monkey virus, known as the simian sarcoma–simian sarcoma associated virus complex (SSV-SSAV). It consists of a virus that transforms fibroblasts *in vitro* (the sarcoma component) and a helper virus for replication (the associated virus component—SSAV), which is usually present in large excess. Subsequently, viruses closely related to SSAV were isolated from Old World gibbon apes with various hematopoietic neoplasms. These viruses, collectively termed the gibbon ape leukemia viruses (GaLV), have most frequently been isolated from animals with lymphosarcomas (Kawakami *et al.,* 1972; Snyder *et al.,* 1976), myelogenous leukemias (Kawakami *et al.,* 1973), and lymphoblastic leukemias (Gallo *et al.,* 1978). However, they have also been isolated from the brains of three clinically normal gibbons, although under unusual conditions (Todaro *et al.,* 1975). GaLVs have spread from animal to animal under natural conditions, and they are able to induce tumors when inoculated into other primates (Wolfe *et al.*, 1971; Parks *et al.*, 1973; Kawakami *et al.*, 1973). They are related to one another by several immunologic criteria (Parks *et al.*, 1973; Gallo *et al.*, 1978; Krakower and Aaronson, 1978; Tronick *et al.*, 1975) and contain related RNA genomes (Todaro *et al.*, 1975; Gallo *et al.*, 1978; Reitz *et al.*, 1979). SSV-SSAV and GaLV are related to each other. Extensive gene sequences homologous to those of the RNAs of GaLV and SSAV have not been detected in the cellular DNA of primates by liquid hybridization studies, showing that these viruses are not endogenous primate viruses (Scolnick *et al.*, 1974; Lieber *et al.*, 1975; Wong-Staal *et al.*, 1975). They were apparently transmitted to primates from mice (Wong-Staal *et al.*, 1975).

Endogenous type-C viruses have been isolated from both Old World and New World monkeys. The baboon endogenous viruses (BaEV) were the first ones isolated (Kalter *et al.*, 1973; Todaro *et al.*, 1974). They are closely related to one another in their biological properties, in the relatedness of their antigenic proteins, and in the similarities of the nucleotide sequences of their RNA genomes, but they are distinct from all other type-C viruses. They are truly endogenous to baboons in that there are BaEV genes in the DNA from all tissues of normal baboons, and there are phylogenetically related genes in the DNA of other primates (Benveniste and Todaro, 1976; Wong-Staal *et al.*, 1976). None of the endogenous viruses have been shown to transform cells *in vitro* or to cause diseases *in vivo*. Their interest is enhanced by reports indicating that BaEV probes can sometimes detect related molecules in some human leukemic cells (Wong-Staal *et al.*, 1976).

The etiology of human cancer cannot yet be fully explained at the molecular level, although some known causes or predisposing factors are involved in some neoplasias and in a few of them a correlation between a particular factor and the incidence of a tumor has been persuasively shown. Considering leukemias and hematopoietic neoplasia, there are associations with some chemicals, and, of course, radiation is known to increase the incidence of leukemia. There is also an increase in leukemia associated with some chromosomal changes. Mendelian genetic factors may also be relevant, as suggested by families with a remarkable incidence of leukemia. Despite these leads, the primary etiology of the vast majority of cases of leukemia in man remains unknown. In every case where the primary etiology of a naturally occurring leukemia in animals is known, it involves an acquired tumor virus, generally, a type-C virus. On this basis an extensive search for the presence of RNA tumor viruses or of their "footprints" in human normal and leukemic tissues was mandatory. Among various families of RNA tumor viruses, the primate type-C viruses have been the most extensively utilized to obtain probes for searching for similar molecules in people. There is substantial evidence now that (1) there has been contact between these agents and man since there are genes in human DNA related to primate type-C viruses, and (2) there is a significantly higher incidence of detection of virus or subviral components in leukemic tissues compared to normal tissues.

As expected from phylogenetic considerations, there exist BaEV-related sequences in normal humans. Unexpected, and more interesting, is the finding of a normal human gene related to SSV-SSAV (Wong-Staal, 1979). One possibility, suggested but not yet confirmed experimentally, is that these sequences in human DNA could represent or include a transforming gene equivalent to the "Src" gene of Rous sarcoma virus (see above).

It is also reported that some human tumors additionally possess acquired, viral-related DNA (Baxt and Spiegelman, 1972; Baxt et al., 1973; Wong-Staal et al., 1976, 1978; Aulakh and Gallo, 1977; Prochownik and Kirsten, 1977). This finding implicates an infectious process in human cancer. The difficulty in forming a hard conclusion is that the experiments cited above for each group of investigators have not been confirmed by other investigators.

However, there are other experiments that support the notion that infection by retroviruses contributes to cancer. Retrovirus proteins are found in human tumors. Reverse transcriptase, an enzyme unique to retroviruses, was detected in human leukemic cells (Gallo et al., 1970) and subsequently shown to be antigenically related to reverse transcriptase of GaLV (Todaro and Gallo, 1973; Gallagher et al., 1974; Chandra et al., 1978). Antigens related to viral structural proteins have also been detected (Sherr and Todaro, 1975; Panem et al., 1978).

Human tumors also carry a cytoplasmic apparatus with biophysical and biochemical properties of retroviruses (Axel et al., 1972; Gallo et al., 1973; Colcher et al., 1974). Particles from leukemic patients contain endogenous reverse transcriptase capable of synthesizing (in an endogenous reaction) cDNA, which hybridizes preferentially to RNA from SSV-SSAV (Miller et al., 1974).

The release of infectious particles following a few (Gallagher and Gallo, 1975; Teich et al., 1975; Nooter et al., 1975; Gabelman et al., 1975; Kaplan et al., 1979) or many (Panem et al., 1975) passages in culture of human cells has been reported, although negative results are the rule. The first report involved a leukemia patient: Two types of virus were released from blood leukocytes and bone marrow cells, one related to the SSV-SSAV group and a second related to BaEV (Teich et al., 1975; Reitz et al., 1976). cDNA synthesized by cytoplasmic particles from fresh uncultured blood leukocytes of this patient contained some sequences related to SSV and others related to BaEV (Reitz et al., 1976; Wong-Staal et al., 1976). All these correlated results are strongly against the possibility of laboratory contamination even if the rarity of the phenomenon partially limits its biologic significance. The release of noninfectious, viruslike particles containing reverse transcriptase or nucleic acid sequences closely related to SSV-SSAV or GaLV have been reported from several laboratories (Mak *et al.*, 1974; Kotler *et al.*, 1973; Kaplan *et al.*, 1977).

The possibility that humans have been exposed to primate retroviruses can also be approached by serologic studies. If primate type-C-related viruses or related endogenous "viral" genetic information plays a role in human leukemia, one might expect an immunologic response in the sera of some humans. Kurth *et al.* 1977; Kurth and Mikschy, 1978; and Snyder *et al.*, 1976 detected humoral antibodies in a significant fraction of humans that precipitate labeled viral antigens of SSAV and BaEV using crude virus. More recently, a purified antigen, the viral envelope protein gp70, was shown to be precipitated (P. Jacquemin, unpublished data). Aoki *et al.* (1976) also reported such positive findings. In their study, antibodies from the sera of leukemic patients were shown to bind to viral envelope antigens on the surface of GaLV-producing gibbon cells, and the antibody activity could be absorbed by disrupted GaLV. Also consistent with previous reports of the presence in leukemic blood cells of reverse transcriptase related to that of the SSV-GaLV group, antibodies reactive with this enzyme have been described in the blood serum of a few leukemic patients (Prochownik and Kirsten, 1977). Finally, since there is a precedent for various tumor cell-bound antibodies that can be eluted from the cell surface, immunoglobulin eluates from the surface of human leukocytes were examined for antibodies that might specifically inhibit purified reverse transcriptase. About 10% of the normals examined were positive (Jacquemin *et al.*, 1978) and the eluted IgG samples reacted most strongly against GaLv RT, as did IgG from the chronic phase of chronic myelogenous leukemia (CML). IgG reactive predominantly with SSAV was observed in 4 out of 10 samples from cases of AML and AMML. Unexpectedly, in nearly every case tested (8 of 9) of CML during the blastic phase of the disease, the eluted IgG specifically and strongly inhibited reverse transcriptase from FeLV.

The several reproducible, often correlated, findings that we have presented indicate that type-C RNA viral genetic information is present in humans. This does not, as yet, permit a complete formulation regarding the possible role of these viral elements in the pathogenesis of human leukemia. Several factors should be considered when interpreting some of the most surprising findings previously mentioned, especially the lack of complete provirus detection in human leukemic DNA and the rarity of virus isolation from leukemic tissues. (1) In the animal systems in which retroviruses have definitively been etiologically related to leukemogenesis, virus replication is usually extensive and the virus is *usually* easy to isolate. However, even in these model systems, there are cases in which virus expression is not readily manifest. For instance, in a significant number of cats with leukemia, FeLV has not been found, and antibodies to FeLV have not been readily detectable (Koshy *et al.,* in preparation). Thus man may not be an exception in lacking abundant replicating viruses; rather the exceptions could be those animals (e.g., cats, gibbons) that for some unexplained reason uniquely fail to control replication of the virus. (2) Detection of any trace of virus (including viral nucleic acids) may be extremely difficult if the processes leading to blood cell transformation are manifold, and if they involve previous alterations necessary to but insufficient for transformation by virus infection. A significant example in this regard may be a few cases of human leukemic recipients of allogeneic bone marrow from HLA identical siblings of the opposite sex. Following successful engraftment, acute leukemia has occurred *de novo* in a few instances in the *transplanted normal* leukocytes as shown by karyotypic characterization, e.g., male karyotype of donor cells in a female leukemic recipient with subsequent development of leukemia in the engrafted male donor cells. These patients were given immunosuppressive therapy, chemotherapy, and radiation therapy before transplantation. Were any of these factors involved in the leukemia? Perhaps whatever caused the original disease might still have been present and, in this regard, viruses must be suspect. (3) Another consideration is that the techniques used to detect viral sequences may not be sufficiently sensitive. This would be a particularly critical factor if leukemic cells are not necessarily the primary target of virus infection and transformation. In this proposal not all the leukemic cells need contain the acquired nucleic acid sequences. This possibility is supported by some new experimental data. Human B lymphocyte cell lines established after exposure of fresh human leukocytes and bone marrow to primate type-C viruses *in vitro* (Markham *et al.,* 1979) consist of a heterogeneous population of infected and uninfected cells (as assayed by *in situ* hybridization; N. Miller, unpublished results), and provirus is detectable in only some of these cell lines. Does this mean a small number of genetically altered cells can secondarily influence the growth of other cells? (4) Finally, integration of fragments of provirus (*incomplete* integrated provirus) provides yet another mechanism by which virus could be sufficient to maintain the transformed phenotype in some cells without ready detection. Evidence for integration of partial proviral integration, in fact, has been presented in the normal cells of a naturally infected gibbon, who died of an intercurrent disease (Wong-Staal *et al.,* 1978).

## IV. DISCUSSION

In this chapter we have reviewed DNA and RNA viruses with oncogenic activity, which may be quite variable in potency and range of effect. In no instance has the mechanism(s) of cellular transformation and neoplastic progression been defined. Particularly little is known about cellular-host factors that are involved in these processes. However, given a susceptible host, certain common factors are emerging with respect to the viruses. In most oncogenic viruses the cell-transforming information is present in a defined portion of the viral genome, sometimes as a single gene. In some oncogenic viruses of both the RNA and DNA classes, evidence is accruing that these "transforming genes" are not essential for the vegetative, replicative, and/or lytic functions of the virus. In general, specific transforming genes have been associated with viruses of greatest oncogenic potency in terms of both the number of virions and the time required for development of transformation. It is not clear if less potent viruses in which transforming genes have not been defined produce transformation by alternative means. However, a possibility that must be recognized is that over time such viruses could acquire such information. The life cycle of these viruses in a susceptible host would favor this possibility. The integration of the viral genome into the DNA of the host cell introduces the possibility of genetic recombination with host cell sequences, which for some retroviruses, at least, has been demonstrated to be the origin of the transforming information. Also, the prolonged association of viral information with the host cell could foster the complementation of genetic variants of the same virus or complementation with the genome of other superinfecting viruses. Certainly, many sources of potentially oncogenic viruses are quite heterogeneous and may contain nonreplicative, defective variants. As repeatedly cited in the text for different viruses, replicative defectiveness is a frequent, and for lytic viruses a necessary, correlate of transforming potential. Since almost all experimental tests most readily detect complete, infectious virus, it is possible that a minor content of defective, transforming variants, the generation of which might be favored by prolonged periods of latency, could be overlooked.

The preceding comments are relevant to our discussion about the search for possible viral involvement in human cancers throughout this chapter. As previously stated, there is no instance in which any human cancer has definitely been linked to a virus etiology. Even in the most suspect cases (by association), i.e., EBV in Burkitt's lymphoma and nasopharyngeal carcinoma, HSV-2 in cervical carcinoma, and certain retroviruses in leukemia, it is assumed that multiple factors are involved and that the virus may not be critical. However, it is important to recognize that the most likely crucial viral elements have not been analyzed. For example, only now are the technical capabilities of virus and gene cloning and analysis being developed to determine whether or not the genome of EBV integrated into the DNA of Burkitt's lymphoma cells contains gene sequences different from other sources of ubiquitous EBV. Up to this point, most experimental work has been advanced on the thesis that the animal systems having abundantly available replicative virus with a quantitatively measurable transforming effect provide the only valid model for viral oncogenesis. As emphasized at various points in the text, there are already indications that this is oversimplified and that extremely sensitive, specific tests able to detect viral gene fragments in perhaps a minor population of host cells are required to rule in or out a role for viruses in human neoplasia. Only then will it be possible to assess accurately the possible interaction of specific viral information with other intrinsic and extrinsic factors affecting tumorigenesis.

## ACKNOWLEDGMENTS

We are grateful to P. Pavlos, K. C. White, and A. Russell for assistance in preparation of this chapter.

## REFERENCES

Andersson-Anvret, M., Forsby, N., Klein, G., and Henle, W. (1977). Relationship between the Epstein-Barr virus and undifferentiated nasopharyngeal carcinoma: Correlated nucleic acid hybridization and histopathological examination. *Int. J. Cancer* **20**, 486-494.

Aoki, T., Walling, M. J., Bushar, G. S., Liu, M., and Hsu, K. C. (1976). Natural antibodies in sera from healthy humans to antigens on surfaces of type C RNA viruses and cells from primates. *Proc. Natl. Acad. Sci. U.S.A.* **73**, 2491-2495.

Aulakh, G. S., and Gallo, R. C. (1977). Rauscher leukemia virus related sequences in human DNA: Presence in some tissues of some patients with hematopoietic neoplasias and their absence in DNA from other tissues. *Proc. Natl. Acad. Sci. U.S.A.* **74**, 353-357.

Axel, R., Gulati, S., and Spiegelman, S. (1972). Particles containing RNA instructed DNA polymerase and virus related RNA in human breast cancers. *Proc. Natl. Acad. Sci. U.S.A.* **69**, 3133-3137.

Baxt, W., and Spiegelman, S. (1972). Nuclear DNA sequences present in human leukemic cells and absent in normal leukocytes. *Proc. Natl. Acad. Sci. U.S.A.* **69**, 3737-3741.

Baxt, W., Yates, A. W., Wallace, H. J., Jr., Holland, J. F., and Spiegelman, S. (1973). Leukemia specific DNA sequences in leukocytes of the leukemic member of identical twins. *Proc. Natl. Acad. Sci. U.S.A.* **70**, 2629-2632.

Benacerraf, B., and Katz, D. H. (1975). The histocompatability-linked immune response genes. *Adv. Cancer Res.* **21**, 121-173.

Benveniste, R. E., and Todaro, G. J. (1974). Evolution of type-C viral genus: Inheritance of exogenous acquired viral genes. *Nature (London)* **242**, 456-459.

Benveniste, R. E., and Todaro, G. J. (1976). Evolution of type-C viral genes: Evidence for an Asian origin of man. *Nature (London)* **261**, 101–108.

Beneniste, R. E., Sherr, C. J., and Todaro, G. J. (1975). Evolution of type-C viral genes: Origin of feline leukemia virus. *Science* **190**, 886–888.

Chandra, P., Balikcioglu, S., and Mildner, B. (1978). Biochemical and immunological characterization of a reverse transcriptase from human melanoma tissues. *Cancer Lett.* **5**, 299–310.

Churchill, A. E., and Biggs, P. M. (1967). Agent of Marek's disease in tissue culture. *Nature (London)* **215**, 528–530.

Colcher, D., Spiegelman, S., and Schlom, J. (1974). Evidence of particle associated RNA directed DNA polymerase and high molecular weight RNA in human gastrointestinal and lung malignancies. *Proc. Natl. Acad. Sci. U.S.A.* **71**, 3304–3308.

Corallini, A., Barbanti-Brodano, G., Portolani, M., Balboni, P. G., Grossi, M. P., Possati, K., Honorati, C., LaPlaca, M., Macroni, A., Caputo, A., Veronesi, U., Orefice, S., and Cardinali, G. (1976). Antibodies to BK virus structural and tumor antigens in human sera from normal persons and from patients with various diseases, including neoplasia. *Infect. Immun.* **13**, 1684–1691.

Crawford, D. H., Rickinson, A. B., Fingerty, S., and Epstein, M. A. (1978). Epstein-Barr (EB) virus genome-containing, EB nuclear antigen-negative B-lymphocyte populations in blood in acute infectious mononucleosis. *J. Gen. Virol.* **38**, 449–460.

Davis, D. B., and Kingsbury, D. T. (1976). Quantitation of viral DNA present in cells transformed by UV-irradiated herpes simplex virus. *J. Virol.* **17**, 788–793.

de-Thé, G., Geser, A., Day, N. E., Tuckei, P. M., Williams, E. H., Beri, D. P., Smith, P. G., Dean, A. G., Bornkamm, G. W., Feorino, P., and Henle, W. (1978). Epidemiological evidence for casual relationship between Epstein-Barr virus and Burkitt's lymphoma from Ugandan prospective study. *Nature (London)* **274**, 756–761.

Dougherty, R. M., and DeStefano, H. S. (1973). Isolation and characterization of a papovavirus from human urine. *Proc. Soc. Exp. Biol. Med.* **146**, 481–487.

Duff, R., and Rapp, F. (1971). Oncogenic transformation of hamster cells after exposure to herpes simplex virus type 2. *Nature (London) New Biol.* **233**, 48–50.

Eddy, B. E., Borman, G. S., Grubbs, G. E., and Young, R. D. (1968). Identification of the oncogenic substance in rhesus monkey-cell cultures as simian virus 40. *J. Virol.* **17**, 65–75.

Epstein, M. A. (1978). An assessment of the possible role of viruses in the etiology of Burkitt's lymphoma. *Prog. Exp. Tumor Res.* **21**, 72–99.

Falk, L., Deinhardt, F., Nonoyama, M., Wolfe, L. G., and Bergholz, C. (1976). Properties of a baboon lymphotropic herpesvirus related to Epstein-Barr virus. *Int. J. Cancer* **18**, 798–807.

Fareed, G. C., and Davoli, D. (1977). Molecular biology of papovavirus. *Annu. Rev. Biochem.* **46**, 471–522.

Fiori, M., and diMayorca, G. (1976). Occurrence of BK virus DNA in DNA obtained from certain human tumors. *Proc. Natl. Acad. Sci. U.S.A.* **73**, 4662–4666.

Frank, A., Andiman, W. A., and Miller, G. (1976). Epstein-Barr virus and nonhuman primates: Natural and experimental infection. *Adv. Cancer Res.* **23**, 171–201.

Frenkel, N., Roizman, B., Cassai, E., and Nahamias, A. (1972). A herpes simplex 2 DNA fragment and its transcription in human cervical cancer tissue. *Proc. Natl. Acad. Sci. U.S.A.* **69**, 3784–3789.

Frenkel, N., Locker, H., Batterson, W., Hayward, G. S., and Roizman, B. (1976). Anatomy of herpes simplex virus DNA. IV. Defective DNA originates from the S component. *J. Virol.* **20**, 527–531.

Fresen, K. ⁻O., Cho, M. ⁻S., and zur Hausen, H. (1978). Recovery of transforming EBV from non-producer cells after superinfection with non-transforming P3HR-1 EBV. *Int. J. Cancer* **22**, 378–383.

Gabelman, N., Waxman, S., Smith, W., and Douglas, S. D. (1975). Appearance of C-type virus-like particles after co-cultivation of a human tumor-cell line with rat (XC) cells. *Int. J. Cancer* **16**, 355–369.

Gallagher, R. E., and Gallo, R. C. (1975). Type-C RNA tumor virus isolated from cultured human acute myelogenous leukemia cells. *Science* **187**, 350–353.

Gallagher, R. E., Todaro, G. J., Smith, R. G., Livingston, D. M., and Gallo, R. C. (1974). Relationship between RNA-directed DNA polymerase (reverse transcriptase) from human acute leukemic blood cells and primates type-C viruses. *Proc. Natl. Acad. Sci. U.S.A.* **71**, 1309–1313.

Gallo, R. C., Yang, S. S., and Ting, R. C. (1970). RNA dependent DNA polymerase of human acute leukemic cells. *Nature (London)* **228**, 927.

Gallo, R. C., Miller, N. R., Saxinger, W. C., and Gillespie, D. (1973). Primate RNA tumor virus-like DNA synthesized endo-enously by RNA-dependent DNA polymerase in virus-like particles from fresh human acute leukemic blood cells. *Proc. Natl. Acad. Sci. U.S.A.* **70**, 3219–3224.

Gallo, R. C., Gallagher, R. E., Wong-Staal, F., Aoki, T., Markham, P., Schetters, H., Ruscetti, F., Valerio, M., Saxinger, W. C., Smith, R. G., Gillespie, D., and Reitz, M. S. (1978). Biochemical and biological studies on a gibbon ape (Hylobates lar) with lymphocytic leukemia. *Virology* **84**, 359–373.

Gardner, S. D., Field, A. M., Coleman, D. V., and Hulme, B. (1971). New human papovavirus (BK) isolated from urine after renal transplantation. *Lancet* **1**, 1253–1257.

Gerber, P. (1972). Activation of Epstein-Barr virus by 5′ bromodeoxyuridine in virus free human cells. *Proc. Natl. Acad. Sci. U.S.A.* **69**, 83–85.

Giraldo, G., Beth, E., Henle, W., Henle, G., Mike, V., Safai, B., Huraux, J. M., McHardy, J., and de-Thé, G. (1978). Antibody patterns to herpesviruses in Kaposi's sarcoma. II. Serological association of American Kaposi's sarcoma with cytomegalovirus. *Int. J. Cancer* **22**, 126–131.

Goldberg, R. J., and Gravell, M. (1976). A search for herpes simplex type 2 markers in cervical carcinoma. *Cancer Res.* **36**, 795–799.

Green, M., and Mackey, J. K. (1978). Are oncogenic human adenoviruses associated with human cancer? Analysis of human tumors for adenovirus transforming gene sequences. *In* "Origins of Human Cancer" (H. H. Hiatt, J. D. Watson, and J. A. Winsten, eds.), pp. 1027–1042. Cold Spring Harbor Lab., Cold Spring Harbor, New York.

Hampar, B., Derge, J. G., Martos, L. M., and Walker, J. L. (1972). Synthesis of Epstein-Barr virus after activation of the viral genome in a "virus negative" human lymphoblastoid cell (Raji) made resistant to 5-bromodeoxyuridine. *Proc. Natl. Acad. Sci. U.S.A.* **69**, 78–82.

Hardy, W. D., Old, L. J., Hess, P. W., Essex, M., and Cotter, S. M. (1973). Horizontal transmission of feline leukemia virus. *Nature (London)* **244**, 266–269.

Hayward, G. S., Frenkel, N., and Roizman, B. (1975). The anatomy of herpes simplex virus DNA: Strain difference and heterogeneity in the location of restriction endonuclease cleavage sites. *Proc. Natl. Acad. Sci. U.S.A.* **72**, 1768–1772.

Hayward, S. D., and Kieff, E. D. (1976). Epstein-Barr virus-specific RNA. I. Analysis of viral RNA in cellular extracts and in the polyribosomal fraction of permissive lymphoblastoid cell lines. *J. Virol.* **18**, 518–525.

Heinonen, O. P., Shapiro, S., Monson, R. R., Hartz, S. C., Rosenberg, L., and Slone, D. (1973). Immunization during pregnancy against poliomyelitis and influenza in relation to childhood malignancy. *Int. J. Epidemiol.* **2**, 229–235.

Henle, W., Henle, G., Zajac, B., Pearson, R., Waubke, R., and Scriba, M. (1970). Differential reactivity of human serums with early antigens induced by Epstein-Barr virus. *Science* **169**, 188-190.

Henle, W., Ho, J. H. C., Henle, G., Chau, J. C. W., and Kivan, H. C. (1977). Nasopharyngeal carcinoma: Significance of changes in Epstein-Barr virus-related antibody patterns following therapy. *Int. J. Cancer* **20**, 663-672.

Jacquemin, P., Saxinger, C., and Gallo, R. C. (1978). Surface antibodies of human myelogenous leukemia leukocytes reactive with specific type-C viral reverse transcriptase *Nature (London)* **276**, 230-236.

Jarrett, W. F. H. (1979). High incidence alimentary carcinoma in cattle associated with an environmental carcinogen and viral papillomas. In "Antiviral Mechanisms in the Control of Neoplasia" (P. Chandra, ed.), Plenum, pp. 39-45. New York.

Jordan, M. C., Rousseau, W. E., Stewart, J. A., Noble, G. R., and Chin, F. D. Y. (1973). Spontaneous cytomegalovirus mononucleosis clinical and laboratory observations in nine cases. *Ann. Intern. Med.* **79**, 153-160.

Jordan, M. C., Shanley, J. D., and Stevens, J. G. (1977). Immunosuppression reactivates and disseminates latent murine cytomegalovirus. *J. Gen. Virol.* **37**, 419-423.

Kaiser-McCaw, B., Epstein, A. L., Kaplan, H. S., and Hecht, F. (1977). Chromosome₁₄ translocation in African and North American Burkitt's lymphoma. *Int. J. Cancer* **19**, 482-486.

Kalter, S. S., Helmke, R. J., Panigel, M., Heberling, R. L., Felsburg, P. J., and Axelrod, L. R. (1973). Observations on apparent C-type particles in baboon (Papio cynocephalus) placentas. *Science* **179**, 1332-1333.

Kaplan, H. S., Goodenow, R. S., Epstein, A. L., Gartner, S., DeCleve, A., and Rosenthal, P. N. (1977). Isolation of type-C RNA virus from an established human histiocytic lymphoma cell line. *Proc. Natl. Acad. Sci. U.S.A.* **74**, 2564-2568.

Kaplan, H. S., Goodenow, R. S., Gartner, S., and Bieber, M. M. (1979). Biology and virology of the human malignant lymphomas. *Cancer* **43**, 1-24.

Katsuki, T., Hinuma, Y., Yamamoto, N., Abo, T., and Kumagai, K. (1977). Identification of the target cells in human B lymphocytes for transformation by Epstein-Barr virus. *Virology* **83**, 287-294.

Kawakami, T. G., Huff, S. D., Buckly, P. M. et al. (1972). C-type virus associated with gibbon lymphosarcoma. *Nature (London), New Biol.* **235**, 170-171.

Kawakami, T. G., Buckly, P. M., McDowell, T. S., et al. (1973). Antibodies to simian C-type virus antigen in sera of gibbon (Hylobates sp.). *Nature (London), New Biol.* **246**, 105-107.

Kidd, J. G., and Rous, P. (1940). Cancers deriving from the virus papillomas of wild rabbits under natural conditions. *J. Exp. Med.* **71**, 469-494.

Klein, G. (1975). Studies on the Epstein-Barr virus genome, and the EBV-determined nuclear antigen in human malignant disease. *Cold Spring Harbor Symp. Quant. Biol.* **39**, 783.

Klein, G. (1975). The Epstein-Barr virus and neoplasia. *N. Engl. J. Med.* **293**, 1353-1356.

Klein, G., Giovanella, B. V., Lindahl, T., Fialkow, P. J., Singh, S., and Stehlin, J. S. (1974). Direct evidence for the presence of Epstein-Barr virus DNA and nuclear antigen in malignant epithelial cells from patients with poorly differentiated carcinoma of the nasopharynx. *Proc. Natl. Acad. Sci. U.S.A.* **71**, 4737-4741.

Klemola, E., von Essen, R., Henle, G., and Henle, W. (1970). Infectious-mononucleosis-like disease with negative heterophile agglutination test. Clinical features in relation to Epstein-Barr virus and cytomegalovirus. *J. Infect. Dis.* **121**, 608-614.

Kotler, M., Weinberg, E., Hospel, O., Olshesky, U., and Becker, Y. (1973). Particles released from arginine-deprived human leukemic cells. *Nature (London), New Biol.* **244**, 197-200.

Kraiselberg, E., Cage, L. P., and Weissbach, A. (1975). Presence of a herpes simplex virus

DNA fragment in L cell clone obtained after infection with irradiated herpes simplex virus 1. *J. Mol. Biol.* **97**, 533–542.

Krakower, J. M., and Aaronson, S. A. (1978). Radioimmunological characterization of RD-114 reverse transcriptase: Evolutionary relatedness of mammalian type-C viral *pol* gene products. *Virology* **86**, 127–137.

Kurth, R., and Mikschy, U. (1978). Human antibodies reactive with purified envelope antigens of primate type-C tumor virus. *Proc. Natl. Acad. Sci. U.S.A.* **75**, 5692–5696.

Kurth, R., Teich, N., Weiss, R., and Oliver, R. T. D. (1977). Natural antibodies reactive with primate type-C viral antigens. *Proc. Natl. Acad. Sci. U.S.A.* **74**, 1237–1241.

Lang, D. J., Cheung, K. S., Schwartz, J. N., Daniels, C. A., and Harwood, S. E. (1976). Cytomegalovirus replication and the host immune response. *Yale J. Biol. Med.* **49**, 45–48.

Lieber, M. M., Sherr, C. J., Todaro, G. J., Benveniste, R. E., Callahan, R., Conn, H. G. (1975). Isolation from the asian mouse *mus caroli* of an endogenous type-C virus. *Proc. Natl. Acad. Sci. U.S.A.* **72**, 2315–2319.

Luborsky, S. W., Chang, C., Pancake, S. J., and Mora, P. T. (1978). Comparative behavior of simian virus 40 T-antigen and of tumor-specific surface and transplantation antigens during partial purification. *Cancer Res.* **38**, 2367–2371.

Luka, J., Lindahl, T., and Klein, G. (1978). Purification of the Epstein-Barr virus-determined nuclear antigen from Epstein-Barr virus-transformed human lymphoid cell lines. *J. Virol.* **27**, 604–611.

McDougall, J. K., Chen, L. B., and Gallimore, P. H. (1978). Transformation *in vitro* by adenovirus type 2. A model system for studying mechanisms of oncogenicity. *In* "Origins of Human Cancer" (H. H. Hiatt, J. D. Watson, and J. A. Winsten, eds.), pp. 1015–1025. Cold Spring Harbor Lab., Cold Spring Harbor, New York.

McDougall, J. K., Galloway, D. A., and Fenoglio, C. M. (1979). *In situ* cytolocial hybridization to detect herpes simplex virus RNA in human tissues. *In* "Antiviral Mechanisms in the Control of Neoplasia" (P. Chandra, ed.), pp. 233–240. Plenum, New York.

Mackey, J. K., Rigden, P. M., and Green, M. (1976). Do highly oncogenic group A human adenoviruses cause human cancer? Analysis of human tumors for adenovirus 12 transforming DNA sequences. *Proc. Natl. Acad. Sci. U.S.A.* **73**, 4657–4661.

Maitland, N. J., and McDougall, J. K. (1977). Biochemical transforming of mouse cells by fragments of herpes simplex virus DNA. *Cell* **11**, 233–241.

Major, E. O., and diMayorca, G. (1973). Malignant transformation of $BKH_{21}$ clone 13 cells by BK virus-A human papovavirus. *Proc. Natl. Acad. Sci. U.S.A.* **70**, 3210–3212.

Mak, T. W., Manaster, I., Howatson, A. F., McCulloch, E. A., and Till, J. E. (1974). Particles with characteristics of leukoviruses in cultures of marrow cells from leukemic patients in remission and relapse. *Proc. Natl. Acad. Sci. U.S.A.* **71**, 4336–4340.

Markham, P. D., Ruscetti, F., Salahuddin, S., Gallagher, R. E., and Gallo, R. C. (1979). Enhanced induction of growth of B lymphoblasts from fresh blood by primate type-C retroviruses (gibbon ape leukemia virus and simian sarcoma virus). *Int. J. Cancer* **23**, 148–156.

Melnick, J. L., Lewis, R., Wimberly, I., Kaufman, R. H., and Adam, E. (1978). Association of cytomegalovirus (CMV) infection with cervical cancer: Isolation of CMV from cell cultures derived from cervical biopsy. *Intervirology* **10**, 115–119.

Miller, N. R., Saxinger, W. C., Reitz, M. S., Gallagher, R. E., Wu, A. M., Gallo, R. C., and Gillespie, D. (1974). Systematics of RNA tumor viruses and virus-like particles of human origin. *Proc. Natl. Acad. Sci. U.S.A.* **71**, 3177–3181.

Minson, A. C., Thouless, M. E., Eglin, R. P., and Darby, G. (1976). The detection of virus DNA sequences in a herpes type 2 transformed hamster cell line (333-8-9). *Int. J. Cancer* **17**, 493–500.

Mizell, M., Toplin, J., and Isaacs, J. J. (1969). Tumor induction in developing frog kidneys by a zonal centrifuge purified fraction of the frog herpes-type virus. *Science* 165, 1134–1137.

Morison, W. L. (1975). Viral warts, herpes simplex and herpes zoster in patients with secondary immune deficiencies and neoplasms. *Br. J. Dermatol.* 92, 625–630.

Moyer, R. C., Moyer, M. P., and Gerodetti, M. H. (1978). Rescue of infectious virus from permissive monkey cells containing simian virus 40 DNA fragments. *J. Virol.* 26, 272–280.

Munyon, W., Kraiselberg, E., Davis, D., and Mann, J. (1971). Transfer of thymidine kinaseless L cells by infection with ultraviolet-irradiated herpes simplex virus. *J. Virol.* 7, 813–820.

Nazerian, K., Solomon, J. J., Witten, R. L., and Burmester, B. R. (1968). Studies on the etiology of Marek's disease. *Proc. Soc. Exp. Biol. Med.* 127, 177–182.

Nilsson, K. (1971). High frequency establishment of human immunoglobin-producing lymphoblastoid lines from normal and malignant lymphoid tissue and peripheral blood. *Int. J. Cancer* 8, 432–442.

Nilsson, K., Giovanella, B. C., Stehlin, J. S., and Klein, G. (1977). Tumorigenicity of human hematopoietic cell lines in athymic nude mice. *Int. J. Cancer* 19, 337–344.

Nooter, K., Aarssen, A. M., Bentvelzen, P., and d'Groot, F. G. (1975). Isolation of an infectious C-type oncornavirus from human leukemic bone marrow cells. *Nature (London)* 256, 595–597.

Olson, C., Gordon, D. E., Robl, M. G., and Lee, P. K. (1969). Oncogenicity of bovine papilloma virus. *Arch. Environ. Health* 19, 827–837.

Olson, C., Miller, L. D., Miller, J. M., *et al.* (1972). Transmission of lymphosarcoma from cattle to sheep. *J. Natl. Cancer Inst.* 49, 1463–1468.

Orth, G., Brietburd, F., Favre, M., and Croissant, O. (1978a). Papillomaviruses: Possible role in human cancer. *In* "Origins of Human Cancer" (H. H. Hiatt, J. D. Watson, and J. A. Winste, eds.), pp. 1043–1068. Cold Spring Harbor Lab., Cold Spring Harbor, New York.

Orth, G., Jablinska, S., Favre, M., Croissant, O., Jarzabek-Chorzelska, M., and Rzesa, G (1978b). Characterization of two types of human papillomaviruses in lesions of epidermodyslplasia verruciformis. *Proc. Natl. Acad. Sci. U.S.A.* 75, 1537–1541.

Padgett, B. L., and Walker, D. L. (1973). Prevalence of antibodies in human sera against JC virus, an isolate from a case of progressive multifocal leukoencephalopathy. *J. Infect. Dis.* 127, 467–470.

Padgett, B. L., Walker, D. L., zuRhein, G. M., Eckroade, R. J., and Dessel, B. H. (1971). Cultivation of papova-like virus from human brain with progressive multifocal leucoencephalopathy. *Lancet* 1, 1257–1260.

Pagano, J. S. (1975). Diseases and mechanisms of persistent DNA virus infection: Latency and cellular transformation. *J. Infect. Dis.* 132, 209–223.

Pagano, J. S., Huang, C. H., and Huang, Y. T. (1976). Epstein-Barr virus genome in infectious mononucleosis. *Nature (London)* 263, 787–789.

Panem, S., Prochownik, E. V., Reale, F. R., and Kirsten, W. H. (1975). Isolation of C-type virions from a normal human fibroblast strain. *Science* 189, 297–299.

Panem, S., Nelson, G., Ordonez, G., Dalton, H., and Soltani, K. (1978). Viral immune complexes is systemic lupus erythematosus: C-type viral complex deposition in skin. *J. Invest. Dermatol.* 71(4); 260–262.

Parks, W. P., Scolnick, E. M., Noon, M. C., *et al.*, (1973). Radioimmunoassay mammalian type-C polypeptides. IV. Characterization of woolly monkey and gibbon viral antigens. *Int. J. Cancer* 12, 129–137.

Pearson, G. R., Johansson, B., and Klein, G. (1978). Antibody-dependent cellular cytotoxicity against Epstein-Barr virus-associated antigens in African patients with nasopharyngeal carcinoma. *Int. J. Cancer* **22**, 120–125.

Pfister, H., and zur Hausen, H. (1978). Seroepidemiological studies of human papilloma virus (HPV-1) infections. *Int. J. Cancer* **21**, 161–165.

Prochownik, E. V., and Kirsten, W. H. (1977). Nucleic acid sequences of primate type-C viruses in normal and neoplastic human tissues. *Nature (London)* **267**, 175–177.

Pyrrhonen, S. (1976). Serological aspects of the immunology of human warts. *In* "Biomedical Aspects of Human Wart Virus Infection" (M. Prunieras, ed.), p. 191. Fondation Merieux, Lyon, France.

Raab-Traub, N., Pritchett, R., and Kieff, E. (1978). DNA of Epstein-Barr virus. III. Identification of restriction enzyme fragments that contain DNA sequences which differ among strains of Epstein-Barr virus. *J. Virol.* **27**, 388–398.

Rabin, H., Neubauer, R. H., and Hopkins, R. F., III (1977). Studies on Epstein-Barr (EPV) like viruses of Old World non-human primates. *In* "Advances in Comparative Leukemia Research 1977" (P. Bentvelzen, J. Hilgers, and D. S. Yohn, eds.), pp. 205–208. Elsevier/North-Holland Biomedical Press, Amsterdam.

Rangan, S. R. S., and Gallagher, R. E. (1979). *Adv. Virus Res.* **24**, 1–123.

Rapp, F., Geder, L., Muraski, D., Lausch, R., Ladda, R., Huang, E. S., and Webber, M. M. (1975). Long-term persistence of cytomegalovirus genome in cultured human cells of prostatic origin. *J. Virol.* **16**, 982–990.

Rawls, W. E., and Adam, E. (1978). Herpes simplex viruses and human malignancies. *In* "Origins of Human Cancer" (H. H. Hiatt, J. D. Watson, and J. A. Winsten, eds.), pp. 1133–1155. Cold Spring Harbor Lab., Cold Spring Harbor, New York.

Reitz, M. S., Miller, N. R., Wong-Staal, F., Gallagher, R. E., Gallo, R. C., and Gillespie, D. H. (1976). Primate type-C virus nucleic acid sequences (woolly monkey and baboon types) in tissues from a patient with actue myelogenous leukemia and in viruses isolated from cultured cells of the same patient. *Proc. Natl. Acad. Sci. U.S.A.* **73**, 2113–2117.

Reitz, M. S., Jr., Wong-Staal, F., Haseltine, W. A., Kleid, D. G., Trainor, C. D., Gallagher, R. C., and Gallo, R. C. (1979). Gibbon Ape Leukemia virus—Hall's Island: New strain of Gibbon Ape leukemia virus. *J. Virol.* **29**(1); 395–400.

Roizman, B., and Kieff, E. (1975). Herpes simplex and Epstein-Barr viruses in human cells and tissues: A study in contrasts. *In* "Cancer: A Comprehensive Treatise" (F. F. Becker, ed.), Vol. 2, pp. 241–322. Plenum, New York.

Rous, P., and Friedewald, W. F. (1944). The effect of chemical carcinogens on virus-induced rabbit papillomas. *J. Exp. Med.* **79**, 511–538.

Rubin, H., Conreliums, A., and Fanshier, L. (1961). The pattern of congenital transmission of an avian leukosis virus. *Proc. Natl. Acad. Sci. U.S.A.* **47**, 1058.

Scolnick, E. M., Parks, W. P., Kawakami, T. G., *et al.* (1974). Primate and murine type-C viral nucleic acid association kinetics: Analysis of model systems and natural tissues. *J. Virol.* **13**, 363–369.

Shah, K. V., Daniel, R. W., and Warszawski, R. (1973). High prevalence of antibodies to BK virus, and SV40 related papovavirus, in residents of Maryland. *J. Infect. Dis.* **128**, 784–787.

Shah, K. V., Daniel, R. W., and Strandberg, J. D. (1975). Sarcoma in a hamster inoculated with BK virus, a human papovavirus. *J. Natl. Cancer Inst.* **54**, 945–950.

Sheldon, P. J., Papamichael, M., Hemsted, E. H., and Holborow, E. J. (1973). Thymic origin of atypical lymphoid cells in infectious mononucleosis. *Lancet* **1**, 1153–1155.

Sherr, C. J., and Todaro, G. J. (1975). Primate type-C virus p30 antigen in cells from humans with acute leukemia. *Science* **187**, 855–857.

Snyder, H. W., Pincus, T., and Fleissner, E. (1976). Specificities of human immunoglobulins reactive with antigens in preparations of several mammalian type-C viruses. *Virology* **75**, 60–73.

Spencer, E. S., and Anderson, H. K. (1970). Clinically evident, non-terminal infectious with herpesviruses and the wart virus in immunosuppressed renal allograft recipients. *Br. Med. J.* **3**, 251–254.

Steel, C. M., Philipson, J., Arthur, E., Gardiner, S. E., Newton, M. S., and McIntosh, R. V. (1977). Possibility of EB virus perferentially transforming a sub-population of human B lymphocytes. *Nature (London)* **270**, 729–730.

Syverton, J. T. (1952). The pathogenesis of the rabbit papilloma to carcinoma sequence. *Ann. N.Y. Acad. Sci.* **54**, 1126–1140.

Takahashi, M., and Yamanishi, K. (1974). Transformation of hamster embryo and human embryo cells by temperature sensitive mutants of herpes simplex virus type 2. *Virology* **61**, 306–311.

Takemoto, K. K., and Mullarkey, M. F. (1973). Human papovavirus, BK strain: Biological studies including antigenic relationship to simian virus 40. *J. Virol.* **12**, 625–631.

Takemoto, K. K., Solomon, D., Israel, M., Howley, P. M., Khoury, G., Aaronson, S., and Martin, M. A. (1978). Search for evidence of papovavirus involvement in human cancer. *In* "Advances in Comparative Leukemia Research 1977" (P. Bentvelzen, J. Hilgers, and D. S. Yohn, eds.), pp. 260–263. Elsevier/North-Holland Biomedical Press, Amsterdam.

Teich, N. M., Weiss, R. A., Salahuddin, S. Z., Gallagher, R. E., Gillespie, D. H., and Gallo, R. C. (1975). Infective transmission and characterization of a C-type virus released by culured human myeloid leukemic cells. *Nature (London)* **256**, 551–555.

Theilen, G. H., Gould, D., Fowler, M. *et al.* (1971). C-type virus in tumor tissue of a woolly monkey (Lagothrix ssp.) with fibrosarcoma. *J. Natl. Cancer Inst.* **47**, 881–889.

Todaro, G. J., and Gallo, R. C. (1973). Immunological relationship of DNA polymerase from acute leukemia cells and primate and mouse leukemia virus reverse transcriptase. *Nature (London)* **244**, 206–209.

Todaro, G. J.; Benveniste, R. E., Callahan, R., *et al.* (1974). Endogneous primate and feline type-C viruses. *Cold Spring Harbor Symp. Quant. Biol.* **39**, 1159–1168.

Todaro, G. J., Lieber, M. M., Benveniste, R. E., Sherr, C. J., Gibbs, C. J., and Gajdusek, D. C. (1975). Infectious primate type-C viruses: Three isolates belonging to a new subgroup from the normal gibbons. *Virology* **67**, 335–357.

Traul, K. A., Stephens, R., Gerber, P., and Peterson, W. D. (1977). Production Epstein-Barr viral infection of the human lymphoblastoid cell line 6410 with release of early antigen inducing and transforming virus. *Int. J. Cancer* **20**, 247–255.

Tronick, S. R., Stephenson, J. R., Aaronson, S. A., and Kawakami, T. G. (1975). Antigenic characterization of type-C virus isolates of gibbon apes. *J. Virol.* **15**(1); 115–120.

Walker, D. L., Padgett, B. L., zuRhein, G. M., Albert, A. E., and March, R. F. (1973). Human papovavirus (JC): Induction of brain tumors in hamsters. *Science* **181**, 674–676.

Weiss, A. F., Portmann, R., Fischer, H., Simon, J., and Zang, K. D. (1975). Simian virus 40 related antigens in three human miningiomas with defined chromosome loss. *Proc. Natl. Acad. Sci. U.S.A.* **72**, 609–613.

Wigler, M., Silverstein, S., Lee, L.-S., Pellicer, A., Cheng, Y. C., and Axel, R. (1977). Transfer of purified herpes virus thymidine kinase gene to cultured mouse cells. *Cell* **11**, 223–232.

Wold, W. S. M., Mackey, J. K., Brachmann, K. H., Takemori, N., Rigden, R., and Green, M. (1978). Analysis of human tumors and human malignant cell lines for BK virus-specific DNA sequences. *Proc. Natl. Acad. Sci. U.S.A.* **75**, 454–458.

Wolfe, L. G., Deinhardt, F., and Theilen, G. H. (1971). Induction of tumors in marmoset monkeys by simian sarcoma virus, type I, (Lagothrix): A preliminary report. *J. Natl. Cancer Inst.* **47**, 1115–1120.

Wong-Staal, F., Gallo, R. C., and Gillespie, D. (1975). Genetic relationship of a primate RNA tumor virus genome to genes in normal mice. *Nature (London)* **256,** 670–672.

Wong-Staal, F., Gillespie, D., and Gallo, R. C. (1976). Proviral sequences of baboon endogenous type-C RNA virus in DNA of leukemic tissues from seven patients with myelogenous leukemia. *Nature (London)* **262,** 190–195.

Wong-Staal, F., Josephs, S., Dalla Favera, F., and Gallo, R. C. (1978). Detection of integrated type-C viral DNA fragments in two primates (human and gibbon) the restriction enzyme-blotting technique. *In* "Modern Trends in Human Leukemia" (R. Neth and K. M. Mannweiler, eds.), Part III. Springer-Verlag, Berlin and New York (in press).

Wong-Staal, F., and Gallo, R. C. (1979). "Molecular Biology of Primate Retroviruses in Viral Oncology" pp. 399–431. Raven Press, New York.

# Chapter Five

# Recombinant DNA Research

## ROBERT B. HELLING

## I. INTRODUCTION

The revelation of the function and general structure of DNA sparked a decade of intensive research in molecular biology that culminated in 1961–1963 with the discoveries of the basis of the genetic code, the coding

LABORATORY SAFETY: THEORY AND PRACTICE
Copyright © 1980 by Academic Press, Inc.
All rights of reproduction in any form reserved.
ISBN 0-12-269980-7

dictionary, punctuation in the code, messenger RNA (mRNA), transcription and translation, and the general nature of controls over gene expression. A second wave of excitement in molecular biology has been generated by the development of new procedures for the manipulation of genes—procedures that allow genes from any organism to be inserted into unrelated organisms where those genes can be probed and changed at will. Lewis Thomas (1978) stated, "This may be the greatest scientific opportunity for biology in this century."

For thousands of years humans have genetically manipulated their domestic animals and cultivated plants. However, by definition, the exchange of genes between higher organisms has been restricted to individuals of the same species. Now the species barrier can be circumvented in placing small fragments of DNA from any source into such cells as the bacteria *Escherichia coli* and *Bacillus subtilis,* the baker's yeast *Saccharomyces cereviseae,* and mammalian cells grown in culture.

Random DNA pieces cannot normally duplicate independently in any cell. Thus desired DNA duplexes are not simply inserted within cells but must be spliced into other small DNA molecules capable of self-replication. These *replicons* are of two sorts: (1) *plasmids,* or small, circular, self-replicating DNA molecules found commonly in prokaryotes and in eukaryotic organelles and (2) *viruses,* such as the phage lambda and the mammalian viruses polyoma and SV40. The faithful reproduction of a DNA segment in a small replicon is often termed *molecular cloning* or the *cloning of genes.* The term *cloning,* on the other hand, often refers to the exact reproduction of a whole organism to form a population of genetically identical organisms; this does not generally involve recombinant DNA technology.

Development of the ability to join a DNA segment to a replicon stems from the discovery of a natural procedure for joining the two ends of the lambda chromosome (Hershey *et al.,* 1963). The chromosome is bounded by short single-strand termini that are complementary, so that, under appropriate conditions of temperature and ionic strength, they pair, thus circularizing the chromosome or joining two different molecules. This finding suggested that the lambda DNA may normally circularize, and such circular molecules were indeed found in infected cells. However, unlike the natural circles, the circles formed *in vitro* by end pairing are nicked in each strand, that is, each complementary strand contains a single break corresponding to the original ends of the noncircular DNA (Fig. 1). A search for an enzyme capable of sealing the single-strand breaks culminated in the discovery by Gellert (1967) of DNA ligase. The use of the two methods—one for the joining of duplex DNA molecules through hybridization of complementary ends and the other, covalent joining, or ligation, of two

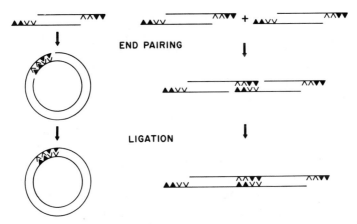

**Fig. 1.**   Joining of complementary ends of DNA molecules. Left; intramolecular joining to form a circular molecule. Right; joining of molecules from different sources.

single strands—led to the *in vitro* construction of a biologically functional plasmid containing DNA from two sources (Cohen *et al.*, 1973).

## II. THE IMPORTANCE OF MOLECULAR CLONING

Articles on molecular cloning and other forms of genetic engineering are commonplace in papers and magazines, and the topic has even found frequent expression through political cartoons and "funnies." This is the result of the tremendous potential for both basic and applied use, and also of the furor over possible hazards (and, on occasion, of the egos of the individuals involved).

The advantages of recombinant DNA technology lie in the ability to make large amounts of any gene and/or its products; to study any gene in a simple, controllable environment; and to extend systems for high-resolution genetic analysis to any organism.

### A. Isolation and Production of Specific Genes and Their Products in High Yield and Purity

The amounts of a specific cloned DNA segment or the corresponding RNA and protein products, or of the products formed under the direction of that protein, can be amplified relative to those amounts in the original cell. This is not an advantage in every instance. For example, one is unlikely to place a globin gene in *E. coli* in order to make large amounts of one of the polypeptides of hemoglobin. However, it has been extremely useful to be able to obtain large amounts of globin DNA by cloning so as to be able to determine its structure (Efstratiadis *et al.,* 1977). Furthermore, most eukaryotic proteins and their corresponding mRNAs are present in small amounts and often are formed only in a limited number of cells during a specific developmental period.

Let us consider the advantage of gene amplification through molecular cloning in more detail. Several hundred to a thousand or more copies of an appropriate plasmid can be formed per cell (Clewell, 1972; Hershfield *et al.,* 1974). This represents an equivalent purification relative to the chromosomal genes of *E. coli* and an overall purification for a unique gene from a eukaryote cloned in that plasmid of perhaps $10^3$ to $10^6$, depending on the complexity of the eukaryotic DNA compared to that of *E. coli.* From these amplified genes it should be possible to make large amounts of the RNA and protein products. Twenty percent or more of the total cellular protein and an equivalent amount of RNA have been shown to be produced from a single operon (*trp*) cloned in a suitable replicon (Hershfield *et al.,* 1974). At these levels of amplification it becomes realistic to think in terms of purifying gram quantities of growth hormone, specific enzymes, antigens, specific immunoglobulins, human insulin, interferon, and almost any other protein.

The requirements for producing large amounts of a gene product go beyond merely placing the gene in the *E. coli* cell. In most cases the gene is unlikely to produce its normal product, at least in high yield (Chang *et al.,* 1975; Meagher *et al.,* 1977), because the controls over transcription, translation, and processing of RNA and protein differ between prokaryotes and eukaryotes, and expression of a given gene can vary even within different strains of the same bacterium. Nevertheless, it is now clear that a eukaryotic gene can often be expressed in prokaryotic cells to give a functional product (Struhl and David, 1977), further engineering can allow a cloned gene to form its product in a substantial yield (Polisky *et al.,* 1976; Itakura *et al.,* 1977), and the availability of a eukaryotic microorganism as the host cell is expected to allow correct expression of most genes from eukaryotes (Hinnen *et al.,* 1978).

## B. Analysis of Gene and Chromosome Structure

Imagine a scientist attempting to determine the nucleotide sequence of a unique gene from among the 2 million gene equivalents of total human DNA. Now contrast that with a second approach in which the same gene is present with only two others on a readily purified plasmid. The near impossibility of the first task is as obvious as the ease of the second.

One might ask why DNA sequencing is important, given the availability of methods for sequencing the corresponding RNA or protein. However, it is obvious that those products must then be purified. Much DNA is transcribed rarely if at all or makes mRNA and protein copies only in specialized cells. Current methods for DNA sequencing are simpler and more rapid than those for RNA or protein (Maxam and Gilbert 1977; Sanger *et al.,* 1977). Furthermore, sequencing the gene products would not provide the structure of regions flanking a structural gene or the identity of adjacent genes, nor would internal DNA sequences not represented in the RNA or protein be found.

Gene cloning has already been used to restrict the models proposed to explain the general structure of the *Drosophila* chromosome (Wensink *et al.,* 1974) and to show that immunoglobulin (Tonegawa *et al.,* 1978) and globin (Tilghman *et al.,* 1978) are made from "split genes." Elegant experiments utilizing cloned DNA have shown that phase variation in *Salmonella* results from a regular, programmed change in the structure of a gene controlling flagellar synthesis. Cells of this bacterium can form either of two types of flagella, but not both at the same time. Zaig *et al.* (1977) cloned the flagellar genes from two *Salmonella* populations exhibiting opposite phases. The cloned segments differed by an inversion ("flip-flop") of a short region of DNA in the controlling gene.

## C. Analysis of Gene Function

The cells of many organisms are difficult to grow in the laboratory. Growth may be extremely slow, may require a complex and expensive medium, or may not even be possible with today's technology. To understand the mechanisms for control of gene expression, it is desirable to have the gene in a simple cell capable of rapid growth in a well-controlled environment and/or to obtain the purified DNA and study its behavior in an *in vitro* transcription and translation system. The extensive knowledge about a few single-celled organisms such as *E. coli* and yeast, and the genetic and physiologic procedures available with these cells enhance their use in the study of gene function. The goal of many geneticists working with cloned eukaryotic genes is to discover how the genes are regulated. Molecular cloning technology will allow scientists to achieve a level of understanding of genes of complex organisms that previously seemed possible only for viruses and prokaryotes.

## D. Manipulation of Genes

Various sophisticated procedures for manipulating genes are available in *E. coli, B. subtilus,* and yeast. These include the use of specific mutagens, powerful selective methods, and several mechanisms for recombination. The procedures can be used to couple genes to new promoters, allowing a high rate of expression (Polisky *et al.,* 1976; Itakura *et al.,* 1977), and to form modified genes with tailormade gene products. For example, the gene for a protein of therapeutic importance might be modified so that the protein has reduced immunogenicity. Imagine the impact on immunology and immunochemistry of having a large number of immunoglobulins available that differed only in a single amino acid at a specified location!

## E. Molecular Genetics for Any Organism

The advantages of recombinant DNA technology lie not only in the results of transferring genes into well-known organisms such as *E coli* but in developing equivalent systems in other organisms heretofore unapproachable at this level. The greatest benefit to be derived from cloning genes in a mammalian cell is the greater understanding of the cell and of viruses that attack it. For example, the availability of a cell line containing a cloned suppressor tRNA gene might allow the development of a conditional lethal genetic system of a type used to great advantage with bacteria and their viruses. It should also be possible, using recombinant techniques with cells growing in culture, to gain a better understanding of many human genetic disorders and to develop possible therapeutic procedures.

## F. Engineering of Cells and Organisms

Development of improved strains of organisms of agricultural, ecological, economic, or medical importance should be possible by the addition of single desired genes or small sets of genes. Commonly cited examples include formation of higher yielding varieties of agriculturally important plants, construction of microorganisms capable of high yields of nitrogen fixation in symbiosis with nonleguminous plants or of varieties of nonleguminous plants that fix nitrogen directly, and development of microorganisms capable of converting wastes to usable end products. The achievement of most of these goals still requires overcoming considerable technical obstacles. Some may be reached in other ways, but it seems quite certain that recombinant DNA procedures will be of great importance to such efforts. No safe or feasible way to correct human disorders by directly adding genes to the human organism is presently available. However, transfer of specific DNA segments to intact animals has been achieved (Cline *et al.*, 1980). The problems and possibilities associated with overcoming genetic disease by genetic engineering have been discussed in detail (Friedmann and Roblin, 1972).

## III. *IN VITRO* CLONING PROCEDURES

Molecular cloning involves the following steps: (1) segmenting the DNA to a relatively small size [generally, the average size is several thousand base pairs; the size of an average gene is about 1000 base pairs (1 kilobase pair or kbp)]; (2) joining the desired segment to a plasmid or virus replicon (*cloning vehicle* or *cloning vector*); (3) inserting into a host cell (*transformation*); (4) selecting the host cell containing the desired DNA.

## A. Cloning Genes in Plasmids

An idealized plasmid for use as a molecular cloning vehicle is depicted in Fig. 2. The plasmid should be small to reduce the likelihood of including undesirable genes and to prevent self-transmission to another cell by conjugation. The yield and purity of cloned genes and the efficiency of ligation and transformation are also greater with smaller DNA molecules. The plasmid contains a replicator locus that allows replication, and a gene $A^+$ to which selection can be applied, so as to be able to isolate cells containing the plasmid from other cells. In the most commonly used plasmids that gene confers resistance to an antibiotic (e.g., ampicillin, tetracycline, kanamycin).

The plasmid is cleaved at a single point, usually by a site-specific nuclease (*restriction endonuclease*). Fragments of the desired DNA are added, and end joining is allowed to take place. The joints are sealed with DNA ligase, and the modified plasmid is inserted into a cell. There the cloned DNA is replicated as a part of the plasmid.

Two additional features of the model plasmid of Fig. 2 are desirable. If the insertion of DNA occurs in a known plasmid gene, the gene will be inactivated (*insertional inactivation*). Cells containing the engineered plasmid can be distinguished from cells containing an unmodified plasmid because the former no longer express the split gene. Presence of a promoter that stimulates transcription across the cloned DNA is also desirable, particularly if the transcription can be initiated or repressed by the experimenter. Thus it might be possible to make the products of a gene that would not otherwise express itself in a foreign cell.

**Fig. 2.** Molecular cloning of DNA in a plasmid. After splitting gene B, the DNA segment to be cloned is inserted at the break point. The dashed arrow depicts transcription from the gene B promoter.

## B. Cloning Genes in Viruses

Procedures for adding genes to most commonly used viruses, such as lambda, differ somewhat from those for plasmids because the viral chromosome is not circular and because the modified chromosome must not be too small or too large to fit into the viral coat. Certain viruses such as the mammalian viruses SV40 and polyoma have circular chromosomes and, if attainment of the finished virus structure is not essential, can be treated essentially as plasmids.

Two approaches to cloning genes into lambda are depicted in Fig. 3. Again the cloning vehicle is cleaved at a specific point, the desired DNA is spliced in, and the DNA is inserted in a sensitive host cell by transformation (Helling and Lomax, 1978).

After infection the modified chromosome replicates and is packaged within a viral coat. Successful packaging in lambda "heads" requires the normal lambda ends, and a total DNA length corresponding to 75–109% of the normal lambda chromosome. As shown in the first model, the lambda DNA is split in nonessential gene A+, and the "foreign" DNA pairs with each of the freed ends. Joining is random but even if rejoining to form a modified chromosome containing an insertion is infrequent, selective procedures can be employed to identify recombinant phage. They will appear as A⁻ instead of A⁺. For example, the appearance of the viral "plaque" in

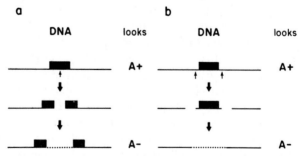

**Fig. 3.** Cloning genes in a virus. Left; genes may be inserted within viral gene A. Right; the cloned genes may replace a segment containing viral gene A. In either case the resulting virus fails to express the normal gene A function.

the usual type of viral assay may be different. The alternative model is quite similar, but instead of inserting foreign DNA at a single cleavage site in the lambda chromosome, the chromosome is cut twice and foreign DNA replaces the middle segment. Again recombinant viruses can be identified because they have lost the A$^+$ gene and so appear as A$^-$.

## C. Segmenting DNA

DNA is normally sheared during routine laboratory operations such as pipeting. However, it is usually advantageous to minimize random breakage and cleave DNA at specific points, using one or more site-specific endonucleases. Over 100 such enzymes have been identified, and about 40 are commercially available. The enzymes are often called restriction enzymes, because, at least in some cases, they restrict entry of intact foreign DNA into cells. Cells containing such a nuclease also contain a site-specific methylase that modifies newly synthesized DNA strands by methylation within the nuclease-sensitive site. Methylation of either strand prevents nuclease activity. DNA from other organisms is generally not methylated in the same sequence, and so is cleaved on entry.

As shown in Table I, all DNA fragments produced by the BamHI en-

TABLE I

**DNA Sequence Specificity of Nucleases Producing Staggered Cuts**[a]

| Enzyme | Specificity | Source |
|---------|-------------|--------|
| BamHI | G\|GATCC | *Bacillus amyloliquefaciens* H |
| EcoRI | G\|AATTC | *Escherichia coli* R413 |
| Hind III | A\|AGCTT | *Hemophilus influenzae* Rd |
| Kpn I | GGTAC\|C | *Klebsiella pneumoniae* OK8 |
| Pst I | CTGCA\|G | *Providencia stuartii* 164 |
| SalGI | G\|TCGAC | *Streptomyces albus* G |

[a] Only one of the two complementary DNA strands is shown, reading from 5′ to 3′. In each case the recognition site is symmetrical and the opposite strand is identical in this site. Specific six base pair sequences such as those shown should occur once in every 4096 nucleotide pairs in DNA containing 50% (G + C). \| designates points of cleavage by the enzyme. The opposite strand is also subject to cleavage as shown for SalGI.

$$5' \ldots \text{G}|\text{TCGAC} \ldots 3'$$
$$3' \ldots \text{C AGCT}|\text{G} \ldots 5'$$

Sal ↓

$$\ldots \text{G} \qquad \text{TCGAC} \ldots$$
$$\ldots \text{CAGCT} \overset{+}{\qquad} \text{G} \ldots$$

zyme have identical single-stranded ends; the same is true for the products of activity of each other enzyme because the sequences cut are symmetrical. The sequence recognized by each enzyme occurs about once in 4100 base pairs, so most DNA fragments produced by the enzyme contain one or more intact genes. Because each sequence is both unique and overlapping, DNA fragments generated from any source by one of the enzymes can be joined through their identical and complementary ends. DNA ligase can be used to seal the union permanently.

## D. Joining DNA

Splicing a piece of DNA to a plasmid or virus cloning vehicle is most simply achieved by joining the complementary ends produced by a restriction enzyme. At low temperature or high salt concentration hybridization of the ends is favored over separation.

Several other joining procedures are available (Fig. 4; Helling 1975; Helling and Lomax, 1979). The "tailing" procedure depends on the use of the enzyme terminal transferase to add nucleotides to the 3′ end of the DNA strand. Tails consisting of only a single type of nucleotide are formed if only one type is present during the transferase reaction. Thus, for example, DNA from a fruitfly or from a snapdragon can be tailed with single-

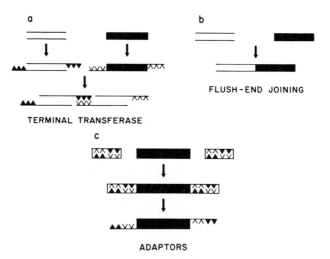

**Fig. 4.** Alternative procedures for joining DNA molecules. A; Terminal transferase adds complementary nucleotides separately to the two types of DNA to be joined. The two DNA preparations are then mixed to allow pairing and ligation. B; T4 ligase will join completely double stranded molecules, although with relatively low efficiency. C; linkers containing sites cleaved by specific nucleases are added as in (B). The single strand complementary ends are then exposed by appropriate nuclease treatment.

stranded polydA and then joined with the DNA of a plasmid or virus to which complementary polydT tails have been added.

Although hybridization through complementary single-strand ends is relatively easy, it is also possible to join completely double-stranded DNA ends by using DNA ligase from phage T4. An elegant combination of "flush end" joining and joining through complementary ends depends on the use of small, chemically synthesized "adapters." The adapters are short, double-strand DNA segments containing the site for cleavage by a specific restriction enzyme. Terminal adapters are added by flush-end ligation. Appearance of single-strand ends follows treatment with the relevant restriction enzyme. The DNA is then available for joining into an appropriate restriction site in a plasmid or virus. The advantage is to allow a DNA sequence to be cloned directly without breaking it up using restriction enzymes, while retaining the ability to recover the DNA from the plasmid or virus for analysis and purification after cloning.

## E. Transformation

Natural procedures for the uptake and integration of DNA into some bacteria have been known since the classic work of Avery *et al.* (1941). However, many bacteria, including *E. coli,* could not be transformed until Mandel and Higa (1970) found that treatment of cells at low temperature with $Ca^{2+}$ allowed the uptake of DNA. Although the mechanism by which such uptake occurs is still unknown, the procedure has been successfully applied to many kinds of bacteria (Benzinger, 1978). Most bacteria can be transformed by one or the other of these methods.

The DNA of the mammalian viruses polyoma and SV40 is infectious to appropriate mammalian cells (at low efficiency). Thus foreign DNA spliced into the viral chromosome can passively enter the nucleus and be maintained as part of that chromosome (Upcroft *et al.,* 1978). Transformation of yeast has recently been achieved (Hinnen *et al.,* 1978), but transformation of higher plants is yet to be demonstrated convincingly.

Thus bits of DNA from a sea urchin, a snail darter, or an orchid joined to a cloning vector can be inserted into *E. coli,* a human cell in culture, or baker's yeast and propagated generation on generation.

## F. Selection of Cells with Cloned Genes

With some proficiency in microbiology and molecular biology, one should be able to clone virtually any gene or short segment of DNA. However, the identification of the individual cell or virus containing the gene of interest may present a formidable problem.

Selection of cells transformed with a plasmid or of a virus with a bit of additional DNA is basically simple. But how does one find the one in a hundred or one in a thousand or one in a million transformed cells or viruses with the desired gene?

Obviously if the added gene can be selected for, the task is simple. For example, the first cloned DNA contained a gene for kanamycin resistance (Cohen *et al.*, 1973). Thus cells containing a plasmid with cloned DNA could be selected directly on medium containing tetracycline (selection for the plasmid) plus kanamycin (selection for a plasmid containing the desired cloned gene). Direct selection for a variety of enzyme activities has been successful, particularly where the genes cloned are from the same or a closely related organism (Clarke and Carbon, 1975).

Several laboratories have devised protocols for identifying cells containing cloned genes making specific proteins by their reaction with specific antibodies (Skalka and Shapiro, 1977; Broome and Gilbert, 1978). With current methods it is feasible to screen thousands of colonies or viral plaques in this way.

In the majority of cases simple selection is not possible. How can one find the clone of interest? In principle, two methods exist: (1) purification of the desired DNA segment prior to cloning (see next section) and (2) identification of the desired cell because it contains DNA capable of hybridizing with appropriate RNA. The latter method is often the method of choice when cloning DNA in an unrelated organism because it does not depend on gene expression or even on cloning an entire gene. It does require that complementary RNA (or DNA) be available for hybridization. Simple procedures for screening thousands of individual colonies or plaques by hybridization have been devised (Kedes *et al.*, 1975; Grunstein and Hogness, 1975; Benton and Davis, 1977; Villarreal and Berg, 1977). A large number of cloned genes have been identified by hybridization with the complementary RNA. A clever procedure based on unavailability for translation of mRNA hybridized with cloned DNA allows one to associate a specific cloned DNA segment with a specific polypeptide even though the function of the latter may be unknown (Paterson *et al.*, 1977; Hastie and Held, 1978).

Large "banks" of cells with cloned DNA from *E. coli,* yeast, and other organisms have been established (Clarke and Carbon, 1976). The availability of a bank to other investigators allows them to focus on identification of desired cells rather than on having to repeat the cloning process.

## G. Selection of Specific Genes

Some genes, such as those for ribosomal RNA, have a base composition quite different from that of the bulk DNA, and so can be separated by density. If mRNA for a specific protein is available, it serves as a template for the enzyme reverse transcriptase, allowing synthesis of a DNA copy. Synthetic sequences for ovalbumin, for $\alpha$- and $\beta$-globin, for insulin, and for other proteins have been cloned. mRNA can also be used to isolate directly the corresponding genes from cellular DNA. The mRNA is bound to an inert support and is allowed to hybridize with denatured DNA; the unbound DNA is then washed away (Shih and Martin, 1974). The single bound strand can be eluted and used in a similar fashion to isolate the opposite DNA strand.

Chemically synthesized DNA can be cloned. A nucleotide sequence corresponding to the amino acid sequence of the hormone somatostatin was synthesized (Itakura *et al.*, 1977). The terminal nucleotides of the synthetic gene served as adapters for joining to termini of a plasmid cleaved by a restriction nuclease. Cells of *E. coli* containing the plasmid synthesized the corresponding polypeptide.

## H. Expression of Cloned Genes

The genetic code for amino acids is universal, so a gene is potentially capable of being expressed to form the same protein in any organism. However, the signal sequences specifying the beginning and end of a gene differ between unrelated organisms, so most eukaryotic genes cloned in bacteria are not expected to form a functional product. RNA from foreign genes may be degraded. Foreign proteins may be rapidly degraded by proteolytic enzymes (Itakura *et al.*, 1977). Furthermore, some eukaryotic proteins reflect posttranscriptional processing of the mRNA or precursor polypeptide. Some eukaryotic genes do form a functional product when cloned in *E. coli,* but it is not clear if eukaryotic genetic punctuation is used (Struhl and Davis, 1977). These cases are exceptional (Chang *et al.*, 1975; Meagher *et al.*, 1977). The utility of yeast as a cell for cloning eukaryotic genes lies in the far greater potential for correct expression of cloned eukaryotic genes than is likely for bacterial hosts. Alternatively, it may be possible to initiate or repress expression of eukaryotic genes in bacterial cells by joining the cloned genes to appropriate prokaryotic punctuation sequences in the cloning vehicle (Polisky *et al.*, 1976; Itakura *et al.*, 1977).

## I. Stability of Cells with Foreign DNA

Few genes chosen at random are expected to be advantageous to a foreign cell or virus. We expect cells containing them to grow more poorly because energy is expended to duplicate the genes and their products. Furthermore, the gene products may interfere directly in normal metabolism.

Consistent with this expectation, lambda phage containing cloned yeast DNA is at a strong competetive disadvantage to normal lambda (Cameron and Davis, 1977). Under most circumstances cells containing plasmids are also selected against. The presence of foreign DNA inserts in the plasmid generally enhances the disadvantage and often interferes with plasmid replication (Adams *et al.*, 1979). Therefore to maintain a cell or virus with a specific cloned segment it is usually necessary to use appropriate selective conditions if available. This is especially important when growing large populations. Several cloned DNA segments have been sequenced and shown to be faithfully replicated for many generations (Helling and Lomax, 1978).

## IV. *IN VIVO* GENETIC MANIPULATION

*In vitro* methods of recombination are having a major impact on molecular biology. *In vivo* procedures now being developed supplement the *in vitro* methods and extend possibilities for genetic manipulation. These procedures include techniques for fusing animal, bacterial, or plant cells and for adding isolated chromosomes to cells (e.g., Miller and Ruddle, 1978).

New techniques for achieving recombination at selected sites in high frequency are also available. Various genetic elements exhibiting the common feature of stimulating genetic translocation or deletion in their vicinity have been discovered (Cohen, 1976; Schwesinger, 1977). These elements include the virus Mu chromosome and specific genetic loci in bacterial sex factors and in resistance factors offering resistance to one or more antibiotics. The translocation catalyzed by these elements differs from classical recombination in not being dependent on extensive nucleotide sequence homology between the two participating DNA molecules or on known enzymes and in the nature of the products. The products represent the addition (or deletion) of DNA segments rather than reciprocal exchange of segments of two homologous chromosomes that remain distinct.

Translocatable elements appear to have played a major role in the construction of the diverse plasmids and viruses found in nature (Reanney, 1976). They may also have been important in the evolution of bacterial chromosomes. Their analogues in higher organisms may play roles in gene regulation and differentiation. The use of translocatable elements in genetic engineering is reviewed in Kleckner *et al.* (1977).

Site-specific endonucleases used in *in vitro* constructions may also drive similar recombinations in the cell. EcoRI endonuclease has been shown to do this during growth of the cell in which it is synthesized (Chang and Cohen, 1977). Both site-specific endonucleases and translocatable elements are likely to be exploited in the near future to allow DNA splicing without *in vitro* manipulation.

The "P" plasmids, so designated because they were found in bacteria of the genus *Pseudomonas,* are of exceptional interest because they are self-transmissible to and compatible with most gram-negative bacteria (Datta *et al.,* 1971). Thus molecular cloning can be achieved in almost any Gram-negative bacterium using a single group P plasmid as a cloning vehicle. Alternatively, cloning can be carried out in one bacterium, and the newly constructed plasmid can be transferred to a different bacterial species for further study.

## V. POSSIBLE HAZARDS ASSOCIATED WITH RECOMBINANT DNA RESEARCH

### A. Microbial Evolution, Ecology, and Gene Exchange

A very small minority of microorganisms is truly pathogenic. The vast majority is beneficial or even essential to most other life. This majority includes scavengers decomposing organic matter and converting inorganic material from one form to another. The photosynthetic bacteria are of great importance in the formation of usable energy. The multiple forms of microbial life reflect the evolution of cells adapted precisely to specific environments. The genetic refinement of organisms to function efficiently in a given environment is the natural consequence of selection over tens of thousands of generations. As a result, every aspect of a microbe seems tailored to its ecological role, and microorganisms introduced from other environs almost never establish themselves in competition with the resident flora. For example, most enteric bacteria fail to survive in water or soil despite the estimated $10^{20}$–$10^{22}$ cells excreted by the human population per day (Davis, 1977). For that reason the presence of coliforms in water is a reliable indicator of recent fecal pollution. Only in a virgin or radically altered environment are intruders able to fluorish outside their normal habitat. It is well known that many bacteria normally considered nonpathogenic can cause transient opportunistic infections in tissues exposed by trauma or in a bowel depleted of its normal flora by antibiotic therapy.

The variants that are subject to natural selection stem from mutation, from recombination of homologous genes, and from acquisition of totally new genes from other kinds of organisms. These latter two processes involve gene exchange, several forms of which are prevalent in the microbial world. Such mating is now believed to be common in most bacteria and to occur between bacteria of quite diverse properties and evolutionary backgrounds. The most common way by which cells receive genes is probably by viruses and self-transmissible plasmids. Many bacteria also undergo transformation in nature.

We can surmise that many potentially dangerous genetic transfer experiments have been performed in nature's laboratory, usually without significant (to humans) result. Many toxins are determined by genes on plasmids or viruses that must repeatedly enter other kinds of cells. Diphtheria toxin is produced by strains of *Corynebacterium* that house a virus with the toxin gene. Production of gas gangrene and botulism by at least some strains of *clostridium* depends on the presence of a virus (Eklund *et al.*, 1974). The ability of *Agrobacterium* to form plant tumors and hairy root disease is plasmid determined (Drummond, 1979; White and Nester, 1980). Virulence in the marine fish pathogen *Vibrio anguillarum* is associated with a plasmid (Crosa, 1980).

Tissue invasiveness by *Yersinia enterocolitica* (Zink *et al.*, 1980) and witches broom in dicotyledenous plants (Murai *et al.*, 1980) also appear to be plasmid determined. Naturally occurring plasmids dictating both toxin-production and antibiotic resistance have been isolated (Gyles *et al.*, 1977). Other metabolic capabilities of considerable ecological significance have also been associated with plasmids, e.g., ability to degrade hydrocarbons or to fix nitrogen (Nuti *et al.*, 1979).

Despite the location of the toxin gene on a mobile replicon, diphtheria is always associated with *Corynebacterium*. Similarly, the causative agents for other kinds of diseases are generally well-characterized specific types of bacteria or virus. It seems quite certain from the preceding observations and from all we know about genetics and evolution that complex properties such as pathogenesis are the result of many genes acting in concert, located in different regions of the genome, and selected over many generations. The possibility that the random addition of a single small segment of DNA from an unrelated organism would convert a cell to a pathogen is equivalent to the expectation that the addition of a small part from a new Porsche would transform a Model T Ford into a modern sportscar. It has in fact been demonstrated that the common *E coli* laboratory strain K12 does not cause clinical disease even after acquiring enterotoxin-determining plasmids from enteropathogenic *E. coli* (Falkow, 1975).

## B. HAZARDS IN RECOMBINANT RESEARCH

Our understanding of the natural microbial world suggests that the accidental construction of a hazardous microorganism is an extremely unlikely possibility. Therefore I assume that no untoward hazard is generally associated with the construction of recombinant organisms. Three sorts of evidence support this conclusion.

First is the record of over 100 years experience with infectious agents in the laboratory. Despite the thousands of individuals who have worked with infectious agents, only a comparative handful of laboratory-acquired infections has been seen (Wedum, 1976). Among those few affected individuals, almost no one transmitted the infectious agent further. The rarity of laboratory-acquired infection among laboratory personnel working with natural infectious agents gives confidence that the accidental spread of organisms carrying recombinant DNA outside the laboratory is extremely unlikely, given standard microbiological technique, uncompromised laboratory workers, and the generally debilitated state of laboratory strains.

Second, microbial recombinations have been carried out in laboratories since the late 1940s without incident. Many of these crosses were between distantly related microorganisms, some of which were pathogens. For example, crosses of *E. coli* with plague bacillus were first made several decades ago. The possibility of obtaining progeny that were recognizably more dangerous than the parents would seem to be greater from such experiments than from the current gene manipulation experiments. Natural mating involves the recombination of many genes, and most complex properties of organisms depend on the smooth interaction of many genes. Nevertheless, I am aware of no experiments in which new types of organisms emerged to cause infection or establish themselves in any environment.

Third, experiments to evaluate possible hazards have found nothing unanticipated in the behavior of recombinant molecules or of the organisms in which they are housed. It is well known that laboratory strains such as *E. coli* K12 survive poorly in the uncompromised gut and that transfer of their genes to normal intestinal microorganisms occurs at a very low frequency (Smith, 1975; Anderson, 1975). Laboratory workers handling *E. coli* K12 strains with plasmids conferring multiple drug resistance fail to show the bacterium or plasmid in the feces (Petrocheilou and Richmond, 1977). Experiments being carried out in containment laboratories in Great Britain and in the United States have revealed no unexpected hazards associated with cloning genes of the DNA tumor virus polyoma (Israel *et al.,* 1979; Chan *et al.,* 1979; Fried *et al.,* 1979).

It is useful to remember that all actions involve some risk, and that many normally harmless or even beneficial microorganisms can be opportunistic pathogens. For this reason individuals compromised by antibiotic or radiation therapy or by disease should not be involved in this work (or any other work handling microbes). Are recombinant cells likely to provide more hazard than the unmodified parent organism? The answer is clearly no. In most cases the modified cells will be less viable and less likely to establish in an undesireable environment. This is because the addition of foreign DNA is almost always detrimental to the host cell, contributing no benefit but consuming energy and possibly inhibiting the organism directly (Helling and Lomax, 1978; Adams et al., 1979). Only in rare cases should an organism containing foreign genes compete successfully with the normal organism. Even in such a rare circumstance, the mere survival of the recombinant is not sufficient to establish disease or change an ecosystem.

I conclude that recombination experiments generally impose no hazard beyond that present in the unmodified organism, and that the vast majority of recombinants present no hazard whatsoever.

Are there any recombinant experiments that might carry hazard? Again the answer seems reasonably clear although in the absence of an adverse effect the answer cannot be absolute.

A hazard is associated with experiments involving pathogens, and their manipulation could conceivably alter pathogenicity undesireably. Possibly cloning of certain toxin genes or genes for pharmacologically active materials could produce a hazardous organism producing those substances. A microbe could be engineered deliberately to be dangerous. (International agreement now precludes the deliberate construction of agents for biological warfare in most countries.) Drug-resistance genes might be disseminated more widely. Possibly antigens from some cloned genes might be allergens.

It is obvious that almost none of these possibilities is really unique to organisms that have been manipulated using recombinant DNA techniques. They are really inherent in the natural organisms with which the work is done. Virtually all hazards can be anticipated by the nature of the genetic donor and of the recipient. Most possible hazards are avoided by cloning genes in E. coli, yeast, or higher organisms. Cloning of a gene whose dissemination would be particularly undesireable (e.g., a toxin gene) should be carried out in a plasmid or virus unlikely to be selected for in nature, or to be readily transferrable to another organism. In the unlikely event of successful infection by a recombinant E. coli, further dissemination is unlikely because E. coli does not normally multiply in the natural world outside the gut, and the infection would be treatable using standard therapy. If pathogens are involved, necessary procedures are usually obvious from the experience of microbiologists in handling the natural organism. To suggest that the hazard associated with molecular cloning experiments is greater than that in other microbiological work is to underestimate the power of natural selection and to ignore the century of experience working with known pathogens.

# REFERENCES

Adams, J., Kinney, T., Thompson, S., Rubin, L., and Helling, R. B. (1979). Frequency-dependent selection for plasmid-containing cells of *Escherichia coli*. *Genetics* **91**, 627–637.

Anderson, E. S. (1975). Viability of, and transfer of a plasmid from, *E. coli* K12 in the human intestine. *Nature (London)* **255**, 502–504.

Avery, O. T., MacLeod, C. M., and McCarty, M. (1944). Studies on the chemical nature of the substance inducing transformation of pneumococcal types. *J. Exp. Med.* **79**, 137–158.

Benton, W. D., and Davis, R. W. (1977). Screening λgt recombinant clones by hybridization to single plaques *in situ*. *Science* **196**, 180–182.

Benzinger, R. (1978). Transfection of *Enterobacteriaceae* and its applications. *Microbiol. Rev.* **42**, 194–236.

Broome, S., and Gilbert, W. (1978). Immunological screening method to detect specific translation products. *Proc. Natl. Acad. Sci. U.S.A.* **75**, 2746–2749.

Cameron, J. R., and Davis, R. W. (1977). The effects of *Escherichia coli* and yeast DNA insertions on the growth of lambda bacteriophage. *Science* **196**, 212–215.

Chang, A. C. Y., Lansman, R. A., Clayton, D. A., and Cohen, S. N. (1975). *Cell* **6**, 231–244.

Chang, S., and Cohen, S. N. (1977). In vivo site-specific genetic recombination promoted by The EcoRI restriction endonuclease. *Proc. Natl. Acad. Sci. U.S.A.* **74**, 4811–4815.

Clarke, L., and Carbon, J. (1975). Biochemical construction and selection of hybrid plasmids containing specific segments of the Escherichia coli genome. *Proc. Natl. Acad. Sci. U.S.A.* **72**, 4361–4365.

Clarke, L., and Carbon, J. (1976). A colony bank containing synthetic Co1E1 hybrid plasmids representative of the entire *E. coli* genome. *Cell* **9**, 91–99.

Clewell, D. B. (1972). Nature of Co1E1 plasmid replication in *Escherichia coli* in the presence of chloramphenicol. *J. Bacteriol.* **110**, 667–676.

Cline, M. J., Stang, H., Mercola, K., Morse, L., Ruprecht, R., Browne, J., and Salser, W. (1980). Gene transfer in intact animals. *Nature (London)* **284**, 422–425.

Cohen, S. N. (1976). Transposable genetic elements and plasmid evolution. *Nature (London)* **263**, 731–738.

Cohen, S. N., Chang, A. C. Y., Boyer, H. W., and Helling, R. B. (1973). Construction of biologically functional bacterial plasmids in vitro. *Proc. Natl. Acad. Sci. U.S.A.* **70**, 3240–3244.

Crosa, J. H. (1980). A plasmid associated with virulence in the marine fish pathogen *Vibrio anguillarum* specifies an iron-sequestering system. *Nature (London)* **284**, 566–568.

Datta, N., Hedges, R. W., Shaw, E. J., Sykes, R. B., and Richmond, M. H. (1971). Properties of an R factor from Pseudomonas aeruginosa. *J. Bacteriol.* **108**, 1244–1249.

Davis, B. D. (1977). Recombinant DNA research. Potential benefits are large, protective methods make risks small. *Chem. & Eng. News* **55**, 27–31.

Drummond, M. D. (1979). Crown gall disease. *Nature (London)* **281**, 343–347.

Efstratiadis, A., Kafatos, F. C., and Maniatis, T. (1977). The primary structure of rabbit ß-globin mRNA as determined from cloned DNA. *Cell* **10**, 571–585.

Eklund, M. W., Poysky, F. T., Meyers, J. A., and Pelroy, G. A. (1974). Interspecies conversion of *Clostridium botulinum* type C to *Clostridium novyi* type A by bacteriophage. *Science* **186**, 456–458.

Falkow, S. (1975). "Infectious Multiple Drug Resistance." Pion Ltd. London.

Friedmann, T., and Roblin, R. (1972). Gene therapy for human genetic disease? *Science* **175**, 949–955.

Gellert, M. (1967). Formation of covalent circles of lambda DNA by E. coli extracts. *Proc. Natl. Acad. Sci. U.S.A.* **57**, 148–155.

Grunstein, M. G., and Hogness, D. S. (1975). Colony hybridization: A method for the isolation of cloned DNAs that contain a specific gene. *Proc. Natl. Acad. Sci. U.S.A.* **72**. 3961-3965.

Gyles, C. L., Palchaudhuri, S., and Maas, W. K. (1977). Naturally occurring plasmid carrying genes for enterotoxin production and drug resistance. *Science* **198**, 198-199.

Hastie, N. D., and Held, W. A. (1978). Analysis of mRNA populations by cDNA·mRNA hybrid-mediated inhibition of cell-free protein synthesis. *Proc. Natl. Acad. Sci. U.S.A.* **75**, 1217-1221.

Helling, R. B. (1975). Eukaryotic genes in prokaryotic cells. *Stadler Symp.* **7**, 15-36

Helling, R. B., and Lomax, M. I. (1978). The molecular cloning of genes—general procedures. *In:* "Genetic Engineering" (A. M. Chakrabarty, ed.), pp. 1-30. CRC Press, West Palm Beach, Fla.

Hershey, A. D., Burgi, E., and Ingraham, L. (1963). Cohesion of DNA molecules isolated from phage lambda. *Proc. Natl. Acad. Sci. U.S.A.* **49**, 748-755.

Hershfield, V., Boyer, H. W., Yanofsky, C., Lovett, M. A., and Helinski, D. R. (1974). Plasmid ColE1 as a molecular vehicle for cloning and amplification of DNA. *Proc. Natl. Acad. Sci. U.S.A.* **71**, 3455-3459.

Hinnen, A., Hicks, J. B., and Fink, G. R. (1978). Transformation of yeast. *Proc. Natl. Acad. Sci. U.S.A.* **75**, 1929-1933.

Itakura, K., Tadaaki, H., Crea, R., Riggs, A. D., Heyneker, H. L., Bolivar, F., and Boyer H. W. (1977). Expression in *Escherichia coli* of a chemically synthesized gene for the hormone somatostatin. *Science* **198**, 1056-1063.

Kedes, L., Chang, A., Houseman, D., and Cohen, S. (1975). Isolation of histone genes from unfractionated sea urchin DNA by subculture cloning in *E. coli. Nature (London)* **255**, 533-538.

Kleckner, N., Roth, J., and Botstein, D. (1977). Genetic engineering in vivo using translocatable drug-resistance elements. New methods in bacterial genetics. *J. Mol. Biol.* **116**, 125-159.

Mandel, M., and Higa, A. (1970). Calcium-dependent bacteriophage DNA infection. *J. Mol. Biol.* **53**, 159-162.

Maxam, A. M., and Gilbert, W. (1977). A new method for sequencing DNA. *Proc. Natl. Acad. Sci. U.S.A.* **74**, 560-564.

Meagher, R. B., Tait, R. C., Betlach, M., and Boyer, H. W. (1977). Protein expression in E. coli minicells by recombinant plasmids. *Cell* **10**, 521-536.

Miller, C. L., and Ruddle, F. H. (1978). Co-transfer of human X-linked markers into murine somatic cells via isolated metaphase chromosomes. *Proc. Natl. Acad. Sci. U.S.A.* **75**, 3346-3350.

Murai, N., Skoog, F., Doyle, M. E. and Hanson, R. S. (1980). Relationships between cytokinin production, present of plasmids, and fasciation caused by strains of *Corynebacterium fascians. Proc. Natl. Acad. Sci. U.S.A.* **77**, 619-623.

Nuti, M. P., Lepidi, A. A., Prakash, R. K., Schilperoort, R. A., and Cannon, F. C. (1979). Evidence for nitrogen fixation (*nif*) genes on endogenous *Rhizibium* plasmids. *(London) Nature* **282**, 533-535.

Paterson, B. M., Roberts, B. E., and Kuff, E. L. (1977). Structural gene identification and mapping by DNA·mRNA hybrid-arrested cell-free translation. *Proc. Natl. Acad. Sci. U.S.A.* **74**, 4370-4374.

Petrocheilou, V., and Richmond, M. H. (1977). Absence of plasmid or *Escherichia coli* K12 infection among laboratory personnel engaged in R-plasmid research. *Gene* **2**, 323-327.

Polisky, B., Bishop, R. J., and Gelfand, D. H. (1976). A plasmid cloning vehicle allowing regulated expression of eukaryotic DNA in bacteria. *Proc. Natl. Acad. Sci. U.S.A.* **73**, 3900-3904.

Reanney, D. (1976). Extrachromosomal elements as possible agents of adaptation and development. *Bacteriol. Rev.* **40**, 552–590.

Sanger, F., Nicklen, S., and Coulson, A. R. (1977). DNA sequencing with chain-terminating inhibitors. *Proc. Natl. Acad. Sci. U.S.A.* **74**, 5463–5467.

Schwesinger, M. D. (1977). Additive recombination in bacteria. *Bacteriol. Rev.* **41**, 872–902.

Seeburg, P. H., Shine, J., Martial, J. A., Baxter, J. D., and Goodman, H. M. (1977). Nucleotide sequence and amplification in bacteria of structural gene for rat growth hormone. *Nature (London)* **270**, 486–494.

Shih, T. Y., and Martin, M. A. (1974). Chemical linkage of nucleic acids to neutral and phosphorylated cellulose powders and isolation of specific sequences by affinity chromatography. *Biochemistry* **13**, 3411–3416.

Skalka, A., and Shapiro, L. (1977). *In situ* immunoassays for gene translation products in phage plaques and bacterial colonies. *Gene* **1**, 65–79.

Smith, H. W. (1975). Survival of orally administered *E. coli* K12 in alimentary tract of man. *Nature (London)* **255**, 500–502.

Struhl, K., and Davis, R. W. (1977). Production of a functional eukaryotic enzyme in *Escherichia coli:* Cloning and expression of the yeast structural gene for imidazole-glycerolphosphate dehydrogenase (his3). *Proc. Natl. Acad. Sci. U.S.A.* **74**, 5255–5259.

Thomas, L. (1978). Hubris in science? *Science* **200**, 1459–1462.

Tilghman, S. M., Curtis, P. J., Tiemeier, D. C., Leder, P., and Weissman, C. (1978). The intervening sequence of a mouse ß-globin gene is transcribed within the 15S ß-globin mRNA precursor. *Proc. Natl. Acad. Sci. U.S.A.* **75**, 1309–1313.

Tonegawa, S., Maxam, A. M., Tizard, R., Bernard, O., and Gilbert, W. (1978). Sequence of a mouse germ-line for a variable region of an immunoglobulin light chain. *Proc. Natl. Acad. Sci. U.S.A.* **75**, 1485–1489.

Upcroft, P., Skolinik, H., Upcroft, J. A., Solomon, D., Khoury, G., Hamer, D. H. and Fareed, G. C. (1978). Transduction of a bacterial gene into mammalian cells. *Proc. Natl. Acad. Sci. U.S.A.* **75**, 2117–2121.

Villarreal, L. P., and Berg, P. (1977). Hybridization in situ of SV40 plaques: Detection of recombinant SV40 virus carrying specific sequences of nonviral DNA. *Science* **196**, 183–195.

Wedum, A. G. (1976). The Detrick experience as a guide to the probable efficacy of P4 microbiological containment facilities for studies on microbial recombinant DNA molecules. *In* "Recombinant DNA Research" (Office of the Director, National Institutes of Health, DHEW publication 1138), pp. 372–376. NIH, Bethesda, Md.

Wensink, P. C., Finnegan, D. J., Donelson, J. E., and Hogness, D. S. (1974). A system for mapping DNA sequences in the chromosomes of Drosophila melanogaster. *Cell* **3**, 315–325.

White, F. E., and Nester, E. W. (1980). Hairy root: plasmid encodes virulence traits in *Agrobacterium rhizogenes. J. Bacteriol.* **141**, 1134–1141.

Zaig, J., Silverman, M., Hilmen, M., and Simon, M. (1977). Recombinational switch for gene expression. *Science* **196**, 171–172.

Zink, D. L., Feeley, J. C., Wells, J. G., Vanderzant, C., Vickery, J. C., Roof, W. D., and O'Donovan, G. A. (1980). Plasmid-mediated tissue invasiveness in *Yersinia enterocolitica. Nature (London)* **283**, 224–226.

# Chapter Six

# Identification, Analysis, and Control of Biohazards in Viral Cancer Research*†

D. L. WEST, D. R. TWARDZIK,
R. W. MCKINNEY, W. E. BARKLEY,
AND A. HELLMAN

## I. INTRODUCTION

Research on the role of viral agents in the etiology of cancer is in a fluid state and consequently undergoes continuing change as new data are developed. In the course of this research it has become apparent that genetic information for RNA tumor viruses is present in the cells of most or all

*This work was supported in part by Contract No. N01-CP6-1021 of the Biological Carcinogenesis Research Program of the National Cancer Institute.

† It is with respect and admiration that the authors dedicate this chapter to the late Dr. Arnold G. Wedum in recognition of his leadership and innumerable contributions, including much of the data in this chapter, to laboratory safety. All persons engaged in microbiological research have benefited from his efforts which earned him justly deserved international recognition.

vertebrate species and that this information is usually maintained in a repressed form in the normal cell. Studies elucidating the influence of factors such as hormones and chemical carcinogens and radiation in activating latent oncogenic viral information and subsequent transformation are currently of great interest.

Some of the most significant studies directed toward developing an understanding of the molecular mechanisms of cancer have been accomplished using RNA tumor viruses in animal model systems. RNA tumor viruses are enveloped viruses that contain an RNA genome and an RNA-dependent DNA polymerase. They have been isolated from reptiles, birds, and various mammals, including nonhuman primates. Reports of their isolation from or the expression of related antigens in man are the subject of current investigations. In susceptible hosts, RNA tumor viruses cause leukemias, sarcomas, and lethal infections; however, in some cases they are also nonpathogenic.

The oncornaviruses are generally classified on the basis of morphology as either type A, B, C, or D. Type A particles are intracellular and primarily localized in the endoplasmic reticulum of some mouse tumor cells. Type B oncornaviruses, of which the mouse mammary tumor virus (MMTV) is the prototype, have eccentrically located nucleoids and demonstrate distinct envelope spikes. Because of their structural protein similarities it has been suggested that type B particles are derived from intracytoplasmic type A particles.

The major interest in many laboratories has been with the type C class of oncornaviruses. They have a centrally located nucleoid and the appearance of budding particles at the plasma membrane. Type C oncornaviruses have been shown to provide helper functions for replication of defective RNA sarcoma viruses and possess structural proteins and other biochemical characteristics that distinguish them from other classes of viruses. Endogenous type C viruses of rodent and primate origin have been isolated from a variety of mammalian species, including Old World monkeys and apes and possibly man. The horizontal transmission of some type C viruses has also been demonstrated particularly in the feline system [e.g., with feline leukemia virus (FeLV) and feline sarcoma virus (FeSV)].

Recently another class of oncornaviruses has been recognized and is now identified as type D. These are distinguished by their bullet-shaped nucleoid and larger particle size. Isolates include the Mason-Pfizer monkey virus (MPMV) and the squirrel monkey retravirus.

The role of DNA viruses as etiologic agents of proliferative cellular diseases in several animal species has been suggested. The possibility that members of this group might be involved in human cancer has not been confirmed; however, studies are continuing to determine whether such an association exists. These studies are pertinent because these agents morphologically transform *in vitro* cells derived from a variety of mammalian species, including human beings. Genetic information homologous to the human isolate BK virus has been detected in several human tumors and similar studies exploring the role of Herpes simplex virus (HSV), Epstein-Barr virus (EBV), and other DNA viruses in neoplastic transformation are in progress. The reader is directed to the many excellent reviews on the subject of viruses and cancer presently available in the literature.

Thus, although a viral etiology for human cancer has not been established, research with the recognized oncogenic viruses can be expected to continue at a significant level.

## II. DETERMINANTS OF LABORATORY HAZARD

The sources of potential hazards associated with the viral oncology research laboratory may be broadly classified as biological, chemical, and physical. The degree of hazard presented by activities in each of these categories will vary among laboratories as a function of the research program emphasis. However, certain basic considerations are applicable to developing procedures and practices for the safe conduct of research in the viral oncology research laboratory.

In assessing the potential hazards associated with work involving known or suspect oncogenic viruses, the same basic approach that is used for estimating the hazards associated with the use of other known microorganisms may be employed. However, the precision with which the assessment is accomplished will be directly dependent on the data that are considered. Listed in Table I are some of the major considerations developed for assessing the risks of infection associated with the laboratory manipulation of known pathogens and which are believed applicable to assessing risks in viral oncology research.

The availability of data regarding the number (Hanson *et al.*, 1967; Pike *et al.*, 1965) and severity of laboratory-acquired infections with a given organism are of value in assessing the risk associated with manipulation of the organism in the laboratory. These data, when integrated with the procedures employed and the activities of the personnel, serve to describe the nature of the laboratory-associated hazard that in turn permits development of control procedures.

In the area of viral oncology research, we do not have data comparable to those available for other laboratory-acquired infections. However, some of the latter are presented in Table II to illustrate the hazards associated with the manipulation of infective agents.

TABLE I

Assessment of Risk of Human Infection

Number and severity of laboratory infections
Modifiers of human susceptibility and resistance:
  vaccines, therapy, antibodies, immunosuppression, sex,
  race, others
Infectious dose for man
Infection from procedure or equipment
Horizontal transmission of disease to cagemates
  by the inoculated animals
Agent excretion in urine, feces, or saliva
Hazards peculiar to the animal species

Although in oncogenic virus research there is no compilation of data comparable to that presented in Table II, there are suggestions that there have been occurrences of laboratory-acquired malignancies. At least four cases, two of them fatal, of an unusual form of malignant lymphoma in reasearch workers have been reported through the press or by discussion among scientists. The only case of laboratory-acquired malignancy of which there is little question was reported in 1926. The case involved the ac-

TABLE II

Laboratory-Acquired Human Infections

| Disease | Infections | References |
|---|---|---|
| Brucellosis (1930–1971) | 398 | Sulkin, 1972 |
| Typhoid (191 in Germany by 1939) | 292 | Draese, 1937–1939 |
| Q fever | 261 | Sulkin, 1972 |
| Tularemia | 227 | Sulkin, 1972 |
| Tuberculosis | 164 | Sulkin, 1972 |
| Typhus | 189 | Sulkin *et al.,* 1969 |
| Hepatitis | 172 | Sulkin, 1972 |
| Venezuelan encephalitis | 138 | Sulkin, 1972 |
| Hemorrhagic fever/renal syndrome [a] | 113 | Kulagin *et al.,* 1962 |
| Coccidioidomycosis (1926–1965) | 108 | Hanel and Kruse, 1967 |
| Psittacosis | 100 | Sulkin, 1972 |

[a] This illustrates the potential human infectivity of microorganisms excreted in urine and feces and aerosolized during routine animal care. In 1961, at the Gamalaya Institute in Moscow, 357 field-collected mouselike rodents, housed in open cages in two animal rooms, were asymptomatic carriers of the virus of hemorrhagic fever. During two months, 113 persons were infected, including 52 in the laboratory building, 17 from other buildings, and 44 of 94 visitors of whom 18 were in the building for only three or four hours.

cidental inoculation of serum from the site of a breast carcinoma. It has been summarized (Gross, 1971) as follows:

> The medical student, Henry Vadan, accidentally punctured his left hand with a syringe with which he aspirated serum from a wound of a patient after a radical mastectomy. After *two* years, a sarcomatous nodule developed at the site of inoculation in the palm of his hand and grew rapidly; the unfortunate student died one year later from metastatic tumors.

In contrast to this dramatic case, results of serologic tests with sera of persons working with oncogenic viruses have been encouragingly negative. On the one hand, this may be a reflection of the lack of pathogenicity of these agents for humans; on the other, it may reflect the need for improved serologic tests. The results may also be a reflection of the quality and safety of the practices and procedures utilized in modern viral oncology research laboratories. A sampling of the results of serologic tests performed with sera of persons working with some oncogenic viruses is presented in Table III.

Although the data presented in Table III indicate that limited infection of humans with the agents listed has occurred, this should not be taken as evidence that infection with oncogenic viruses has not occurred, or in particular circumstances may not occur. In regard to the latter, it is important to consider factors that may modify the susceptibility of humans to infection.

It has long been recognized that the susceptibility of humans and laboratory animals to infection may be modified by a variety of treatments or conditions. These include stress (Riley, 1975), tissue injury (Thomas *et al.,* 1974), immunosuppression (Dent, 1972; Kersey *et al.,* 1973; Penn,

**TABLE III**

Seroconversion in Laboratory Workers

| Agent | Findings |
|---|---|
| Marek's disease and herpesvirus of turkeys | 0/100 (Sharma *et al.,* 1972) |
| Murine leukemia | 1/4 (Fink, 1974); 0/71 (Charman *et al.,* 1974); 0/400 p-30 (Aaronson and Hellman, 1978) |
| Feline leukemia/sarcoma | 0/132 (Sarma *et al.,* 1970); 0/400 p-30 (Aaronson and Hellman, 1978) |
| SSV-1 | 0/400 p-30 (Aaronson and Hellman, 1978) |
| M-7 baboon/RD 114 | 0/400 p-30 (Aaronson and Hellman, 1978) |
| Shope rabbit papilloma | 4/12 (Rogers, 1966) |
| SV40 | 15/37 (Horvath, 1972); 31/102 (Chumakova *et al.,* 1963) |
| Herpesvirus saimiri | 0/20 (Wolfe *et al.,* 1971a) |

1974), pregnancy (Rabin, 1971; Munroe, 1969; Harvey, 1964), and the sex of the host (Mason and McKay, 1974; Cutler *et al.,* 1974; Karande *et al.,* 1975). Of these, the last three are of direct concern in the development of biohazard controls. The need for considering the health status of the laboratory worker is reflected in the policy statement made in the National Cancer Institute Safety Standards for Research Involving Oncogenic Viruses (U.S. Department of Health, Education, and Welfare, 1974b).

. . . persons with reduced immunological competency, pregnant women, and patients under treatment with steroids or cytotoxic drugs shall receive a medical evaluation before work in areas where oncogenic viruses are used.

In addition to the foregoing considerations, there is a recognized variation in susceptibility that is dependent on the route of exposure. However, one obvious generalization is that the more direct the route of exposure (e.g., injection), the more likely it is that infection will result. This generalization is modified by a number of variables, but it does serve as a basic tenet in identifying the sources of potential hazards and for developing practices and procedures.

The routes of exposure to be considered in the potential infection of laboratory workers include direct inoculation, ingestion, and inhalation. Affecting the rate of infection by any one of the routes is the variation in susceptibility exhibited by humans exposed by the same route to the same quantity of material. Thus it is necessary to utilize procedures and practices that minimize or, it is hoped, eliminate exposure for the most susceptible person.

In any discussion of susceptibility the question of the pathogenicity of the infecting organism must be considered concurrently. To this extent, susceptibility of the host and pathogenicity of the organism for that host cannot be clearly separated. A case in point is the fact that humans demonstrate an increased susceptibility to tumors transplanted from genetically related donors (Gross, 1971). Several cases of human cancer after transplantation of kidneys from tumor-bearing donors have been reported. Based on these observations, it is essential that laboratory workers *do not* manipulate cell cultures derived from their own tissues or those of close relatives (Charman *et al.,* 1974).

Modification of the host range of a virus by the formation of a pseudotype is of value in the study of oncogenesis. However, the procedure frequently results in expanding the host range of the agent. It has been suggested that a defective human cancer virus genome might be rescued by providing a feline or murine leukemia virus envelope. However, this could result in a pseudotype that would be a significant hazard to humans. The role of the virus envelope as a determinant of the host range is reflected in several publications pertinent to the hazards to humans (Hanafusa and Hanafusa, 1966; Fischinger and O'Connor, 1970; Huebner *et al.*, 1970; Boettiger *et al.*, 1975).

Closely associated with the question of the susceptibility to infection is the recognition of the wide variation in the number of microorganisms that constitute an infective dose for humans. Among various microorganisms the number ranges from one to several million. In general, the number is relatively large and this, coupled with the fact that repeated small doses by any route do not have a cumulative infective effect, contributes to reducing the hazard potential. Thus control of procedures that may otherwise result in the presentation of infective dose quantities is effective in reducing the potential hazards associated with work involving pathogenic or oncogenic agents.

In addition to the biological hazards present in the cancer research laboratory, there are hazards associated with chemical agents and physical factors. The control of these hazards is an integral part of a biohazards control program, since some of them may contribute to enhancing the effects of exposure to a biologic agent. On the other hand, some of the materials are significant independent hazards and require specific action for their control.

An increased potential hazard is now evident as research programs with oncogenic viruses are expanded to include chemical carcinogens. The consequences of combining the effects of an oncogenic virus and chemical carcinogens have been demonstrated in a variety of *in vitro* and *in vivo* systems. These studies serve to illustrate the potential for modifying biohazards for laboratory personnel by effecting a change in the research program. Unlike biologic agents, chemical carcinogens do not usually produce an immediate measurable tumorigenic reaction; they tend to be cumulative and may be retained in the body for long periods of time; their presence in host tissues is not readily detected; and certain combinations may be synergistic.

A major physical hazard present in the research laboratory is that resulting from radioactive materials. The hazard is similar to that associated with chemicals, because the effects of isotope exposure are cumulative and long term. In addition to the physical hazards associated with isotopes there are those that may result in injury. These range from simple cuts from broken glass to permanent injury or even death from fire, explosions, or equipment failure. The physical hazards must be considered as part of the total biohazard problem, since they may contribute to the occurrence of an accident involving biohazardous material.

Thus the determinants of hazard in the laboratory will vary from one laboratory to another as a function of the procedures and practices being conducted. What does not vary is the principle that hazards cannot be viewed independently but must be considered collectively in providing for the safety of the personnel.

## III. CRITERIA FOR TUMOR VIRUS CLASSIFICATION

In developing a classification for any microorganism that reflects the risks associated with manipulation of the organism, it is often necessary to utilize data that are not directly related to humans but that provide the basis for making a best judgment decision concerning the potential risks. This is the situation in viral oncology research in the assignment of a risk classification for oncogenic viruses.

At the present time, the classifications of low, moderate, and high risk are used in categorizing oncogenic viruses (U.S. Department of Health, Education and Welfare, 1974b). The viruses included in the low-risk category are those that are not otherwise classified, whereas the high-risk classification is reserved for a virus proved to induce cancer in humans. Such viruses will retain the high-risk classification until their complete hazard potential can be established. Since there are no viruses currently classified as high risk, consideration will be given to the criteria employed in the classification of moderate-risk oncogenic viruses. These criteria are not absolute but are subject to modification as new data are developed; however, currently they provide a basis for considering the risk potential of oncogenic viruses. The criteria are listed in Table IV.

In addition to the listed criteria, one must also consider the extent to which there is prior experience with the virus that may reflect the absence of hazard. The presently recognized moderate-risk viruses are listed in Table V.

As previously indicated, the criteria are not absolute; as a consequence,

**TABLE IV**

**Criteria for Moderate-Risk Oncogenic Viruses**

A.  Suspected oncogenic virus isolate from man
B.  Virus that transforms human cells *in vivo*, as evidenced by a morphological and/or functional alteration that is transferred genetically
C.  Virus that produces cancer without the aid of experimental host modification either (1) in a subhuman primate of any age or (2) across another mammalian species barrier in juvenile or adult animals
D.  A genetic recombinant between an animal oncogenic virus and a microorganism infectious for man, until its oncogenic potential for man is determined
E.  Any concentrated oncogenic virus or infectious transforming viral nucleic acid

TABLE V

**Moderate-Risk Oncogenic Viruses**

Ribonucleic acid (RNA)
    Feline leukemia
    Feline sarcoma
    Woolly monkey fibrosarcoma
    Gibbon ape lymphosarcoma
Deoxyribonucleic acid (DNA)
    *Herpesvirus saimiri*
    *Herpesvirus ateles*
    Yaba virus
    Epstein-Barr virus
    Nondefective Ad2-SV40 Hybrid
RNA and/or DNA virus isolates from man with possible oncogenic potential

there are some exceptions. In order to illustrate these exceptions each of the criteria will be discussed separately.

The rationale for criterion A is self-evident in that any candidate human cancer virus must be considered to present a potential risk until such time as data are required that justify reclassification. The assignment of the moderate-risk classification to candidate agents may be questioned, since proof that the virus is etiologically associated with cancer in humans would result in classifying the agent as high risk. Despite this possibility, the moderate-risk classification is justified in relation to the risk/benefit ratio associated with the research, since it is recognized that in the initial phases of the work with a candidate agent the quantity and concentration of the virus would be small, the procedures used would be relatively conventional and not result in creating hazards greater than those associated with recognized pathogens of moderate hazard, and importantly, rigid control in handling the agent would be achieved by virtue of the fact that only a few persons would be involved in the work.

Use of criterion B in defining moderate-risk viruses is not fully reliable. Although human cells do not exhibit spontaneous transformation (Macpherson, 1970), the fact that such cells undergo other changes during serial *in vitro* passage and that oncogenic viruses may be modified by serial *in vivo* passages complicates interpretation of transformation studies.

Difficulties in establishing a correlation between cell transformation and oncogenic potential has led to the conclusion that viral transformation of cells is not always a sufficient or a necessary condition for the transformed cell to induce tumor formation, and as yet there are not generally valid *in vitro* criteria for predicting tumorgenicity (Eagle *et al.,* 1970). In an attempt to resolve this problem it has been suggested that cultures that are genotypically transformed *and* induce tumor formation be identified as a malignant transformed cell, in contrast to genotype transformation for cells that acquire abnormal genotypic characteristics but do not induce tumor formation (Freeman and Huebner, 1973; Pontén, 1971). Sanford (1974) has suggested that the term *transformed* be limited in use to describe those cells that replicate *in vivo* as a malignant neoplasm or be qualified to reflect the nature of the change.

When the characteristics of the low- and moderate-risk viruses are evaluated against criterion B, it is of interest to note that three low-risk agents appear to meet the criteria, whereas only two moderate-risk agents do not (Table VI). It is to be noted that only five of the 13 oncogenic viruses (seven low risk and six moderate risk) that have been shown either to alter the morphology of human cells *in vitro* or to affect a permanent transformation have been tested to determine whether the transformed cells induce tumor formation in animals. These are Simian virus 40 (SV40), CELO (avian adenovirus), Rous sarcoma virus (RSV), Simian sarcoma virus (SSV-1), and Epstein-Barr virus (EBV).

The data summarized in Table VI serve to illustrate the difficulty of using genetic transformation of a human cell as a criterion for classifying a virus as a moderate-risk agent. However, until there are additional data that clarify the significance of transformation in relation to tumorgenicity, the criterion should be retained as at least one of the criteria for classifying low- and moderate-risk oncogenic viruses.

The results of studies dealing with animal host range of oncogenic viruses are summarized in Tables VII and VIII for low- and moderate-risk viruses, respectively. As may be seen there are disagreements between results obtained for particular viruses; thus the exact significance of these results in terms of defining human risk is not clear. However, the data obtained with subhuman primates deserve serious consideration; that is particularly true of those obtained in studies with the gibbon, chimpanzee, rhesus monkey, and baboon. The significance of these data is emphasized by close correlation to man of the taxonomic–phylogenetic relationship (Nahmias, 1974) and the homology of the nucleic acid sequences of tissue cellular DNA (Benveniste and Todaro, 1974a,b; Gilden *et al.*, 1974; Kohne, 1970).

The study by Harvey *et al.* (1974) relating to the immune responsiveness of selected subhuman primates is of interest in considering these data. Harvey and his co-workers found that the immune response of the baboon closely resembled that of humans. The squirrel monkey was similar to the baboon, whereas the cebus and marmoset showed several abnormalities.

It is also important to bear in mind that the studies represented by the data in Tables VII and VIII were performed under different protocols; thus the variation in results may be a reflection of differences in inocula, routes of administration, etc.

Despite some of the difficulties in assessing the significance of the animal data, it is reasonable to conclude that the propagation of oncogenic viruses in cells derived from subhuman primates does present a hazard to humans and that this may be reduced by the use of other host cells.

The potential hazards associated with manipulation of genetic recombinants developed with an oncogenic virus and a microorganism infectious for man (e.g., the hybrids produced with nondefective adenovirus 2 and Simian virus 40) have not been defined. Therefore it is considered appropriate that such recombinants be classified as a moderate risk until their oncogenic potential for man can be determined.

A similar situation prevails for the area of research involving recombinant DNA technology. Although this newer technology has potential application to viral oncology research, the possible hazards have not been established. Therefore such research should be conducted in accordance with the requirements set forth in the *Guidelines for Research Involving Recombinant DNA Molecules* issued by the National Institutes of Health (U.S. Department of Health, Education and Welfare, 1978a).

The definition provided in criterion E does not per se relate to *an* oncogenic virus; rather, it reflects the need to handle concentrated preparations of any oncogenic virus or transforming nucleic acid under the same conditions as prescribed for moderate-risk agents. The rationale for this criterion is predicated on the knowledge that initiation of infection is dependent on the quantity of infective material presented to the host and the route of administration.

The oncogenic potential of infectious viral nucleic acid is also a matter for consideration in evaluating the potential hazards associated with viral oncology research. This is reflected in the results of studies that demonstrated the infection of host cells with DNA isolated from cells infected or transformed with type C oncornaviruses. This mode of infection bypasses early virus–host interaction and removes restrictions that previously limited the host range (Scolnick and Baumgarner, 1975; Hsu *et al.,* 1978). Other findings also indicate the need for caution in the handling of viral nucleic acid. Studies by Burnett and Harrington (1968) demonstrated that the nucleic acid of simian adenovirus SA7 was more tumorigenic than the complete virus. The potency of nucleic acid is also indicated in studies by Graham *et al.* (1974), who reported that the viral nucleic acid of adenovirus 5 transformed rate cells *in vitro* and that noninfectious sheared adenovirus 5 nucleic acid fragments, representing approximately 5% of the viral genome, would initiate and maintain transformation of human and hamster cells *in vitro*. The transformed hamster cells were shown to be oncogenic when inoculated into newborn hamsters.

Taken together the data referable to the criteria for classifying oncogenic

**TABLE VI**

Oncogenic Viruses that Transform Human cells in Vitro[a]

| Virus | Cell-free virus produced in human cell line in vitro | Prolonged life of, or permanent, viral-transformed-in vitro human cell line[j] | Tumor induction[k] by inoculation of human-cells transformed in vitro |
|---|---|---|---|
| **Low Risk** | | | |
| Simian virus 40 | Yes (1–3) | Yes (3, 30–32) | No: hamster (41, 42) |
| Simian virus 40[b] | Yes (4, 5) | | No: mouse, monkey (32, 43) |
| Simian virus 40 | | | Yes: nodules, man |
| CELO (avian adeno) | No (6) | Yes (3, 6)[i] | Yes: hamster[h] (6) |
| Human adeno 12 | Yes (3,7–9) | Yes (3, 8); No (7, 31, p. 101; 33) | ND |
| Def. Adeno 7-SV40 | Yes (10) | Yes (3, 10) | ND |
| Rous sarcoma | No (11, p. 150; 12) | Yes (3); No (11[h], 31, P. 135; 13) | Yes: chick, rat[i] (13) |
| | No (13) | | Yes: hamster[c,h] (13) |
| Murine sarcoma | Yes (13, 14, 15) | No (11); Difficult (34)[i] | ND |
| Murine sarcoma | Yes (16) | Yes (22, 3); No (13)[i] | ND |
| EK (SV40-PML 1) | Yes (17) | "Partial transformation"(17) | ND |
| DAR (SV40-PML 2) | Yes (17) | "Partial transformation" (17) | ND |
| Human papilloma | ND | "Transformed (35, p. 398)[i] | ND |
| **Moderate risk** | | | |
| Feline sarcoma | Yes (3, 18, 19) | Yes 3, 36, 37)[d,i] | ND |
| Herpesvirus saimiri | Yes (3, 19, 20) | No transformation[d] | ND |
| Herpesvirus ateles | Yes (19) | Transforms in vitro[d] (38)[i] | ND |
| Simian sarcoma-VI | Yes (21–23) | "Transformed" (23, 39) | Yes: marmoset[i] (44) |

| | | No morphological alteration[i] | |
|---|---|---|---|
| Gibbon ape (GALV) | Yes (21, 24) | | ND |
| Yaba | Yes (25) No (26) | Yes (3); No (31, p. 45) | ND |
| Herpes simplex 2 | Yes (3, 27f) | Yes (3, 27) | Under study (27) |
| Epstein-Barr | Yes (3, 28, 29) | Yes (3, 28, 29, 40) | Yes: mouse (28) |

[a] Note: ND = no data. Numbers in parentheses indicate references:

[b] Titers vary with cell lines, time of cultivation and cultural conditions (30, pp. 93, 94, 142).

[c] In cortisone-conditioned 3-methylchloanthrene-treated weanling hamsters.

[d] "Morphologically transformed."

[e] Letter, F. Deinhardt, 26 March 1974.

[f] Only in cell culture passages 37 and 38.

[g] "Confirmation needed" (31, 45).

[h] Injected into the cheek pouch.

[i] "RNA tumor viruses . . . not a single case of an *established* (transformed cell) line has been described" (31, p. 184). No permanent avian- or mouse-human cell line (11, p. 150).

[j] There is no agreement on the definition of an infinite growth transformation (established cell line). Seventy population doublings and 150 passages both have been suggested (31).

[k] The word "induction" as used here and elsewhere in this report is not intended as a differentiation between tumor formation caused by released complete virus and tumor formation resulting from transplantation of cells containing subviral oncogenic units (31, p. 18).

[l] Newborn.

[m] References: (1)Butel et al., 1972; (2) Eddy, 1964; (3) Rauscher, 1971; (4) Fogh et al., 1973; (5) Rabson et al., 1962b; (6) Anderson et al., 1969 (7) McAllister et al., 1972; (8) Sultanian and Freeman, 1966; (9) Trentin et al., 1968; (10) Black and Todaro, 1965; (11) Hellman et al., 1973; (12) Jensen et al., 1964a; (13) Sekiyo, 1971; (14) Boiron et al., 1969; (15) Klement et al., 1971; (16) Aaronson and Todaro, 1970; (17) Narayan and Weiner, 1974; (18) Jarret et al., 1972; (19) Melendez et al., 1972b; (20) Ablashi et al., 1971; (21) Kawakami et al., 1972; (22) Larson et al., 1973; (23) Wolfe et al., 1972a; (24) Larson et al., 1974; (25) Levinthal and Shein, 1964; (26) Ambrus et al., 1963; (27) Darai and Munk, 1973; (28) Deal et al., 1971; (29) Gerber et al., 1969; (30) Fogh et al., 1973; (31) Pontén, 1971; (32) Stiles et al., 1975; (33) Todaro and Aaronson, 1968; (34) Todaro and Meyer, 1974; (35) Black, 1968; (36) Chan et al., 1974; (37) Sarma et al., 1970; (38) Falk et al., 1974a; (39) Aaronson, 1973; (40) Epstein and Probert, 1973; (41) Ashkenazi and Melnick, 1963; (42) Eagle et al., 1970; (43) Jensen et al., 1964; (44) Wolfe and Deinhardt, 1972; (45) Rowson and Mahy, 1967.

viruses do not provide for a clear definition of the risks to humans associated with the manipulation of the agents. However, the lack of such data cannot be used in mitigating the potential risk. In an approach to providing a schema that reflects the risks, a gradation of risks to humans from oncogenic viruses is presented in Table IX.

As may be seen, there is not a clear separation between the low- and moderate-risk viruses. Thus the classification of the virus should not be used as the sole determinant of the potential risks. Consideration must be given to the procedures and practices to be employed in use of the virus and the environment in which the work will be conducted. Also included in these considerations must be the provisions for maintaining the integrity of the research materials. A discussion of these is provided in the subsequent sections.

## IV. CONTAINMENT SYSTEMS AND CONTROL PRACTICES

A strategy frequently utilized to protect workers from occupational exposure to hazardous materials is comprised of three elements: one to control the hazardous material at the source to prevent release into the worker's environment, one to minimize the consequences of accidental release of material to the environment, and another to protect the worker from contact with the material in the event he is in a contaminated environment. This strategy is implemented by the design of engineering controls and the selection of control practices for the specific situation at hand. If to this strategy for occupational safety are added further safeguards to protect the surrounding community and general environment, most biomedical research programs can be conducted safely. Safe conduct of biomedical research generally, and viral oncology research specifically, is dependent on an awareness by the laboratory staff of the potential hazards associated with their work, the staff's discipline in good laboratory practices, the availability and proper use of appropriate safety equipment, and a properly designed and constructed research facility.

Although it is axiomatic that facilities are essential for the conduct of research, facilities alone cannot create a research environment that is safe. A safe research environment is primarily dependent on the competence and dedication of the laboratory staff in performing good laboratory practice. Engineering controls (or barrier systems, in laboratory vernacular) complement, rather than substitute for, good laboratory practice. A well-designed facility, however, can facilitate good laboratory practice and can offer ad-

**TABLE VII**

**Low-Risk Oncogenic Viruses that Cross Animal Species Barriers**

| Virus | Result of injecting other animals | | References |
|---|---|---|---|
| | Tumor | Species other than that of viral origin | |
| Adenovirus 1,2,5,6-human | No | Hamster[n] | (1-3) |
| Adenovirus 2,7,12-human | No | Monkey: rhesus, cynomolgus; marmoset | (2, 4) |
| Adenovirus, various[k] | Yes | Hamster,[n] mouse,[n] rat,[n] mastomys[n] | (4-6) |
| Sheep papilloma | Yes | Only regressive tumors in sheep, hamsters[n] | (7) |
| Avian leukosis | No | Man; (lamb, guinea pig, mouse)[y,n] | (8,9, p. 210; 10) |
| CELO (avian adeno) | No | Chicken, mouse, rabbit | (5) |
| CELO (avian adeno) | Yes | Hamster[n] | (5, 11, 12) |
| Herpesvirus, turkeys | No | <19 days old: rhesus, cynomolgus, bonnet | (13) |
| Marek's disease | No | Marmoset,[a,n] rhesus[a] | (13, 14) |
| Marek's disease | No | <19 days old: rhesus, cynomolgus | |
| Shope rabbit fibroma | No | Guinea pig[w] | (15, 16) |
| Rous sarcoma | Yes | Lamb,[n] lemming, reptiles, drosophila | (17, 18) |
| Rous sarcoma-SR | No | Baboon,[a] squirrel M. >4 yr. old, rat[a] | (4, 19, 20) |
| Rous sarcoma-SR | Yes | Rat,[n] rat,[d] rat,[c] hamster,[n] mouse[a] | (19, 21) |
| Rous sarcoma-SR | Yes | Squirrel monkey <303 days old, tumors[r] in 50% | (20) |
| Rous sarcoma | Yes | Baboon[n] <9 months old[i]; (dog, rabbit)[i,m,p] | (4, 22) |
| Rous sarcoma | Yes[r] | Galagos and Old World monkeys | (4) |
| Rous sarcoma-SR | Yes[p] | Marmoset,[b] rhesus[m,c or h] | (4, 23, 24) |
| Rous sarcoma-Mill Hill | No | Rat, mouse, hamster, rabbit | (25, 26) |
| Rous sarcoma-SR | Yes | G. pig,[n] rabbit,[r] (rat, mouse, hamster, G. pig)[p] | (19, 25-27) |
| Rous sarcoma | Yes | Duck, turkey, pigeon, goose, pheasant | (17, 24, 28) |
| Bovine rhinotracheitis | Yes | Hamster,[b] by viral transformed hamster cells | (29) |
| Bovine lymphosarcoma | No | Monkey, mouse, G. pig, hamster, rat | (30, 31) |
| Bovine lymphosarcoma | Yes | Lamb | (31, 32) |
| Squirrel fibroma | Yes | Rabbit, woodchuck[x] | (33) |

*(cont.)*

# TABLE VII (CONT.)

| Virus | Result of injecting other animals | | References |
|---|---|---|---|
| | Tumor | Species other than that of viral origin | |
| Murine leukemia | No | Rat,[a] hamster,[n] G. pig,[n] cat,[g] rhesus[n,i] | (34–38) |
| Murine leukemia, sarcoma | No | Rabbit,[x] marmoset, galagos, chick[e] | (4, 39) |
| Murine leukemia | Yes | Rat,[n] hamster[n] | (35, 37, 38, 40, 41) |
| Murine sarcoma | Yes | Hamster,[n,w] rat,[n] G. pig[n,w] | (23, 42–46) |
| Deer, sheep, fibroma | Yes | Hamster[n] at site of inoculation | (47) |
| Bovine papilloma | Yes | Mouse,[x] hamster,[b] horse[a,r]; No: rat[q] | (47–49) |
| Human papilloma (warts) | Yes | Man, monkey (rare), dog? | (50, 51) |
| Polyoma-murine | No | Mouse,[a] rat,[w,a] cat,[n] dog,[n] chick | (21, 52, 53) |
| Polyoma-murine | Yes | Rat,[n] hamster,[n] rabbit,[n,r] ferret,[n] G. pig | (47, 53–55) |
| Polyoma-murine | Yes | Mouse,[c,n] mastomys,[n] hamster 4–5 mo. | (21, 52, 56) |
| Mason-Pfizer Monkey | No | Rhesus monkey[n] | (57–59) |
| Herpes simplex type 1 | No[j] | Rabbit, G. pig, mouse, hamster, c. rat, monkey | (60–62) |
| Herpes simplex type 1 | Yes | Hamster[n] by HVS1 transformed h. cells | (63, 64) |
| Simian virus 40 | No | Female spider monkey[c] | (65) |
| Simian virus 40 | No[m] | Monkey,[a,n] (mouse, rat, G. pig, rabbit)[r] | 4, 66–68) |
| Simian virus 40 | Yes | Hamster,[c,n,w,a] mastomys[n] | (21, 69–71) |
| Simian virus 40 | Yes | Mouse,[i] rat[n] | (72, 73) |
| Papovavirus BK (human) | Yes | Hamster,[a,n] | (9, p. 283; 74, 75) |
| JC, EK, DAR (from PML) | Yes | Hamster[n] | (76,77) |
| EK, DAR | No | Mouse[n] | (76) |
| Cytomegalovirus | Yes | Hamster[n] by virus-transformed h. cells | (64) |

Note: CELO = chicken-embryo-lethal-orphan virus; the only known oncogenic avian adenovirus. PML = progressive multifocal leukoencephalopathy.

[a] Adult.
[b] All ages.
[c] Adult immunosuppressed.

[d] In adults only after several serial passages in newborn rats.

[e] Embryo

[f] Intracerebral inoculation.

[g] Germ-free.

[h] Pregnant.

[i] Immunosuppressed.

[j] No tumor, but infection, virus in tissues, and often overt disease.

[k] Human including Ad7 and 12, simian, canine, bovine.

[m] Tested in marmosets?

[n] Newborn.

[p] Usually progressive oncogenesis.

[q] Bovine tumors regress unless inoculation is into the brain (78).

[r] Tumors usually regress.

[s] Pathological lymphoid and reticulum-cell hyperplasia that appeared to be progressive.

[w] Weanling.

[x] Age not stated.

[y] References: (1) Lewis et al., 1974; (2) McAllister et al., 1972; (3) Trentin et al., 1962; (4) Rabin, 1971; (5) Merkow and Slifkin, 1973; (6)Rondhuis, 1973; (7) Gibbs et al., 1975; (8) Davis, 1962; (9) Hellman et al., 1973; (10) Hilleman, 1968; (11) Anderson et al., 1969; (12) Stenbach et al., 1973; (13) Sharma et al., 1973; (14) Sharma et al., 1972; (15) Chaproniere and Andrewes, 1957; (16) Olsen and Yohn, 1974; (17) Loan, 1974; (18) Pontén, 1971; (19) Ahlstrom and Forsby, 1962; (20) Rabin and Cooper, 1971; (21) Allison et al., 1967; (22) Hanguenau et al., 1971; (23) Deinhardt et al., 1972; (24) Munroe, 1969; (25) Zilber, 1964; (27) Ahlstrom et al., 1962; (28) Zilber, 1961; (29) Michalski and Hsiung, 1975; (30) Barthold et al., 1976; (31) Olson et al., 1973; (32) Hoss and Olson, 1973; (33) Huebner, 1963; (34) Cohen et al., 1974; (35) Ioachim et al., 1973; (36) McCullough et al., 1971; (37) Mirand and Grace, 1962; (38) Rich et al., 1965; (39) Rich, 1968; (40) Maruyama and Dmochowski, 1973; (41) Moloney, 1962; (42) Forni et al., 1975; (43) Harvey, 1964; (44) Huebner, 1967; (45) Heubner et al., 1966; (46) Kelloff et al., 1970; (47) Barthold and Olson, 1974; (48) Boiron et al., 1964; (49) Robl and Olson, 1968; (50) Jablonska et al., 1972; (51) Rowson and Mahy, 1967; (52) Eddy, 1969; (53) Flocks et al., 1965; (54) Eddy et al., 1959; (55) Rowe, 1961; (56) Barski et al., 1962; (57) Fine et al., 1974; (58) Fine et al., 1975; (59) Rangan, 1974; (60) Katzin et al., 1967; (61) Nahmias and Dowdle, 1968; (62) Nahmias et al., 1970; (63) Duff and Rapp, 1973; (64) Rapp, 1974; (65) Cole and Shah, 1974; (66) Butel et al., 1972; (67) Eddy, 1964; (68) Rabson et al., 1965; (69) Diamondopoulos, 1973; (70) Rabson et al., 1972a; (71) Takemoto et al., 1968; (72) Allison and Taylor, 1967; (73) Ashkenazi and Melnick, 1963; (74) Portolani et al., 1975; (75) Shah et al., 1974; (76) Narayan and Weiner, 1974; (77) Walker et al., 1973; (78) Olson et al., 1969.

183

**TABLE VIII**

**Moderate-Risk Oncogenic Viruses that Cross Animal Species Barriers**

| Virus | Result of injecting other animals | | References |
|---|---|---|---|
| | Tumor | Species other than that of viral origin | |
| Feline leukemia | No | Mouse, rat, mouse" | (1, 2) |
| Feline sarcoma-ST | No | Talapoin monkey," baboon | (3, 4) |
| Feline sarcoma-ST | yes' | Sheep,'," pig," dog," rabbit," | (4–6) |
| Feline sarcoma-ST | Yes' | Squirrel monkey,^b 3 species Macaca" | (4, 6, 7) |
| Feline sarcoma-ST | Yes | Rat," regress and then reappear; dog" | (8, 9) |
| Feline sarcoma-ST | Yes^p | Marmoset,^b sheep^a | (6, 10, 11) |
| Feline sarcoma-GA | No | Rabbit, hamster, mouse, porpoise, baboon | (8, 12) |
| Feline sarcoma-GA | Yes' | Dog^a | (13) |
| Feline sarcoma-GA | Yes^p | Marmoset,^b dog^{j,n} | (11, 13) |
| Feline sarcoma, -ST or -GA | Yes | Mouse^e | (3) |
| Woolly monkey SSV-1 | No | Squirrel monkey," Galago" | (7, 14, 15) |
| Woolly monkey SSV-1 | Yes | Marmoset," relatively benign | (7, 16) |
| Woolly monkey SSV-1 | Yes^p | Marmoset^{n,a} | (15, 17) |
| Gibbon ape^x | No | Squirrel monkey," marmoset | (14, 18) |
| Gibbon ape^x | No | Pig," cat" | (18) |
| *Herpesvirus ateles* | No^h | Black spider monkey^a and others | (19–22) |
| *Herpesvirus ateles* | No | Rabbit, squirrel monkey, African grivet^a | (20–22) |
| *Herpesvirus ateles* | No | Owl monkey,^a mouse, goat | (19, 22, 23) |
| *Herpesvirus ateles* | Yes | Marmoset^a | (21, 22, 24) |
| *Herpesvirus saimiri* | No^h | Squirrel monkey^a and others | (19, 25) |
| *Herpesvirus saimiri* | No | Mouse," chick embryo, dog, goat | (19, 24, 25) |
| *Herpesvirus saimiri* | No | Simians: galagos, talapoin, rhesus | (7, 24, 26) |
| *Herpesvirus saimiri* | No | Baboon, chimpanzee, stumptailed | (24) |
| *Herpesvirus saimiri* | No^i | Green monkey, capuchin | (7, 24) |
| *Herpesvirus saimiri* | No | Hamster, but fatal disease" | (19, 25) |
| *Herpesvirus saimiri* | Yes^p | simians: marmose,^b owl, spider | (24, 26, 26) |
| *Herpesvirus saimiri* | Yes | Cinnamon ringtail, rabbit; No: rabbit (316) | (24, 28) |
| Herpes simplex type 2^x | No | Hamster," rabbit | (29–31) |
| Herpes simplex type 2^x | No | Cebus monkey, marmoset, baboon | (30, 32) |

| | | | |
|---|---|---|---|
| Herpes simplex type 2[x] | Yes | Hamster,[n,w,c] mouse,[a,c] mouse? | (30, 33–35) |
| Yaba | No | (Dog, rabbit, G. pig, rat, hamster, mouse)[a,n] | (36, 37) |
| Yaba | No | Simians: marmoset, squirrel, vervet, mona, sooty | (36) |
| Yaba | Yes[r] | Rhesus, cynamolgus, A. green, pigtail, man | (36) |
| Epstein-Barr | No | Gibbon, A. green, chimpanzee, rhesus | (38–42) |
| Epstein-Barr[d] | Yes[p] | Marmoset,[j,n] (hamster, rat, mouse[i])[n] | (40, 43–46) |
| Nondefective Ad2-SV40 | Yes | Hamster[a] by inoculum[c] | (47) |
| E. coli-SV40 hybrid | ND | ND | |

Note: ND = no data.

[a] Adult.
[b] All ages.
[c] Virus-transformed cells used as inoculum.
[d] In marmosets, only EBV derived from human infectious mononucleosis induced tumors (40, 45, 48, 49).
[e] Young mice immunosuppressed by antithymocyte serum, inoculated with FeSV-transformed baboon testis cells.
[f] Fetal.
[g] Changed to Candidate Human Isolate, 3 October 1974.
[h] Infection and virus in tissues without detectable disease.
[i] Sometimes yes.
[j] Juvenile.
[n] Newborn.
[p] Progressive.
[r] Tumors regress.
[w] Weanling.
[x] Viral-induced leukemia in the gibbon ape reported by T. Kawakami at the 9th Annual Joint Working Conference, NCI, VCP, November 1974.

[y] References: (1) Jarrett, 1970; (2) McCullough et al., 1971; (3) Melnick et al., 1971; (4) Rabin et al., 1972; (5) Pearson et al., 1973; (6) Theilen et al., 1970; (7) Rabin, 1971; (8) Jarrett et al., 1970; (9) Maruyama and Dmochowski, 1973; (10) Hall et al., 1975; (11) Wolfe et al., 1972b; (12) Gardner, 1971; (13) Gardner et al., 1970; (14) Rabin, 1974; (15) Wolfe and Deinhardt, 1972; (16) Wolfe et al., 1971b; (17) Neubauer et al., 1974; (18) Snyder et al., 1973; (19) Deinhardt et al., 1974b; (20) Falk, 1974; (21) Hunt et al., 1972; (22) Laufs and Melendez, 1973; (23) Melendez et al., 1972a; (24) Melendez et al., 1972b; (25) Melendez et al., 1969; (26) Wolfe et al., 1971a; (27) Melendez et al., 1970; (28) Daniel et al., 1974; (29) Nahmias et al., 1970; (30) Sever, 1973; (31) Walker et al., 1973; (32) Palmer et al., 1976; (33) Boyd and Orme, 1975; (34) Duff and Rapp, 1971; (35) Rapp, 1974; (36) Ambrus et al., 1963; (37) Espana, 1971; (38) Editorial, 1974; (39) Henle and Henle, 1973; (40) Shope et al., 1973; (41) Werner et al., 1972; (42) Werner et al., 1973; (43) Adams et al., 1971; (44) Deal et al., (45) Falk et al., 1974b; (46) Southam et al., 1969; (47) Lewis et al., 1974; (48) Deinhardt et al., 1947a; (49) Werner et al., 1975.

**TABLE IX**

**Gradation of Risk to Humans from Oncogenic Viruses**

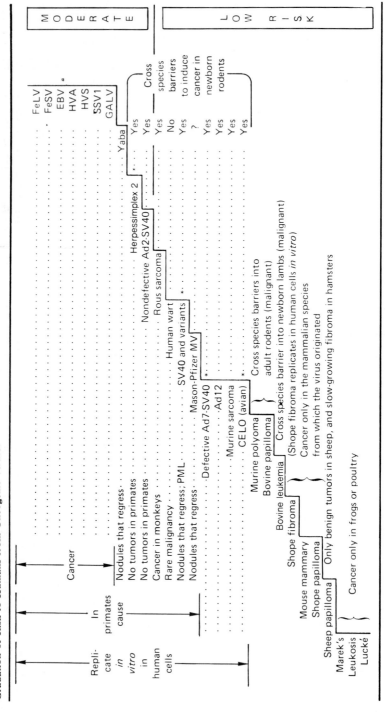

[a] Notes: *, Transform human cells *in vitro*. ?, Possibly due to a murine endogenous virus in nude mice. PML-progressive multifocal leukoencephalopathy associated with three SV40-like viruses.

ditional safeguards to ensure the protection of persons and the environment outside the laboratory.

The following discussion deals with the engineering controls and control practices that have evolved during the era of infectious disease microbiology and that have been further refined during the recent years of viral oncology research.

## A. Barrier Systems

From an analysis of 3921 laboratory-associated infections, Pike (1976) reported that only 703 infections (18%) resulted from recognized accidents. In 827 infections (21%) it was only known that the victim worked with the agent, and in 767 infections (20%) the cause was unknown or not indicated. These two categories represent 41% of the cases analyzed. The sources that accounted for the balance of the infections were animals for ectoparasites (658 cases), clinical specimens (287 cases), discarded glassware (46 cases), human autopsy (75 cases), intentional infection (19 cases), aerosols (522 cases), and other sources (16 cases). Of the 703 recognized accidents, contact with infectious material resulting from aerosol-associated events such as spills and sprays was the identified cause for 188 (27% of 703). In addition to the 188 aerosol-associated infections, aerosols were identified as the probable sources of 522 other infections. Accordingly, 710 of all infections analyzed by Pike may have resulted from aerosols.

It is generally assumed that the majority of unexplained laboratory infections (e.g., the 41% in Pike's study) are the consequence of the release of undetected aerosols that normally occur during routine laboratory operations. In such instances, exposure results from the inhalation of respirable aerosol particles as well as from contact contamination from splatter droplets and settling aerosol particles. This contact contamination can subsequently be transferred to the skin and absorbed through cuts or abrasions or may be transferred to the mouth. Stern *et al.* (1974), in describing aerosol production associated with clinical laboratory procedures, suggest that the surface contamination resulting from aerosol splatter and settling particles can be a more significant source of laboratory contamination than the aerosol per se.

Barrier systems are used to confine aerosols of hazardous agents and the resulting contact contamination at the source during manipulation of the agent by workers or processing of the agent with laboratory equipment. The primary barrier is interposed between the agent and the personnel. It is intended that the primary barrier provide protection to the individual manipulating the agent, as well as to other persons in the laboratory, to

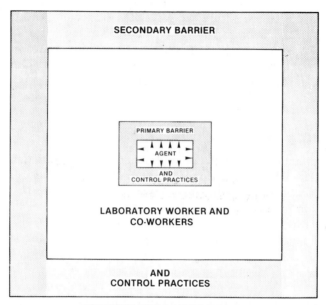

Fig. 1. The roles of primary and secondary barriers. (Reprinted by permission of the American Society of Heating, Refrigerating and Air-Conditioning Engineers, Inc., from ASHRAE TRANSACTIONS, copyright 1978.)

personnel in other areas of the building, and to the environment (see Fig. 1). Examples of primary barriers are ventilated hoods and safety cabinets and glove box enclosures. Architectural and engineering features of the laboratory can form a secondary barrier to protect personnel in other areas of the building and the environment from exposure to hazardous agents that may be released because of failure of primary barriers or because of laboratory accidents occurring outside a primary barrier. It is to be recognized that the secondary barrier does not reduce the risk of exposure for personnel inside the laboratory where material may be released.

From a safety standpoint, primary and secondary barriers serve to protect personnel and the environment from hazardous research materials. However, as barriers to the movement of contamination, some of these systems also have proved useful in maintaining the purity of research materials. Contamination detrimental to the research mission may include ubiquitous fungal or bacterial contamination from outdoors, indoors, or even laboratory personnel. The potential of cross contamination of research materials in the laboratory can also be minimized by barrier systems.

**Fig. 2.**   Physical barrier.

## 1. Primary Barriers: Design Concepts, Capabilities, and Limitations

There are four elemental types of barriers that can be assembled in various combinations and configurations to produce a barrier device or system. The four elemental types are physical, air, filtration, and destruction or inactivation barriers.

A physical barrier is an impervious surface through which material cannot move. Physical barriers are often arranged to form an enclosure to separate contaminated space from contaminant-free space. They are typified by the solid sides, view panels, gaskets and sealants, and gloves of a laboratory glove box (see Fig. 2). The materials of construction should be selected to withstand wear and the corrosive action of laboratory gases, liquids, and decontaminants and also to contribute to the maintenance of the integrity of the barrier at all anticipated pressures. Gaskets, sealants, and gloves can be damaged or become worn during installation or use, or they may deteriorate with age. Consequently, these parts of a barrier system should be checked for their integrity immediately following initial installation and periodically thereafter. Physical barriers, when intact, are considered absolute barriers in that their effectiveness in confining contamination is total and complete.

An air barrier is created by maintaining a flow of air with relatively uniform direction and velocity. The air inflow through the face or work opening of a biological safety cabinet, as shown in Fig. 3, is an example of an air barrier. The directional flow of air is maintained at a velocity great enough to prohibit the movement of contamination against the direction of airflow. The flow must also overcome the effects of eddy currents that always accompany turbulent flow and disruption of the normal flow by personnel activity near the face of the cabinet. The integrity of an air barrier can be compromised by aerosols released from high-energy operations, by disruption of the airflow by personnel activity near the cabinet, or by mechanical failure of the air-moving equipment. Contact contamination, as may be found on a laboratory worker's gloved hands that are withdrawn from the work area of a cabinet can be transported across an air barrier. For these reasons, air barriers are considered only partial barriers.

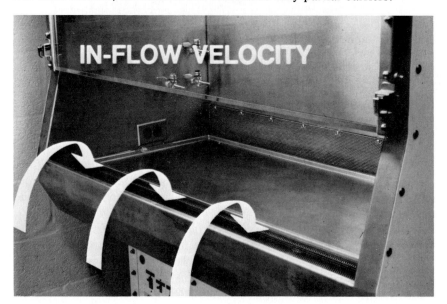

**Fig. 3.**   Air barrier.

Filtration barriers are used to filter or remove particulate contamination from air that is drawn into or exhausted from a primary barrier system as well as to filter air recirculated within a device. There are numerous types of filters, and it is possible that one type can be manufactured to meet various performance requirements.

Filters most often used in laboratory primary barriers are ultra-high-efficiency filters known as High-Efficiency Particulate Air (HEPA) filters. HEPA filters are efficient in removing at least 99.97% of particles 0.3 μm in diameter. Because of the mechanics of particle filtration, particles of larger *and* smaller sizes are removed with even greater efficiency. In addition, the efficiency of a particulate filter may increase slightly as the filter medium becomes loaded with contaminants; the efficiency of an intact filter will not decrease with use. However, as the filter becomes loaded, the resistance to air movement through the filter increases, with the result that the rate of airflow will decrease. Therefore filters must be replaced periodically to assure the design conditions of airflow and the performance of the barrier system are maintained. The manufacturer of the system should be consulted for its criteria for filter maintenance. Because filters are susceptible to damage during manufacture, shipping, installation and maintenance, a primary barrier system should not be used until the integrity and efficiency of the filters have been verified following installation, relocation, or maintenance of the system.

Particulate filters do not remove gaseous constituents of air. Either wet collectors or adsorption systems (e.g., beds of activated charcoal) can be used to remove gaseous contaminants, the latter being the one employed most frequently in the laboratory setting. The performance characteristics of adsorptive media air purifiers are less defined than those of particulate filters, since performance at a moment in time can be effected by a variety of factors, including ambient temperature, relative humidity, chemical constituents of the air (other than the contaminant of interest) and their concentrations, and extent to which the adsorptive capacity of the media has been depleted.

Destruction or inactivation barriers represent the fourth elemental type of barrier for physical containment and control of biohazardous contamination. A destruction barrier destroys the contamination that passes through it; an incinerator that destroys airborne contaminants in the air exhausted from a safety cabinet is such a device. Closely allied to destruction barriers are those that inactivate, in a biological sense, viable contaminants yet do not necessarily affect the vehicle of the contaminant. Inactivation barriers are usually employed to facilitate decontamination of equipment or supplies transferred into or out of primary barrier systems. These include double-door sterilizers (employing either steam or gas sterilant), double-door fumigation or spray chambers, and dunk baths filled with disinfectants. In using one of these devices, it is necessary for the laboratory worker to understand its principle of operation and its limitations; only practices proven to be effective should be followed.

## 2. Primary Barriers: Biological Safety Cabinets

Although numerous models of biological safety cabinets incorporating the elemental types of barriers are marketed by manufacturers, three classes of cabinets (Classes I, II, and III), each distinguished by its design and containment capabilities, have become widely recognized. In the early 1970s, the National Cancer Institute formalized design and performance specifications for laminar flow biological safety cabinets; these specifications were subsequently incorporated into a design standard (National Sanitation Foundation, 1976). More recently, the National Institutes of Health has formalized specifications for all three classes of cabinets (U.S. Department of Health, Education and Welfare, 1978b). Recommended applications and minimum performance specifications for biological safety cabinets are summarized in Tables X and XI.

**a. Class I Cabinets.** The Class I cabinet is the simplest form of a biological safety cabinet and consists of an enclosure with a front view panel and a full-width work opening (see Fig. 4a). Room air drawn into the cabinet through the work opening prevents airborne contaminants inside the cabinet from escaping into the room (see Fig. 4b). The air flows across the work area, over and under a back wall baffle, and out through a HEPA filter to an exhaust duct and blower. The air from the cabinet is discharged either into a building air exhaust system or directly to the outdoors. The airflow, inward through the unrestricted work opening, should have a minimum face velocity of 75 ft/min (normal to the work opening).

Optional modes of operation are available for the Class I cabinet. A front closure panel with access ports through which the operator's arms can be extended into the cabinet, can be placed over the work opening (see Fig. 4c). This reduces the amount of open area and raises the face velocity of the air through the access ports. Another option is to attach arm-length rubber gloves to the access ports of the closure panel, and in this mode, the cabinet serves as a glove box (see Fig. 4d) but does not provide containment equivalent to a Class III system (see Section IV,A,2,c). In this last mode the amount of air flowing into the cabinet is very small; the air pressure within the cabinet should be at least 0.5 in. water gauge (W.G.) below that of the laboratory.

When used in either of these optional modes, equipment and materials can be introduced and removed through a hinged front view panel or a side

**TABLE X**

**Applications of Biological Safety Cabinets in Microbiological Research**

| Biological safety cabinet | | | Research uses/applications | | | |
|---|---|---|---|---|---|---|
| Type | Work opening | Face velocity (ft/min) | Oncogenic viruses[a] | Chemical carcinogens[b] | Etiologic agents[c] | Recombinant DNA[d] |
| Class I | Front panel not in place | 75 | Low and moderate | No | CDC 1-3 | P1-P3 |
| | Front panel in place without gloves | 150 | Low and moderate | Yes | CDC 1-3 | P1-P3 |
| | Front panel in place with gloves | NA | Low and moderate | Yes | CDC 1-3 | P1-P3 |
| Class II | | | | | | |
| Type A | Fixed height, usually 10 in. | 75, minimum | Low and moderate | No | CDC 1-3 | P1-P3 |
| Type B | Sliding sash provides opening adjustable from 8 to 20 in. for introduction and removal of equipment and materials. To obtain proper face velocity, experimentation should be done with 8-in. opening. | 100 at 8-in. sash opening | Low and moderate | Yes in low dilution and volatility | CDC 1-3 | P1-P3 |
| Class III | No direct opening. Access is through double-door sterilizer and decontaminant dunk bath. | NA | Low, moderate, and high | Yes | CDC 1-4 | P1-P4 |

[a] U.S. Department of Health, Education and Welfare (1974b).
[b] U.S. Department of Health, Education and Welfare (1975).
[c] U.S. Department of Health, Education and Welfare (1976b).
[d] U.S. Department of Health, Education and Welfare (1978b).

**TABLE XI**

**Recommended Minimum Performance Specifications of Biological Safety Cabinets**

| Cabinet | | Face velocity, ft/min | Velocity profile | Negative pressure (in., w.g.) | Permissible leak rate | Exhaust filter efficiency |
|---|---|---|---|---|---|---|
| Class I | Open front | 75 | NA | NA | NA | 99.97% for 0.3 $\mu$m particles |
| | Front panel without gloves | 150 | NA | NA | NA | " |
| | Front panel with gloves | NA | NA | $\triangle$p > 0.5 | NA | " |
| Class II | Type A | 75 | [a] | NA | < 1 × 10⁻⁴ cm³/sec at 2 in. w.g. pressure | " |
| | Type B | 100 | [a] | NA | NA | " |
| CLASS III | | NA | NA | $\triangle$p > 0.5 | < 1 × 10⁻⁵ cm³/sec at 3 in. w.g. pressure | [b] |

[a] Dependent on National Sanitation Foundation (NSF) certification in accordance with NSF Standard 49.

[b] Both HEPA filters must be certified to have a filtration efficiency of 99.97% for 0.3 $\mu$m particles. When an incinerator is used in lieu of the second HEPA filter, the incinerator must be capable of destroying all spores of *Bacillus subtilis* when challenged at a concentration of $10^5$ spores per cubic foot of air.

air lock, if provided. Service cocks and piping for the usual laboratory utilities, water service, cup sinks, and drains can be provided; however, gas should not be used if the cabinet is operated with the gloves attached.

Class I cabinets of usual configuration (Fig. 4a) easily accommodate many routine laboratory operations such as pipetting, blending, and sonicating. Cabinets can be custom built to contain larger items of equipment, such as lyophilizers and ultracentrifuges (see Fig. 4e).

Since room air is drawn through the work opening and across the work area, the Class I cabinet does not provide protection to the experimental materials from airborne contamination that is ubiquitous or of laboratory or personnel origin. Even when the cabinet is used with the closure panel and attached gloves, protection of materials within the cabinet cannot be guaranteed because the air that does not enter the cabinet is not filtered.

The limitations of air barriers, discussed in Section V,A,1, apply to the Class I cabinet. The protective function of the cabinet can be compromised

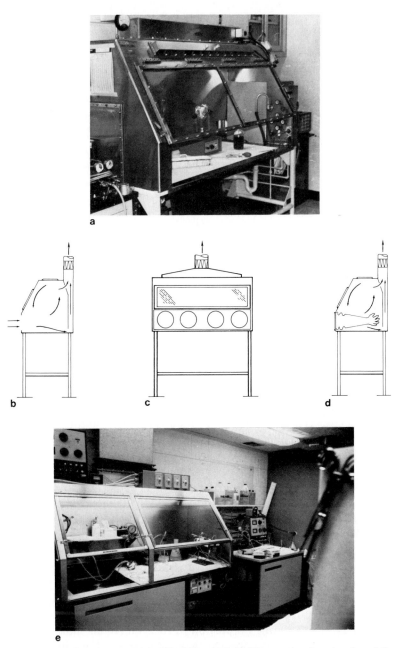

**Fig. 4.**   Class I cabinet: (a) with full-width opening, (b) cutaway view showing airflow at work opening, (c) with front closure panel with access ports, (d) cutaway view showing closure panel and attached gloves, (e) custom built for special application.

by any disturbance of the air barrier. This can be caused by the operator suddenly withdrawing his hands from the cabinet or by air currents such as those caused by personnel traffic near the work opening, by the closing or opening of a nearby room door, or by the discharge of air from a ventilation supply grille located near the cabinet. Except when the cabinet is used in the mode involving attached arm-length gloves, the operator's hands and arms are not protected from contact contamination. Control of contact contamination is dependent on the use of gloves and other protective clothing and of procedures for their decontamination and/or removal.

Class I cabinets are suitable for general research with low- and moderate-risk oncogenic viruses (U.S. Department of Health, Education and Welfare, 1974b), etiologic agents assigned to classes 1, 2, and 3 by the Center for Disease Control (U.S. Department of Health, Education and Welfare, 1974a), and recombinant DNA materials requiring P1 to P3 level of physical containment (U.S. Department of Health, Education and Welfare, 1978a). Chemical carcinogens and low-level radioactive materials and volatile solvents can be used in Class I cabinets, *provided* (1) the face velocity is 100 ft/min or a higher value that may be specified by an applicable regulation, (2) careful evaluation reveals that concentrations of the compounds will not reach dangerous levels or cause problems of decontamination of the cabinet or associated exhaust system, and (3) quality of the effluent meets emission regulations that might apply.

**b. Class II Cabinets.** The Class II cabinet, commonly known as a laminar flow biological safety cabinet, is the cabinet with the most sophisticated design. The Class II cabinet utilizes an air barrier at the work opening to provide both containment of contamination inside the cabinet and protection of the experimental equipment and materials from contamination by extraneous matter. Airborne contaminants in the cabinet are prevented from escaping by an air barrier formed by (1) room air flowing into the work opening and (2) HEPA filtered air flowing down from an overhead grille inside the cabinet. These two flows are drawn into a narrow grille that extends the length of the front edge of the work surface, immediately inside the work opening (see Fig. 5a and 5b). This air barrier also prevents airborne contaminants in room air from entering the work space of the cabinet, since the room air flowing into the cabinet through the work opening is drawn into the front grille. Air drawn into the front grille passes through a HEPA filter and is recirculated into the cabinet work space through the overhead grille. A portion of this filtered air maintains the air barrier at the work opening and the remainder passes down onto the work surface and is drawn from the work space through grilles at the back edge of the work surface. The HEPA filtered air from the overhead grille flows

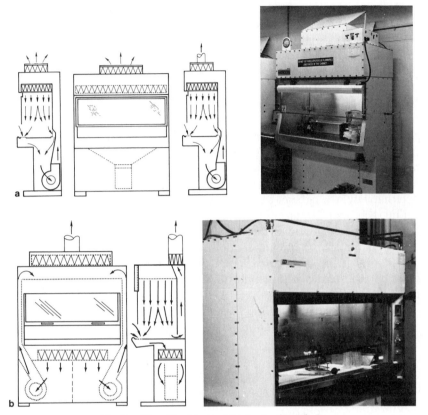

**Fig. 5.** Class II cabinets: (a) type A, (b) type B.

downward with uniform velocity and minimum turbulence or back mixing. This unidirectional flow of filtered air minimizes lateral movement of aerosolized contamination within the cabinet and at the same time purges the work space. The result is a clean-air work environment. A portion of the air that is drawn into the grilles at the front and back of the work surface is exhausted from the cabinet after passing through a HEPA filter.

There are two types of Class II cabinets: Type A (formerly known as Type 1) and Type B (formerly known as Type 2). These differ principally in the vertical dimensions of the work opening; the proportion of air recirculated; the velocity of the airflows at the opening and downward to the work surface; the manner of discharge of exhaust air; and the pressure of the contaminated air plenums relative to the room.

The Type A cabinet (see Fig. 5a) has a work opening of fixed vertical dimension, usually from 8 to 10 in. A relatively high proportion, approximately 70%, of the air drawn into the front and back grilles is recirculated through the work space; approximately 30% is exhausted. The average velocity of the air flowing out of the overhead grille and downward to the work surface is usually 75 ft/min. The average velocity of room air into the cabinet through the work opening is at least 75 ft/min. A blower, an integral part of the cabinet, forces portions of the contaminated air through the recirculation filter and the exhaust filter; the plenum between the blower and these two filters is contaminated and at positive pressure relative to the room. The air exhausted from the cabinet can be handled in either of two manners. In one approach it can be discharged directly into the laboratory room, where it mixes with the room air. This method minimizes the impact on the ventilation system serving the room, unless the heat produced by the operation of the cabinet cannot be accommodated. Alternatively, the cabinet exhaust can be connected to a duct via a thimble fitting (National Sanitation Foundation, 1976) and discharged either into a building exhaust system or directly to the outdoors. This mode of discharge is often required for removal of odor or excessive heat. When the exhaust is ducted out of the laboratory room, the general ventilation for the room must be adjusted to compensate for the room air exhausted through the cabinet.

The Type B cabinet (see Fig. 5b) has a vertical sliding sash that enables the laboratory worker to adjust the work opening from 8 to 18 in., although the recommended work height is 8 in. A relatively low proportion, usually 30–50%, of the air drawn into the front and back grilles is recirculated through the work space; 50–70% is exhausted. The average velocity of air flowing out of the overhead grille and downward to the work surface is usually 50 ft/min, whereas the average velocity of the room air into the cabinet is 100 ft/min. This face velocity is attained when the sash is lowered to a work opening of 8 in. and is reduced as the sash is raised. A pair of blowers in the base of the Type B cabinet draws air through the front grille near the work opening and then through a HEPA filter. The blowers force filtered air up through plenums along the side of the cabinet and downward through the overhead grille above the work surface. Air is drawn from the work area through the rear gille, and via a plenum through a HEPA filter by an exhaust blower that is located outside the cabinet and that is not usually an integral part of the cabinet. The contaminated plenums between the inlet grilles and the HEPA filters are at negative pressure relative to the room. The exhaust air from the external blower may be discharged either to a building exhaust system or directly to the outdoors. The room ventilation system must be adjusted to compensate for the room air exhausted through the cabinet.

Because of their sophisticated design and the requirement for delicate balance of air flows to achieve the design performance, Class II cabinets are not readily modified to special configurations. However, some manu-factureres of Type A cabinets do offer hinged front-view panels and side access doors to facilitate introduction and removal of equipment and materials. Service cocks and piping for usual laboratory utilities may be provided; water drains cannot be provided.

Although sophisticated in design, the Class II cabinet utilizes an air bar-rier at the work opening to control movement of airborne contamination. The limitations of air barriers discussed in general terms in Section IV,A,1, and with reference to Class I cabinets in Section IV,A,2a, apply to both types of Class II cabinets. The double protective function (i.e., protection of the laboratory environment and of the environment of the work space within the cabinet) of the Class II cabinet can be compromised by disturb-ance of the air barrier. In addition to the conditions that can disturb the air barrier of both Class I and Class II cabinets, the barrier of the Class II cabinet can be disturbed by an imbalance of airflows that may be caused by mechanical failure, dirty filters, or blockage of the air-intake grilles that ex-tend along the front and back of the work surface. Furthermore, the uniform, downward flow of air over the work surface can be disturbed by overcrowding the cabinet interior with equipment and supplies or by con-vection currents from heat sources (e.g., bunsen burners). Contact con-tamination on hands or equipment and supplies can traverse an air barrier in either direction: The external contamination can enter the cabinet by contact to jeopardize the experiment just as easily as hazardous con-taminants can escape the cabinet by surface contact.

If the limitations of air barriers and the operation of Class II cabinets are understood by laboratory workers, they will be able to operate the cabinets according to design intent and achieve a high degree of environmental con-trol. Good laboratory practices to be used in conjunction with Class II cab-inets are discussed in Section IV,B.

Class II cabinets are suitable for procedures with low-risk oncogenic viruses that have the likelihood of generating aerosols (U.S. Department of Health, Education and Welfare, 1974b) and for general research pro-cedures with moderate-risk oncogenic viruses (U.S. Department of Health, Education and Welfare, 1974b), etiologic agents assigned to classes 1 to 3 by CDC (U.S. Department of Health, Education and Welfare, 1974a), and recombinant DNA materials requiring P1 to P3 level of physical contain-ment (U.S. Department of Health, Education and Welfare, 1978a). Be-cause Class II cabinets recirculate a large fraction of the air flowing through them, they are not generally suitable for procedures involving the use of solvents, toxic chemicals, or radioactive materials. The Type B cabinet, however, is recommended by the National Cancer Institute for work with dilute preparations of chemical carcinogens of low volatility

(U.S. Department of Health, Education and Welfare, 1976a), *provided* the face velocity meets the requirements of applicable regulations and the quality of the effluent meets emission regulations that might apply.

Laminar flow clean benches in which filtered air is forced across the work area *toward* the laboratory worker and *out* into the room should not be used in biomedical laboratories. Such clean benches were developed to facilitate the clean assembly of electronic and spacecraft components and the performance of operations requiring extreme cleanliness, or even sterility. However, since the operation of these benches results in the exposure of the laboratory worker to materials that may be released or become airborne during the work, the benches should not be used for work with biologic material, since even sterile or other seemingly innocuous biologic material may contain potentially allergenic substances.

**c. Class III Cabinets.** The Class III cabinet (Figs. 6a, 6b), commonly known as a glove box, is an hermetically sealed enclosure for confining extremely hazardous research materials. Absolute containment is provided by physical barriers; stainless steel and glass are the usual materials of construction. Air barriers, which provide only partial containment, are not utilized. Besides protecting personnel and the environment from research materials, the Class III cabinet protects the experiment from extraneous matter.

Ventilation is provided by drawing air into the cabinet through a HEPA filter and exhausting it through two HEPA filters in series or one HEPA filter and an incinerator. Class III cabinets should be ventilated at a rate to provide at least 10 changes of cabinet volume per hour, or at a rate sufficient to limit to 10°F (5.6°C) the temperature rise due to internal heat load, whichever is greater. Cabinets should be maintained under negative air pressure of at least 0.5 in. water gage relative to the laboratory room. The exhaust fan, usually not an integral part of the cabinet, should provide an air velocity of at least 100 ft/min through any glove port in the event a glove is accidentally detached. The air exhausted from Class III cabinets is not ducted to other building exhaust systems, but is discharged directly outdoors.

Operations within the cabinets are conducted through attached rubber gloves (Fig. 6b). Materials are introduced and removed through an attached double-door sterilizer or dunk bath filled with liquid disinfectant (Fig. 6a). Service cocks and piping for the usual laboratory utilities can be provided; however, flammable gas should not be used because of explosion hazards. Water service, cup sinks, and drains can also be provided, but liquid wastes must be drained to a holding tank for sterilization prior to being released to the building sewer system. The modular design of Class III

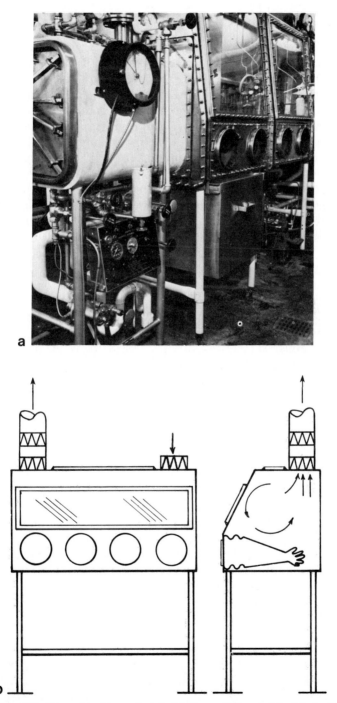

**Fig. 6.** Class III cabinet: (a) with attached double-door sterilizer and dunk bath, (b) showing attached gloves.

cabinets facilitate the enclosure of most types of laboratory equipment, including refrigerators, freezers, incubators, centrifuges, animal cages, etc.

Class III cabinets provide the highest level of personnel and environmental protection from exposure to research materials. Since work is conducted through attached rubber gloves rather than an air barrier the Class III cabinet provides protection from vapor and aerosol exposure and from splatter and contact contamination. The protection of the Class III cabinet, however, can be compromised by puncture of the gloves, breakage of seals, or conditions that create positive pressure in the cabinet.

Class III cabinets are suitable for all research procedures with high-risk oncogenic viruses (U.S. Department of Health, Education and Welfare, 1974b), etiologic agents assigned to class 4 by CDC (U.S. Department of Health, Education and Welfare, 1974a), and recombinant DNA materials requiring P4 level of physical containment (U.S. Department of Health, Education and Welfare, 1978a). The cabinets may also be used for research with highly toxic chemicals and chemical carcinogens *provided* the effluents are treated to meet whatever emission regulations might apply. Flammable solvents should not be used in these cabinets unless a careful evaluation has been made to determine that concentrations do not reach dangerous levels.

Despite the apparent attractiveness of Class III cabinets, they have several inherent disadvantages that have limited their usage to critical operations. These disadvantages include (1) very high cost of construction, (2) very poor work accessibility, (3) explosion hazards from chemicals, and (4) poor air circulation inside the cabinet that may allow cross contamination of experimental materials within the cabinet.

## 3. Biological Safety Cabinets: Selection Criteria

The factors that should be considered in the selection of a biological safety cabinet for a research program include (1) the hazard classification of the agent(s) to be used in the research; (2) the chemical and physical nature of other research materials involved; (3) the need for protection of the laboratory personnel, the environment, and the research materials; (4) the extent to which hazardous aerosols are involved; and (5) the suitability of the cabinet configuration for the research operations and equipment.

## 4. Biological Safety Cabinets: Certification Requirements

In an unpublished survey of NCI contractor and grantee laboratories, it was found that 65% of all tested laminar flow biological safety cabinets failed to pass leak tests of the filtration system. Subsequent testing programs conducted by and for the NCI show that newly installed cabinets consistently fail to meet design performance criteria. The most frequent problems encountered include improper balance of air flows and damage to filters, filter housings, and air plenum seals. For these reasons it is the recommended practice to test each safety cabinet (1) after it has been purchased and installed *but* before it is used; (2) after it has been moved, relocated, or serviced; and (3) at least annually. The cabinets should be tested and certified as meeting design performance specifications for filter integrity and efficiency, airflow velocities, integrity of the enclosure around the work space and contaminated air plenums, and relative air pressure of work space or air plenums.

### 5. Secondary Barriers: Laboratory Features

As stated in the introduction to Section V, the laboratory room can form a secondary barrier to protect personnel in other areas of the building and the environment from exposure to research materials released because of absence or failure of primary barriers or because of laboratory accidents occurring outside the primary barrier. The secondary barrier is not intended to reduce the risk of exposure for personnel inside the laboratory where the release of contamination occurs.

The secondary barrier can be thought of as an envelope consisting of the architectural and engineering features of the laboratory and its supporting mechanical systems. The secondary barrier may include control features such as materials and methods of construction that facilitate cleaning and prevent accumulation of contamination; pest- and vector-proof construction techniques; protection of utility distribution systems from contamination; treatment of liquid and air effluents to remove contaminants; treatment of recirculated ventilation air to remove contaminants; separation of functional areas of the laboratory or building; and air pressure gradients to maintain migration and infiltration of air from noncontaminated areas to areas that might be contaminated.

The secondary barrier is an integral part of the research facility. Although a research facility often contains several laboratories (i.e., several individual secondary barrier units), it is convenient to consider the features of a secondary barrier in terms of facility design and construction. Facilities differ in their design for accommodating biomedical research involving potentially hazardous materials and can be identified as one of three types: (1) the basic facility, (2) the containment facility, and (3) the maximum containment facility.

**a. The Basic Facility.** The basic facility is not characterized by special engineering features. Conventional design approaches are adequate for this facility; however, certain important design objectives should be achieved.

The arrangement of space within the basic facility should complement the functional needs of the research program while avoiding the intermingling of incompatible functions. For example, areas known to be sources of general contamination such as animal rooms and waste staging areas should not be adjacent to media processing areas or tissue culture laboratories. Public areas and general offices to which nonlaboratory staff require frequent access should be separated from spaces that primarily support research functions.

The surface finishes of floors and walls should facilitate cleaning and withstand detergent cleaning solutions. Bench tops should be impervious to liquids and resistant to acid, alkali, organic solvents, and moderate heat. Laboratory furniture should be sturdy and voids should be accessible for cleaning. Hand washing facilities should be available in each laboratory. Doors to main laboratory rooms should have self-closing devices. Exhaust air from ventilated safety cabinets should be discharged to the outdoors. Although it is generally undesirable to do so, filtered exhaust air from Class I and Class II Type A biological safety cabinets may be returned to the laboratory environment provided that the exhaust HEPA filters have been certified by scan testing procedures. An autoclave should be installed in the basic facility in which research with infectious materials is conducted. The autoclave should be located either within the individual laboratory module where such work is conducted or in a designated waste staging area.

Conditioned air for comfort ventilation can be provided to the basic facility in a conventional manner. Mechanically supplied air may be recirculated within the laboratory environment. In the design of new facilities, however, nonrecirculating air-handling systems are generally recommended, since this provides the flexibility that would be needed to upgrade the basic facility to a containment facility.

**b. The Containment Facility.** The containment facility has special engineering features that make it possible for laboratory workers to handle moderately hazardous materials without endangering others in the building or the environment. The unique features that distinguish the containment facility from the basic facility are the provisions for access control and a specialized ventilation system.

The containment facility may be an individual laboratory room, a suite of rooms within a building, or an entire building. In any case, the containment facility is separated by a controlled access zone from areas open to the public and other laboratory persons who do not work within the containment facility. Various arrangements of space can be used to achieve separation as shown in Figs. 7a, 7b, and 7c. Figure 7a illustrates three approaches to separating individual containment laboratories from a common-use cor-

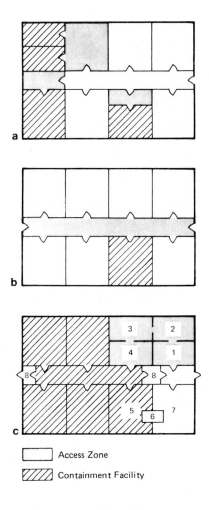

Access Zone

Containment Facility

1 Clean Clothing Change Room
2 Drying Room
3 Shower Room
4 Contaminated Clothing Change Room
5 Contaminated Waste Handling Room
6 Double Door Autoclave
7 Washing Room
8 Air Lock

**Fig. 7.** Containment facilities showing various arrangements of space.

ridor. Figure 7b illustrates how a corridor can serve as the access zone for an individual containment laboratory. This approach is acceptable but undesirable unless strict access control can be ensured. Figure 7c shows the access zone for a containment laboratory suite as a change room and shower facility. Access to this containment facility is by passage from the clean-clothing change room through the drying room, shower room, and "contaminated-clothing" change room. This traverse is reversed for egress. In this example the airlocks are used only for the passage of equipment, materials, or supplies into the containment facility.

The ventilation system supporting the containment facility should be capable of controlling the movement of air between functional areas. The system should be balanced so that there is infiltration of air into each laboratory from its adjacent corridor. It is recommended that the infiltration rate be at least 50 ft³/min. The containment facility may be served by the same ventilation system that serves areas outside the containment facility, provided the air exhausted from the containment facility is not recirculated and the air balance can be maintained. The air exhausted from the containment facilities should be discharged to the outdoors clear of occupied buildings and supply air intakes. The general exhaust air (i.e., the air exhausted from the rooms) can be discharged to the outdoors without filtration or other treatment.

Each laboratory of the containment facility should be capable of accommodating a ventilated safety cabinet. The cabinet exhaust air may be discharged either directly to the outdoors through an individual duct and exhaust fan or through the general exhaust system of the containment facility. In the latter case it is important that the exhaust system be designed and operated in a manner that prohibits interference between the air balance of the containment facility and the operation of the ventilated safety cabinet. Pressurization of the exhaust duct should be avoided.

Surface finishes of walls, floors, and ceilings should be impervious to liquids and should be readily cleanable. The laboratories should be constructed so that they can be sealed to permit space decontamination. If windows are provided they should be secured in the closed position. Horizontal surfaces that may serve as dust collectors (e.g., exposed air ducts and utility pipes, tops of wall cabinets, etc.) should be kept to a minimum. If suspended ceilings are installed to conceal air ducts and utility pipes, they should be constructed of plaster or drywall. All ceiling joints should be taped and sealed, and the ceiling should be finished with epoxy, phenolic, or polyurethane paint.

A foot, elbow, or automatically operated hand washing facility should be provided near the exit area of each laboratory. All doors of the containment facility should be self-closing.

An autoclave should be located within the containment facility if the facility is used for research involving hazardous microbiologic materials. With appropriate procedural controls it is possible to locate the autoclave outside of the containment facility, providing it is located within the same building.

**c. The Maximum Containment Facility.** The design objective of the maximum containment facility is to create a facility that will allow the safe conduct of research involving biologic agents that are extremely hazardous to the laboratory worker or that may cause serious epidemic disease. The distinguishing characteristic of the maximum containment laboratory is the provision for special secondary barrier features that prevent the escape of hazardous materials to the environment. These secondary barrier features serve to isolate the laboratory area from the surrounding environment and include

1. Monolithic walls, floors, and ceilings in which all penetrations, such as air ducts, electrical conduits, and utility pipes, are sealed to ensure the physical isolation of the laboratory area.
2. Air locks through which supplies and materials can be brought safely into the facility.
3. Contiguous clothing-change and shower rooms through which personnel enter and exit the facility.
4. Double-door autoclaves to sterilize and safely remove wastes and other materials from the facility.
5. A biowaste treatment system for sterilization of liquid wastes.
6. A separate ventilation system that maintains negative air pressures and directional air flow within the facility.
7. A treatment system to decontaminate exhaust air before discharge into the atmosphere.

Although the maximum containment facility is generally a separate building, it may be constructed as an isolated area within a building. The perimeter wall partitions of the facility should be installed the full height from finished floor to the under surface of the floor or roof above. If windows are installed in the perimeter partitions, they should be secured in the closed position and the frames sealed; the glazing should be safety glass. Perimeter doors should be insect- and rodent-proof. Wall, floor, and ceiling construction joints, utility pipe and duct penetrations, and electrical conduits and other passages should be sealed to assure isolation of the laboratory environment. The surface finishes should be selected on the basis of their providing a monolithic surface barrier. Epoxy, phenolic, and polyurethane finishes have proved satisfactory for this purpose.

The clothing change and shower rooms are contiguous to the perimeter structure of the facility. They are generally arranged so that the clean-clothing change area is separated from the laboratory zone by an air lock or shower area. Personnel egress from the laboratory zone must be through the shower area to the clean-clothing change room. Air locks for movement of materials, supplies, and equipment into the facility are also a part of the perimeter structure. A double door autoclave is located so that either the interior or exterior door frame is sealed to the perimeter barrier wall. It is preferable to make the interior door frame contiguous with the barrier wall so that autoclave maintenance can be performed outside the laboratory zone.

The maximum containment facility is ventilated by its own supply and exhaust systems. The systems are balanced so that the air pressure within the facility can be maintained lower than that outside the perimeter walls and so that air flow is from areas of least hazard potential to areas of greatest hazard potential.

The air-handling system should provide an air supply consisting of 100% outdoor air on a year-round basis. The system should provide separate branch supply and exhaust air ducts to each space to permit proper air balance. The supply and exhaust fans should be interlocked to prevent reversal of air flow in event of exhaust fan failure.

The general exhaust air is filtered by passage through HEPA filters before being discharged to the outdoors. The air filters should be located as near to the laboratory space as possible to minimize the length of potentially contaminated air ducts. The filter plenums should be designed to facilitate (1) testing of filters after installation and (2) in-place decontamination before filter removal and replacement.

Mechanical systems should be designed so that maintenance of building machinery, piping, and controls can be performed from outside the laboratory environment.

Liquid effluents from the maximum containment facility should be collected and decontaminated before disposal into the sanitary sewers. Effluents from laboratory sinks, cabinets, floors, and autoclaves should be sterilized by heat treatment. Liquid wastes from the shower room may be decontaminated with chemical disinfectants. The wastes from toilets may be discharged directly into the sanitary sewers.

## 6. Secondary Barriers: Facility Requirements for Research Programs

The National Cancer Institute has developed guidelines for the safe handling of cancer viruses (U.S. Department of Health, Education and

Welfare, 1974b) and chemical carcinogens (U.S. Department of Health, Education and Welfare, 1975); the Center for Disease Control has developed guidelines for work with etiologic agents of infectious disease (U.S. Department of Health, Education and Welfare, 1974a); and the National Institutes of Health has promulgated guidelines for research employing recombinant DNA techniques (U.S. Department of Health, Education and Welfare, 1978a). These documents describe good laboratory practice and facility recommendations (or requirements) appropriate for research involving the respective materials.

The basic facility is appropriate for supporting research involving low-risk cancer viruses (U.S. Department of Health, Education and Welfare, 1974b) and etiologic agents assigned to classes 1 and 2 by CDC (U.S. Department of Health, Education and Welfare, 1974a). The containment facility should be used, however, if the research involves either aerosol exposure studies in experimental animals or large-scale production of any of these agents. The basic facility is also appropriate for conducting work that employs recombinant DNA techniques designated for P1 and P2 levels of containment (U.S. Department of Health, Education and Welfare, 1978a).

The containment facility is recommended for research involving moderate-risk cancer viruses (U.S. Department of Health, Education and Welfare, 1974b), most chemical carcinogens (U.S. Department of Health, Education and Welfare, 1975), and etiologic agents assigned to class 3 by CDC (U.S. Department of Health, Education and Welfare, 1974a). Consideration should be given to the use of the maximum containment facility for aerosol studies and large-scale production activities involving these agents. The containment facility is appropriate for conducting research involving recombinant DNA techniques designated for the P3 level of containment (U.S. Department of Health, Education and Welfare, 1978a).

The maximum containment facility is appropriate for research involving etiologic agents assigned to class 4 by CDC (U.S. Department of Health, Education and Welfare, 1974a), for conducting work that employs recombinant DNA techniques designated as requiring the P4 level of containment (U.S. Department of Health, Education and Welfare, 1978a), and would be used for work with oncogenic viruses proved to be etiologic of cancer in humans (U.S. Department of Health, Education and Welfare, 1974b).

The recommended or required use of facilities in support of biomedical research are summarized below.

Cancer Viruses
    Low-risk viruses        Basic facility
    Moderate-risk viruses    Containment facility
    High-risk viruses      Maximum containment facility

Chemical Carcinogens        Containment facility
Organisms containing recombinant DNA molecules
   P1 classification        Basic facility
   P2 classification        Basic facility
   P3 classification        Containment facility
   P4 classification        Maximum containment facility

Etiologic agents
   Class 1 agent            Basic facility
   Class 2 agent            Basic facility
   Class 3 agent            Containment facility
   Class 4 agent            Maximum containment facility

## B. Control Practices

As previously indicated, the safe conduct of research in viral oncology is dependent on the exercise of good laboratory practices by the personnel. The manner in which procedures are accomplished and the operating practices serve to control the hazards presented to the personnel and the environment.

Although the operating conditions will differ between laboratories, there are basic practices that apply to all viral oncology research activities. These include the role of the management, medical surveillance, and essential personnel practices.

Critical to the development and implementation of safe practices and procedures is the absolute commitment by the institution management or administration to require and support a biohazards safety program. This participation is essential to ensuring that the institution satisfies their moral and legal obligations to themselves, the research personnel, the community, and the quality and integrity of the research. These obligations are emphasized by the continuing growth of the litigious attitude in today's society and the increase in government regulatory actions in response to the demands of the society. Although it is anticipated that all persons engaged in biomedical research would recognize the value of a safety program, on merit alone it is necessary that the management or administration of an institution assume an active role in establishing and maintaining such a program.

A part of the administration's participation in a biohazards safety program in viral oncology research is the establishment of an appropriate medical surveillance program for the research personnel. The basic purpose of such a program is the identification of persons who may have or develop a medical condition that places them at increased risk to the adverse effects of the agents or materials to which they are exposed. Although there are data (Table III), albeit limited, that suggest that human infection has not occurred with a number of oncogenic viruses, it cannot be assumed that infection will not occur in persons with abnormal physical or physiologic conditions. Therefore medical surveillance of persons engaged in viral oncology research should be accomplished by performance of an appropriate medical evaluation prior to exposure to an oncogenic virus and at regular intervals thereafter. At a minimum, regular medical counseling should occur to determine whether there has been a change in the employee's health status that would place that person at increased risk. Included in the latter are persons with reduced immunologic competency, pregnant women, and persons under treatment with steroids or cytotoxic drugs. It is recommended that the periodic monitoring include the collection of a serum sample from each person engaged in work involving oncogenic viruses. Serum samples should also be collected following accidental exposure to an oncogenic agent and at intervals thereafter. These latter samples and those collected at the regular intervals may be subjected to testing but should also be stored under appropriate conditions for future reference.

The key person in the implementation of a safety program is the senior investigator, who is responsible for establishing practices and procedures that minimize the hazards associated with the research and, importantly, is also responsible for ensuring the conformance of his staff to these practices and procedures. An integral part of the investigator's role in gaining this conformance is the development of a positive attitude toward laboratory safety on the part of his staff, since the success of a safety program requires the active participation of each person in the laboratory.

The maintenance of safety in laboratory operations requires that each person comply with the established practices and procedures and accomplishes the work in a manner that does not create hazards.

The control of potential hazards in the viral oncology research evolves from the development and practices and procedures that minimize the potential for exposure of personnel to inhalation, ingestion, or self-inoculation of the agent used.

Minimizing exposure by inhalation requires the control of all procedures that may result in the generation of aerosols. These procedures range from those used in direct manipulation of the agent (e.g., pipetting, sonication, or homogenization) to those involved with the cleaning of glassware and maintenance of the laboratory.

The control of these potential sources of aerosols does not present a significant problem. Aerosol generating processes such as pipetting or sonication may be accomplished in a primary barrier unit such as a Class II cabinet, whereas those associated with the handling of glassware are minimized by the use of appropriate decontamination procedures. It is preferable to treat glassware immediately after use and prior to transport within the facility; this practice serves to reduce the potential hazards in the event of an accidental spill. The reduction of potentially infective aerosols generated by maintenance procedures is accomplished by prompt and proper decontamination of materials spilled on work surfaces or the floor and by use of "wet" procedures in routine cleaning.

The control of exposures by ingestion is accomplished by application of a strict prohibition against all mouth pipetting. There should be no exceptions to this practice, since to do so would permit personnel to rationalize the potential risks. This prohibition should be augmented by others that are directed toward the control of any practice with the potential for ingestion of infective materials.

Although the control of potential sources of exposure by inhalation or ingestion is relatively easy to establish and monitor, such is not the case for what is probably the most significant source—self-inoculation. Obviously, the degree of risk to the laboratory staff of self-inoculation can be reduced by minimizing the use of the syringe and needle as a laboratory device. However, the risk remains unaltered for the individual using procedures in which there is no alternative to the use of a syringe and needle. In these situations, only due care will avoid self-inoculation. Related to self-inoculation is exposure through cuts or abrasions. However, control of exposure can be achieved by the use of personal protective equipment.

In summary, the implementation of procedures for the control of hazards in the viral oncology research laboratory requires not the employment of sophisticated equipment but a commitment on the part of the administration of the institution and the research staff to conduct the research using safety practices. Development of the appropriate practices is contingent on the assessment of the hazards associated with a particular research program.

## ACKNOWLEDGMENT

The authors gratefully acknowledge the contribution of Mr. Leon Idoine in providing editorial assistance in the review of the text and acquisition of reference materials.

# REFERENCES

Aaronson, S. A. (1973). Biological characterization of mammalian cells transformed by a primate sarcoma virus. *Virology* **52,** 562-567.

Aaronson, S. A., and Hellman, A. (1978). Unpublished Observation.

Aaronson, S. A., and Todaro, G. J. (1970). Transformation and virus growth by murine sarcoma viruses in human cells. *Nature (London)* **225,** 458-459.

Ablashi, D. V., Armstrong, G. R., Heine, V., and Manaker, R. A. (1971). Propagation of *Herpesvirus saimiri* in human cells. *J. Natl. Cancer Inst.* **47,** 241-244.

Adams, R. A., Hellerstein, E. E., Pothier, L., Foley, G. E., Lazarus, H., and Stuart, A. B. (1971). Malignant potential of a cell line isolated from the peripheral blood in infectious mononucleosis. *Cancer* **27,** 651-658.

Ahlstrom, C. G., and Forsby, N. (1962). Sarcomas in hamsters after injection with Rous chicken tumor material. *J. Exp. Med.* **115,** 839-852.

Ahlstrom, C. G., Jonsson, N., and Forsby, N. (1962). Rous sarcoma in mammals. *Acta Pathol. Microbiol. Scand., Suppl.* **154,** 127-133.

Ahlstrom, C. G., Bergman, S., and Ehrenberg, B. (1963). Neoplasms in guinea pigs induced by an agent in Rous chicken sarcoma. *Acta Pathol. Microbiol. Scand.* **58,** 177-190.

Allison, A. C., and Taylor, R. B. (1967). Observation on thymectomy and carcinogenesis. *Cancer Res.* **27,** 703-707.

Allison, A. C., Chesterman, C., and Baron, S. (1967). Induction of tumors in adult hamsters with SV-40. *J. Natl. Cancer Inst.* **38,** 567-572.

Ambrus, J. L., Feltz, E. T., Grace, J. T., and Owens, G. (1963). A virus-induced tumor in primates. *Natl. Cancer Inst. Monogr.* **10,** 447-458.

Anderson, J., Yates, V. J., Jasty, V., and Mancini, L. O. (1969). The *in vitro* transformation by an avian adenovirus (CELO). III. Human amnion cell cultures. *J. Natl. Cancer Inst.* **43,** 575-580.

Ashkenazi, A., and Melnick, J. L. (1963). Tumorigenicity of simian papovavirus SV-40 and of virus-transformed cells. *J. Natl. Cancer Inst.* **30,** 1227-1265.

Barski, G., Chlap, Z. B., Gotliev-Stematsky, T., and Charlier, M. (1962). Neoplasms and hemorrhagic disease induced in adult hamsters with polyoma virus. *Nature (London)* **193,** 298-299.

Barthold, S. W., and Olson, C. (1974). Papovavirus-induced neoplasia. *In* "Handbook of Laboratory Animal Science" (E. C. Melby and N. H. Altman, eds.), Vol. II, pp. 70-75. CRC Press, Cleveland, Ohio.

Barthold, S. W., Baumgartener, L. E., and Olson, C. (1976). Lack of infectivity of bovine leukemia (C-type) virus to rats. *J. Natl. Cancer Inst.* **56,** 643-644.

Benveniste, R. E., and Todaro, G. J. (1974a). Evolution of type C viral genes. I. Nucleic acid from baboon type C virus as a measure of divergence among primate species. *Proc. Natl. Acad. Sci. U.S.A.* **71,** 4513-4518.

Benveniste, R. E., and Todaro, G. J. (1974b). Evolution of C-type viral genes: Inheritance of exogenously acquired viral genes. *Nature (London)* **252,** 456-459.

Black, P. H. (1968). The oncogenic DNA viruses: A review of *in vitro* transformation studies. *Annu. Rev. Microbiol.* **22,** 391-426.

Black, P. H., and Todaro, G. J. (1965). *In vitro* transformation of hamster and human cells with the adeno 7-SV-40 hybrid virus. *Proc. Natl. Acad. Sci. U.S.A.* **54,** 374-381.

Boettiger, D., Love, D. N., and Weiss, R. A. (1975). Virus envelope markers in mammalian tropism of avian RNA tumor viruses. *J. Virol.* **15,** 108-114.

Boiron, M., Levy, J. P., Thomas, M., Friedmann, J. C., and Bernard, J. (1964). Some properties of bovine papilloma virus. *Nature (London)* 201, 423–424.

Boiron, M., Bernard, C., and Chuat, J. C. (1969). Replication of mouse sarcoma virus Moloney strain (MSV-V) in human cells. *Proc. Am. Assoc. Cancer Res.* 10, 8.

Boyd, A. L., and Orme, T. W. (1975). Transformation of mouse cells after infection with with ultraviolet irradiation-inactivated Herpes simplex type 2. *Int. J. Cancer* 16, 526–538.

Burnett, J. P., and Harrington, J. A. (1968). Simian adenovirus SA7 DNA: Chemical, physical, and biological studies. *Proc. Natl. Acad. Sci. U.S.A.* 60, 1023–1029.

Butel, J. S., Tevethia, S. S., and Melnick, J. L. (1972). Oncogenicity and cell transformation on papovavirus SV-40: The role of the viral genome. *Adv. Cancer Res.* 15, 1–15.

Chan, E. W., Schiop-Stansly, P. E., and O'Connor, T. E. (1974). Mammalian sarcoma-leukemia viruses. I. Infection of feline, bovine, and human cell cultures with Snyder-Theilen feline sarcoma virus. *J. Natl. Cancer Inst.* 52, 473–481.

Chaproniere, D. M., and Andrewes, C. H. (1957). Cultivation of rabbit myxoma and fibroma viruses in tissues of nonsusceptible hosts. *Virology* 4, 351–365.

Charman, H. P., Kim, N., White, M., and Gilden, R. V. (1974). Failure to detect antibodies cross-reactive with mouse leukemia virus group specific antigen in human sera. *J. Natl. Cancer Inst.* 52, 1409–1413.

Chumakova, M. Y., Chumakov, M. P., Elbert, L. B., Avgustinovich, G. I., Ralf, N. M., Voroshilova, M. K., Taranova, G. B., and Tapupere, V. O. (1963). Demonstration of antibody to simian virus (SV40) in human sera. Serological evidence of infection from monkeys and kidney cultures. *Vopr. Virusol.* 8, 457–459.

Cohen, M. H., Bernstein, A. D., and Levine, P. H. (1974). Hematological and serological effects of Rauscher leukemia virus and Epstein-Barr virus on immunosuppressed newborn subhuman primates. *Oncology* 29, 353–363.

Cole, G. A., and Shah, K. V. (1974). Experimental simian virus 40 infection of normal and immunosuppressed spider monkeys. *Acta Virol. (Engl. Ed.)* 18, 65–69.

Cutler, S. J., Scotto, J., Devesa, S. S., and Connelly, R. R. (1974). Third National Cancer Survey—An overview of available information. *J. Natl. Cancer Inst.* 53, 1565–1575.

Daniel, M. D., Melendez, L. V., Hunt, R. D., Anver, M., Fraser, C. E. O., Barahona, H, King, N. W., and Baggs, R. B. (1974). *Herpesvirus saimiri.* VII. Induction of malignant lymphoma in New Zealand white rabbits. *J. Natl. Cancer Inst.* 53, 1803–1807.

Darai, G., and Munk, K. (1973). Human embryonic lung cells abortively infected with herpes virus homines type 2 show some properties of cell transformation. *Nature (London), New Biol.* 241, 268–269.

Davis, O. S. (1962). Pathological findings in mammals inoculated with avian visceral lymphomatosis tumor triturates. *World's Poult. Congr. Proc., 12th, 1962* Sydney Sect. Pap., pp. 309–314. (summaries in Suppl., pp. 54–55).

Deal, D. R., Gerber, P., and Chisari, F. V. (1971). Heterotransplantation of two human lymphoid cell lines transformed *in vitro* by Epstein-Barr virus. *J. Natl. Cancer Inst.* 47, 771–780.

Deinhardt, F., Wolfe, L., Northrup, R., Margzynska, B., Ogden, J., McDonald, R., Falk, L., Shramek, G., Smith, R., and Deinhardt, J. (1972). Induction of neoplasms by viruses in marmoset monkeys. *J. Med. Primatol.* 1, 24–50.

Deinhardt, F., Falk, L. A., and Wolfe, L. G. (1974a). Transformation of nonhuman primate lymphocytes by Epstein-Barr virus. *Cancer Res.* 34, 1241–1244.

Deinhardt, F. W., Falk, L. A., and Wolfe, G. (1974b). Simian herpesvirus and neoplasia. *Adv. Cancer Res.* 19, 167–205.

Dent, P. B. (1972). Immunodepression by oncogenic viruses. *Prog. Med. Virol.* 14, 1–35.

Diamondopoulos, G. T. (1973). Induction of lymphocytic leukemia, lymphosarcoma, reticulum cell sarcoma, and osteogenic sarcoma in the Syrian golden hamster by oncogenic DNA simian virus 40. *J. Natl. Cancer Inst.* **50,** 1347-1365.

Draese, K. D. (1937-1939). Uber laboratoriumsinfektionene mit typhus bazillen and anderen bacterien. *Arch. Hyg.* **121,** 232-291.

Duff, R., and Rapp, F. (1971). Properties of hamster embryo fibroblasts transformed *in vitro* after exposure to ultraviolet-irradiated herpes simplex virus type 2. *J. Virol.* **8,** 469-477.

Duff, R., and Rapp, F. (1973). Oncogenic transformation of hamster embryo cells after exposure to inactivated herpes simplex virus type 1. *J. Virol.* **12,** 209-217.

Eagle, H., Foley, G. E., Koprowski, H., Lazarus, H., Levine, E. M., and Adams, R. A. (1970). Growth characteristics of virus-transformed cells. *J. Exp. Med.* **131,** 863-879.

Eddy, B. E. (1964). Simian virus 40 (SV-40): An oncogenic virus. *Prog. Exp. Tumor Res.* **4,** 1-26.

Eddy, B. E. (1969). Polyoma virus. *In* "Polyoma Virus, Rubella Virus" (S. Gard, C. Hallover, and K. F. Meyer, eds.), pp. 1-114. Springer-Verlag, Berlin and New York.

Eddy, B. E., Steward, S. E., Kirchstein, R. L., and Young, R. D. (1959). Induction of subcutaneous nodules in rabbits with the SE poly virus. *Nature (London)* **183,** 766-767.

Editorial (1974). Oncogenicity of E. B. virus: Leads from New-world Primates. *Lancet* **1,** 123-124.

Epstein, M. A., and Probert, M. (1973). Virology and serology--new approaches to identification of an etiologic agent. *Natl. Cancer Inst. Monogr.* **36,** 73-77.

Espana, C. (1971). A pox disease of monkeys transmissible to man. *In* "Medical Primatology II" (E. I. Goldsmith and J. Moor-Jankowski, eds.), pp. 694-708. Karger, Basel.

Falk, L. A. (1974). Oncogenic DNA viruses of non-human primates. A review. *Lab. Anim. Sci.* **24,** 182-192.

Falk, L., Wright, J., Wolfe, L., and Deinhardt, F. (1974a). *Herpesvirus ateles:* Transformation *in vitro* of marmoset splenic lymphocytes. *Int. J. Cancer* **14,** 244-251.

Falk, L. A., Wolfe, L. G., Deinhardt, F., Paciga, J., Dombos, L., Klein, G., Henle, W., and Henle, G. (1974b). Epstein-Barr virus: Transformation of non-human primate lymphocytes *in vitro*. *Int. J. Cancer* **13,** 363-376.

Fine, D. L., Pienta, R. J., Malan, L. B., Kubicek, M. T., Bennett, D. G., Landon, J. C., Valerio, M. G., West, D. M., Fabrizio, D. A., and Chopra, H. C. (1974). Biologic characteristics of transformed rhesus foreskin cells infected with Mason-Pfizer monkey virus. *J. Natl. Cancer Inst.* **52,** 1135-1142.

Fine, D. L., Landon, J. C., Pienta, R. J., Kubicek, M. T., Valerio, M. G., Loeb, W. F., and Chopra, H. C. (1975). Responses of infant rhesus monkeys to inoculation with Mason-Pfizer monkey virus materials. *J. Natl. Cancer Inst.* **54,** 651-658.

Fink, M. A. (1967). Discussion. *In* "Carcinogenesis: A Broad Critique" (W. J. Burdette, ed.), pp. 47. Williams & Wilkins, Baltimore, Maryland.

Fischinger, P. J., and O'Connor, T. E. (1970). Productive infection and morphologic alteration of human cells by a modified sarcoma virus. *J. Natl. Cancer Inst.* **44,** 429-438.

Flocks, J. S., Weis, T. P., Kleinman, D. C., and Kirsten, W. H. (1965). Dose response studies to polyoma virus in rats. *J. Natl. Cancer Inst.* **35,** 259-272.

Fogh, J. *et al.* (1973). "Transformation of Tissue Culture Cells by SV-40 Viruses," p. 193. MSS Information Corp., New York.

Forni, G., Rhim, J. S., Pickeral, S., Shevach, E. M., and Green, I. (1975). Antigenicity of carcinogen and viral-induced sarcomas in inbred and random bred guinea pigs. *J. Immunol.* **115,** 204-210.

Freeman, A. E., and Huebner, R. J. (1973). Problems in interpretation of experimental evidence of cell transformation. *J. Natl. Cancer Inst.* **50**, 303–306.

Gardner, M. B. (1971). Current information of feline and canine cancers and relationships or lack of relationship to human cancer. *J. Natl. Cancer Inst.* **46**, 281–290.

Gardner, M. B., Rongey, R. W., Arnstein, P., Estes, J. D., Sarma, P., Huebner, R. J., and Rickard, C. G. (1970). Experimental transmission of feline fibrosarcoma to cats and dogs. *Nature (London)* **226**, 807–809.

Gerber, P., Whang-Peng, J., and Monroe, J. H. (1969). Transformation and chromosome changes induced by Epstein-Barr virus in normal human leukocyte cultures. *Proc. Natl. Acad. Sci. U.S.A.* **63**, 740–747.

Gibbs, E. P. J., Smale, C. J., and Lawman, M. J. P. (1975). Warts in sheep. Identification of a papilloma virus and transmission of infection to sheep. *J. Comp. Pathol.* **85**, 327–334.

Gilden, R. V., Oroszlan, S., and Hatanaka, M. (1974). Comparison and evolution of RNA tumor virus components. *In* "Viruses, Evolution and Cancer—Basic Considerations" (E. Kurstak and K. Maramorosch, eds.), pp. 235–257. Academic Press, New York.

Graham, F. L., van der Eb, A. J., and Heijneker, H. L. (1974). Size and location of the transforming region in human adenovirus type 5 DNA. *Nature (London)* **251**, 687–691.

Gross, L. (1970). "Oncogenic Viruses," 2nd ed. Pergamon, Oxford.

Gross, L. (1971). Transmission of cancer in man. Tentative guidelines referring to the possible effects of inoculation of homologous cancer extracts in man. *Cancer* **23**, 785–788.

Hall, J. G., Scollay, R. G., Birbeck, M. S. C., and Theilen, G. H. (1975). Studies on FeSV induced sarcomata in sheep with particular reference to the regional lymphatic system. *Br. J. Cancer* **32**, 639–659.

Hanafusa, H., and Hanafusa, T. (1966). Determining factor in the capacity of Rous sarcoma virus to induce tumors in mammals. *Proc. Natl. Acad. Sci. U.S.A.* **55**, 532–538.

Hanel, E., Jr., and Kruse, R. H. (1967). "Laboratory-acquired Mycoses," Misc. Publ. 28 (AD-665-376). Fort Detrick, Frederick, Maryland.

Hanguenau, F., Rabotti, G. F., Lyon, G., and Moraillon, A. (1971). Gliomas induced by Rous sarcoma virus in the dog—an ultrastructural study. *J. Natl. Cancer Inst.* **46**, 538–559.

Hanson, R. P., Sulkin, S. E., Buescher, E. L., Hammon, W. M., McKinney, R. W., and Work, T. H. (1967). Arbovirus infections of laboratory workers: Extent of problem emphasizes the need for more effective measures to reduce hazards. *Science* **158**, 1283–1286.

Harvey, J. J. (1964). An unidentified virus which causes the rapid production of tumors in mice. *Nature (London)* **204**, 1104–1105.

Harvey, J. S., Felsburg, P. J., Heberling, R. L., Kniker, W. T., and Kalter, S. S. (1974) Immunological competence in non-human primates: Differences observed in four species. *Clin. Exp. Immunol.* **16**, 267–278.

Hellman, A., Oxman, M. N., and Pollack, R. (1973). "Biohazards in Biological Research." Cold Spring Harbor Lab., Cold Spring Harbor, New York.

Henle, W., and Henle, G. (1973). Evidence for an oncogenic potential of the Epstein-Barr virus. *Cancer Res.* **33**, 1419–1423.

Hilleman, M. R. (1968). Cells, vaccines, and pursuit of precedent. *Prog. Med. Virol.* **10**, 348–354.

Horvath, L. B. (1972). SV-40 neutralizing antibodies in the sera of man and experimental animals. *Acta Virol. (Engl. Ed.)* **16**, 141–146.

Hoss, H. E., and Olson, C. (1973). Infectivity of bovine C-type (leukemia) virus for sheep and goats. *Am. J. Vet. Res.* **35**, 633–637.

Hsu, I., Yang, W., Tennant, R., and Brown, A. (1978). Transfection of Fv-1 permissive and restrictive mouse cells with integrated DNA of murine leukemia viruses. *Proc. Natl. Acad. Sci. U.S.A.* **75**, 1451-1455.

Huebner, R. J. (1963). Tumor virus study systems. *Ann. N.Y. Acad. Sci.* **108**, 1129-1148.

Huebner, R. J. (1967). *In vitro* methods for detection and assay of leukemia viruses. *Carcinog., Collect. Pap. Annu. Symp. Fundam. Cancer Res., 20th, 1966* pp. 23-27.

Huebner, R. J., Hartley, J. W., Rowe, W. P., Lane, W. T., and Capps, W. I. (1966). Rescue of the defective genome of Moloney sarcoma virus from a noninfectious hamster tumor and the production of pseudotype sarcoma virus with various murine leukemia viruses. *Proc. Natl. Acad. Sci. U.S.A.* **56**, 1164-1169.

Huebner, R. J., Todaro, G. J., Sarma, P., Hartley, J. W., Freeman, A. E., Peters, R. L., Whitmire, C. E., Meier, H., and Gilden, R. V. (1970). "Switched off" vertically transmitted C-type RNA tumor viruses as determinants of spontaneous and induced cancer: A new hypothesis of viral carcinogenesis. *Colloq. Int. C.N.R.S.* **183**, 33-57.

Hunt, R. D., Melendez, L. V., Garcia, F. G. and Trum, B. F. (1972). Pathologic features of *Herpesvirus ateles* lymphoma in cotton-topped marmosets (Saguinus oedipus). *J. Natl. Cancer Inst.* **49**, 1631-1639.

Ioachim, H. L., Keller, S. E., Gimovsky, M. L., and Shepherd, S. (1973). Immunity to viral leukemia in rats and the route of transmission from mother to offspring. *Cancer Res.* **33**, 537-550.

Jablonska, S., Dabrowski, J., and Jakubowicz, K. (1972). Epidermodysplasia verruciformis as a model in studies on the role of papovaviruses in oncogenesis. *Cancer Res.* **32**, 583-589.

Jarrett, O. (1970). Evidence for the viral etiology of leukemia in domestic animals. *Adv. Cancer Res.* **13**, 39-62.

Jarrett, O., Laird, H. M., and Hay, D. (1970). Growth of feline leukemia virus in human, canine and porcine cells. *Bibl. Haematol. (Basel)* **36**, 387-392.

Jarrett, O., Laird, H. M., and Hag, H. (1972). Restricted host range of a feline leukemia virus. *Nature (London)* **238**, 220-221.

Jensen, F. C., Girardi, A. J., Gilden, R. V., and Koprowski, H. (1964a). Infection of human and simian tissue cultures with Rous sarcoma virus. *Proc. Natl. Acad. Sci. U.S.A.* **52**, 53-59.

Jensen, F. C., Koprowski, H., Pagano, J. S., Pontén, J., and Ravdin, R. G. (1964b). Autologous and homologous implantation of human cells transformed *in vitro* by simian virus 40. *J. Natl. Cancer Inst.* **32**, 917-937.

Karande, K. A., Taskar, S. P., and Ranadive, K. J. (1975). Activation of murine leukemia virus under different physiological conditions. *Br. J. Cancer* **31**, 434-442.

Katzin, D. S., Connor, J. D., Wilson, L. A., and Sexton, R. S. (1967). Experimental herpes simplex infection in the owl monkey. *Proc. Soc. Exp. Biol. Med.* **125**, 391-398.

Kawakami, T. G., Huff, S. D., Buckley, P. M., Dungworth, D. L., and Snyder, S. P. (1972). C-type virus associated with gibbon lymphosarcoma. *Nature (London) New Biol.* **235**, 170-171.

Kelloff, G., Huebner, R. J., Chang, N. H., Lee, Y. K., and Gilden, R. V. (1970). Envelope antigen relationships among three hamsters-specific sarcoma viruses and a hamster-specific helper virus. *J. Gen. Virol.* **9**, 19-26.

Kersey, J. H., Spector, B. D., and Good, R. A. (1973). Primary immunodeficiency diseases and cancer: the immunodeficiency registry. *Int. J. Cancer* **12**, 333-347.

Klement, V., Freedman, M. H., McAllister, R. M., Nelson-Rees, W. A., and Huebner, R. J. (1971). Differences in susceptibility of human cells to mouse sarcoma virus. *J. Natl. Cancer Inst.* **47**, 65-73.

Kohne, D. E. 1970. Evolution of higher-organism DNA. *Q. Rev. Biophys.* **3**, 327–375.

Kulagin, S. M., Fedorova, N. I., and Ketiladze, E. S. (1962). Laboratory outbreak of hemorrhagic fever with a renal syndrome: Clinico-epidemiological characteristics. *Zh. Mikrobiol., Epidemiol. Immunobiol.* **33**, 121–126.

Larson, D. L., Garan, C. E., Harewood, K. R., Wolff, J. S., Schidlovsky, G, and Mayyasi, S. A. (1973). Biochemical and biological characterization of simian sarcoma virus. Type 1 (SSV-1) propagated in various mammalian cell lines. *Am. Soc. Microbiol., Abstr. 73rd Annu. Meet. Vol. 177*, p. 224.

Larson, D. L., Ahmed, M., Garon, D. E., Cravet, A., and Mayyasi, S. A. (1974). Antigenic and structural components of simian sarcoma virus (SSV-1) grown in human cells. *Proc. Am. Assoc. Cancer Res.* **15**, 54.

Laufs, R., and Melendez, L. V. (1973). Oncogenicity of *Herpesvirus ateles* in monkeys. *J. Natl. Cancer Inst.* **51**, 599–608.

Levinthal, J. M., and Shein, H. M. (1964). Propagation of a simian tumor agent (Yaba virus) in cultures of human and simian renel cells as detected by immunofluorescence. *Virology* **23**, 268–270.

Lewis, A. M., Rabson, A. S., and Levine, A. S. (1974). Studies of nondefective adenovirus 2-simian virus 40 hybrid viruses. X. Transformation of hamster kidney cells by adenovirus 2 and the nondefective hybrid viruses. *J. Virol.* **13**, 1291–1301.

Loan, R. W. (1974). Avian oncornaviruses. *In* "Handbook of Laboratory Animal Science" (E. C. Melby and N. H. Altman, eds.), Vol. II, pp. 22–38. CRC Press, Cleveland, Ohio.

McAllister, R. M., Gilden, R. V., and Green, M. (1972). Adenoviruses in human cancer. *Lancet* **1**, 831–833.

McCullough, B., Hoover, G. A., and Hardy, W. D. (1971). Susceptibility of the germ-free cats to Rauscher murine leukemia virus. *Am. J. Vet. Res.* **32**, 2077–2079.

Macpherson, I. (1970). The characteristics of animal cells transformed *in vitro*. *Adv. Cancer Res.* **13**, 169–215.

Maruyama, K., and Dmochowski, L. (1973). Cross-species transmission of mammalian RNA tumor viruses. *Tex. Med.* **69**, 65–75.

Mason, T. J., and McKay, F. (1974). "U.S. Cancer Mortality by County: 1950–1969," DHEW Publ. NIH-74-615, p. 729. Supt. of Documents, US Govt. Printing Office, Washington, D.C.

Melendez, L. V., Daniel, M. D., Garcia, F. G., Fraser, C. E. O., Hunt, R. D., and King, N. W. (1969). *Herpesvirus saimiri.* I. Further characterization studies of a new virus from the squirrel monkey. *Lab Anim. Care* **19**, 372–377.

Melendez, L. V., Daniel, M. D., Hunt, R. D., Fraser, C. E. O., Garcia, F. G., King, N. W. and Williamson, M. E. (1970). *Herpesvirus saimiri:* Further evidence to consider this virus as the etiologic agent of a lethal disease in primates which resembles a malignant lymphoma. *J. Natl. Cancer Inst.* **44**, 1175–1181.

Melendez, L. V., Castellanos, H., Barahona, H. H., Daniel, M. D., Hunt, R. D., Fraser, C. E. O., Garcia, F. G., and King, N. W. (1972a). Two new herpesviruses from spider monkeys (*Ateles geoggroyi*). *J. Natl. Cancer Inst.* **49**, 233–237.

Melendez, L. V., Hunt, R. D., Daniel, M. D., Fraser, C. E. O., Barahona, H. H., King, N. W., and Garcia, F. G. (1972b). *Herpesvirus saimiri* and *ateles*—their role in malignant lymphomas of monkeys. *Fed. Proc. Fed. Am. Soc. Exp. Biol.* **31**, 1643–1650.

Melnick, J. L., Altenburg, B., Arnstein, P., Mirkovic, R., and Tevethia, S. S. (1973). Transformation of baboon cells with feline sarcoma virus. *Intervirology* **1**, 386–398.

Merkow, L. P., and Slifkin, M., eds. (1973). "Progress in Experimental Tumor Research," Vol. 18. Karger, Basel.

Michalski, F., and Hsiung, G. D. (1975). Malignant transformation of hamster cells following infection with a bovine herpesvirus (infectious bovine rhinotracheitis virus). *Proc. Soc. Exp. Biol. Med.* **148**, 891–896.

Mirand, E. A., and Grace, J. T. (1962). Induction of leukemia in rats with Friend sarcoma. *Virology* **17**, 364–366.

Moloney, J. B. (1962). The murine leukemias. *Fed. Proc., Fed. Am. Soc. Exp. Biol.* **21**, 19–31.

Munroe, J. S. (1969). Subhuman primates in experimental viral oncology. *Ann. N.Y. Acad. Sci.* **162**, 556–567.

Nahmias, A. J. (1974). The evolution (evovirology) of herpesviruses. *In* "Viruses, Evolution and Cancer: Basic Considerations" (E. Kurstak and K. Maramorosch, eds.), pp. 605–624. Academic Press, New York.

Nahmias, A. J., and Dowdle, W. R. (1968). Antigenic and biologic difference in herpesviruses hominis. *Prog. Med. Virol.* **10**, 110–159.

Nahmias, A. J., Naib, Z. M., Josey, W. E., Murphy, F. A., and Luce, C. F. (1970). Sarcomas after inoculation of newborn hamsters with herpesvirus hominis type 2 strains. *Proc. Soc. Exp. Biol. Med.* **134**, 1065–1069.

Narayan, O., and Weiner, L. P. (1974). Biological properties of two strains of simian virus 40 isolated from patients with progressive multifocal leukoenocophalopathy. *Infect. Immun.* **10**, 173–179.

National Sanitation Foundation (1976). "NSF Standard No. 49 for Class II (Laminar Flow) Biohazard Cabinetry." NSF, Ann Arbor, Michigan.

Neubauer, R. H., Rabin, H., Wallen, W. C., Cicmanec, L., and Valerio, M. G. (1974). Oncogenicity of simian sarcoma virus, type 1 (SSV-1) in adult marmosets. *Proc. Am. Assoc. Cancer Res.* **15**, 35 (Abstr. 139).

Olsen, R. G., and Yohn, D. S. (1974). Tumor-forming poxviruses. *In* "Handbook of Laboratory Animal Science" (E. C. Melby and N. H. Altman, eds.), Vol. II, pp. 76–84. CRC Press, Cleveland, Ohio.

Olson, C., Gordon, D. E., Robl, M. G., and Lee, K. P. (1969). Oncogenicity of bovine papilloma virus. *Arch. Environ. Health* **19**, 827–837.

Olson, C., Miller, L. D., and Miller, J. M. (1973). Role of C-type virus in bovine lymphosarcoma. *Bibl. Haematol. (Basel)* **39**, 198–205.

Palmer, A. E., London, W. T., Nahmias, A. J., Naib, Z. M., Tunca, J., Fuccillo, D. A., Ellenberg, J. H., and Sever, J. L. (1976). A preliminary report on investigation of oncogenic potential of Herpes simplex virus type 2 in Cebus monkeys. *Cancer Res.* **36**, 807–809.

Pearson, L. D., Snyder, S. P., and Aldrich, C. D. (1973). Oncogenic activity of feline fibrosarcoma virus in newborn pigs. *Am. J. Vet. Res.* **34**, 405–410.

Pike, R. M. (1976). Laboratory-associated infections: Summary and analysis of 3921 cases. *Health Lab. Sci.* **13**, 105–114.

Pike, R. M., Sulkin, S. E., and Schulze, M. L. (1965). Continuing importance of laboratory-acquired infections. *Am. J. Public Health* **55**, 190–199.

Penn, I. (1974). Occurrence of cancer in immune deficiencies. *Cancer* **34**, Suppl., 858–866.

Pontén, J. (1971). Spontaneous and virus-induced transformation in cell culture. *Virol. Monogr.* **8**.

Portolani, M., Brodano, G. B., and LaPlaca, M. (1975). Malignant transformation of hamster kidney cells by BK virus. *Virology* **15**, 420–422.

Rabin, H. (1971). Assay and pathogenesis of oncogenic viruses in non-human primates. *Lab. Anim. Sci.* **21**, 1032–1049.

Rabin, H. (1974). Oncornaviruses of non-human primate origin. *In* "Handbook of Laboratory Animal Science" (E. C. Melby and N. H. Altman, eds.), Vol. II, pp. 39–45. CRC Press, Cleveland, Ohio.

Rabin, H., and Cooper, R. W. (1971). Tumor production in squirrel monkeys (*Saimiri scireus*) by Rous Sarcoma Virus. *Lab. Anim. Sci.* **21**, 705–711.

Rabin, H., Theilen, G. H., Sarma, P. S., Dungworth, D. L., Nelson-Rees, W. A., and Cooper, R. W. (1972). Tumor induction in squirrel monkeys by the S-T strain of feline sarcoma virus. *J. Natl. Cancer Inst.* **49**, 441–450.

Rabson, A. S., O'Connor, G. T., Kirschstein, R. L., and Branigan, (1962a). Papillary ependymomas produced in *Rattus* (Mastomys) *Natalensis* inoculated with vacuolating virus (SV-40). *J. Natl. Cancer Inst.* **29**, 765–787.

Rabson, A. S., Malmgren, R. A., O'Connor, O. T., and Kirschstein, R. L. (1962b). Simian vacuolating virus (SV-40) infection in cell cultures derived from adult human thyroid tissue. *J. Natl. Cancer Inst.* **29**, 1123–1145.

Rabson, A. S., Kirchstein, R. L., and Megallis, F. W. (1965). Autologous implantion of rhesus monkey cells "transformed" *in vitro* by simian virus 40. *J. Natl. Cancer Inst.* **35**, 981–991.

Rangan, S. R. S. (1974). C-type oncogenic viruses of nonhuman primates. *Lab. Anim. Sci.* **24**, 193–203.

Rapp, F. (1974). Herpesviruses and cancer. *Adv. Cancer Res.* **19**, 265–302.

Rauscher, F. J. (1971). Major opportunities for determination of etiologies and prevention of cancer in man. *In* "Recent Advances in Human Tumor Virology and Immunology" (W. Nakahara, K. Nishioka, T. Hirogama, and Y. Ito, eds.), pp. 3–24. Univ. of Tokyo Press, Tokyo.

Rich, M. A. (1968). Virus-induced murine leukemia. *In* "Experimental Leukemia" (M. A. Rich, ed.), pp. 15–49. Appleton, New York.

Rich, M. A., Geldner, J., and Meyers, P. (1965). Studies on murine leukemia. *J. Natl. Cancer Inst.* **36**, 523–536.

Riley, V. (1975). Mouse mammary tumors: Alteration of incidence as apparent function of stress. *Science* **189**, 465–467.

Robl, M. G., and Olson, C. (1968). Oncogenic action of bovine papilloma in hamsters. *Cancer Res.* **28**, 1596–1604.

Rogers, S. (1966). Shope papilloma virus: A passenger in man and its significance to the potential control of the host genome. *Nature (London)* **212**, 1220–1222.

Rondhuis, P. R. (1973). Induction of tumors in hamsters with a bovine adenovirus strain (serotype 8). *Arch. Gesamte Virusforsch.* **41**, 147–149.

Rowe, W. P. (1961). The epidemiology of mouse polyoma virus infection. *Bacteriol. Rev.* **25**, 18–31.

Rowson, K. E. K., and Mahy, B. W. J. (1967). Human papova (wart) virus. *Bacteriol. Rev.* **31**, 110–131.

Sanford, K. K. (1974). Biologic manifestations of oncogenesis *in vitro*. A critique. *J. Natl. Cancer Inst.* **53**, 1451–1485.

Sarma, P. S., Basker, J. F., Vernon, L., and Gilden, R. V. (1970). Feline leukemia and sarcoma viruses: Susceptibility of human cells to infection. *Science* **168**, 1098–1100.

Scolnick, E., and Baumgarner, S. (1975). Isolation of infectious xenotropic mouse type C virus by transfection of a heterologous cell with DNA from a transformed mouse cell. *J. Virol.* **15**, 1293–1296.

Sekiyo, S. (1971). Transformation of human cells by virus and chemical carcinogen *in vitro*. *Cancer* **28**, 789–797.

Sever, J. L. (1973). Herpesvirus and cervical cancer studies in experimental animals. *Cancer Res.* **33**, 1509–1510.

Shah, K. V., Daniel, R.W., and Warszawski, R. (1974). Sarcoma in a hamster inoculated with human papovavirus, BK virus. *Fed. Proc., Fed. Am. Soc. Exp. Biol.* **33**, 753; in more detail: *J. Natl. Cancer Ins.* **54**, 945–950. (1975).

Sharma, J. M., Witter, R. L., Shramek, G., Wolfe, L. G., Burmester, B. R., and Deinhardt, F. (1972). Lack of pathogenicity of Marek's Disease virus and herpesvirus of turkeys in marmoset monkeys. *J. Natl. Cancer Inst.* **49**, 1191–1197.

Sharma, J. M., Witter, R. L., Burmester, B. R., and Landon, J. C. (1973). Public health implications of Marek's disease virus and herpesvirus of turkeys. *J. Natl. Cancer Inst.* **51**, 1123–1127.

Shope, T., Dechairo, D., and Miller, G. (1973). Malignant lymphoma in cottontop marmosets after innoculation with Epstein-Barr virus. *Proc. Natl. Acad. Sci. U.S.A.* **70**, 2487–2491.

Snyder, S. P., Dungworth, D. L., Kawakami, T. G., Callaway, E., and Lau, D. T. L. (1973). Lymphosarcomas in two gibbons (*Hylobates lar*) with associated C-type virus. *J. Natl. Cancer Inst.* **51**, 89–94

Southam, C. M., Burchenal, J. H., Clarkson, B., Tanzi, A. T., Mackey, R., and McComb, V. (1969). Heterotransplantation of human cell lines from Burkitts tumors and acute leukemia into newborn rats. *Cancer* **23**, 281–299.

Stenbach, W. A., Anderson, J. P., McCormick, K. J., and Trentin, J. J. (1973). Induction of tumors in the liver of hamsters by an avian adenovirus (CELO). *J. Natl. Cancer Inst.* **50**, 963–970.

Stern, E. L., Johnson, J. W., Vesley, D., Halbert, M. M., Williams, L. E., and Blume, P. (1974). Aerosol production associated with clinical laboratory procedures. *Am. J. Clin. Pathol.* **62**, 591–600.

Stiles, C. D., Desmond, W., Sato, G., and Saier, M. H. (1975). Failure of human cells transformed by simian virus 40 to form tumors in athymic nude mice. *Proc. Natl. Acad. Sci. U.S.A.* **72**, 4971–4975.

Sulkin, S. E. (1972). Department of Microbiology, Southwestern Medical School, Dallas, Texas (revised, unpublished data on overt infections).

Sulkin, S. E., and Pike, R. M., and Hammon, W. M., and Sather, G. E. (1969). Prevention of laboratory infections; and Arboviruses. *In* Diagnostic Procedures for Viral and Reickettsial Infections'' (E. H. Lennette and N. J. Schmidt, eds.), 4th ed., pp. 66–78, 238–241. Am. Public Health Assoc., New York.

Sultanian, I. V., and Freeman, G. (1966). Enhanced growth of human embryonic cells infected with adenovirus 12. *Science* **154**, 665–667.

Takemoto, K. K., Ting, R. C. Y., Ozer, H. L., and Fabisch, P. (1968). Establishment of a cell line from an inbred mouse strain for viral transformation studies: Simian virus 40 transformation and tumor production. *J. Natl. Cancer Inst.* **41**, 1401–1409.

Theilen, G. H., Snyder, S. P., Wolfe, L. G., and Landon, J. C. (1970). Biological studies with viral induced fibrosarcomas in cats, dogs, rabbits and non-human primates. *Bibl. Haematol. (Basel)* **36**, 393–400.

Thomas, W. R., Aw, E. J., Papadimitrioium, J. M., and Simons, P. J. (1974). Effect of tissue injury on tumorigenesis induced by murine sarcoma virus (Harvey). *J. Natl. Cancer Inst.* **53**, 763–766.

Todaro, G. J., and Aaronson, S. A. (1968). Human cell strains susceptible to focus formation by human adenovirus type 12. *Proc. Natl. Acad. Sci. U.S.A.* **61**, 1272–1278.

Todaro, G. J., and Meyer, C. A. (1974). Transformation assay for murine sarcoma viruses using a simian virus 40-transformed human cell line. *J. Natl. Cancer Inst.* **52**, 167–171.

Trentin, J. J., Yabe, Y., and Taylor, G. (1962). The quest for human cancer viruses. *Science* **137**, 835–841.

Trentin, J. J., Van Hoosier, G. L., and Samper, L. (1968). The oncogenicity of human adenoviruses in hamsters. *Proc. Soc. Exp. Biol. Med.* **127**, 683–689.

U.S. Department of Health, Education and Welfare (1974a). "Classification of Etiologic Agents on the Basis of Hazard," 4th ed. Public Health Service, Center for Disease Control, USDHEW, Washington, D.C.

U.S. Department of Health, Education and Welfare (1974b). (1974b). "National Cancer Institute Safety Standards for Research Involving Oncogenic Viruses," DHEW Publ. No. (NIH) 78-790. Public Health Service, National Institutes of Health, USDHEW, Bethesda, Maryland.

U.S. Department of Health, Education and Welfare (1975). "National Cancer Institute Safety Standards for Research Involving Chemical Carcinogens," DHEW Publ. No. (NIH) 76-900. Public Health Service, National Institutes of Health, USDHEW, Bethesda, Maryland.

U.S. Department of Health, Education and Welfare (1976a). "Selecting a Biological Safety Cabinet." Public Health Service, National Institutes of Health, National Cancer Institute, Office of Research Safety, USDHEW, Washington, D.C.

U.S. Department of Health, Education and Welfare (1976b). "Classification of Etiologic Agents on the Basis of Hazard." Public Health Service, Center for Disease Control, USDHEW, Atlanta, Georgia.

U.S. Department of Health, Education and Welfare (1978a). Recombinant DNA research: Guidelines. *Fed. Regist.* **43**(247), 60108–60131.

U.S. Department of Health, Education and Welfare (1978b). "National Institutes of Health Laboratory Safety Monograph: A Supplement to the NIH Guidelines for Recombinant DNA Research." Public Health Service, National Institutes of Health, USDHEW, Bethesda, Maryland.

Walker, D. L., Padgett, B. L., Qurhein, G. M., Albert, A. E., and Marsh, R. F. (1973). Human papovavirus (JC): Induction of brain tumors in hamsters. *Science* **181**, 674–676.

Werner, J., Henle, G., Pinto, C. A., Haff, R. F., and Henle, W. (1972). Establishment of continuous lymphoblast cultures from leukecytes of gibbons (*Hylobates Lar*). *Int. J. Cancer* **10**, 557–567.

Werner, J., Pinto, C. A., Haff, R. F., Henle, W., and Henle, G. (1973). Responses of gibbons to inoculation of Epstein-Barr virus. *J. Infect. Dis.* **126**, 678–681.

Werner, J., Wolf, H., Apodaca, J., and zur Hausen, H. (1975). Lymphoproliferative disease in cotton-top marmoset after inoculation with infectious mononucleosis-derived Epstein-Barr virus. *Int. J. Cancer* **15**, 1000–1008.

Wolfe, L. G., and Deinhardt, F. (1972). Oncornaviruses associated with spontaneous and experimentally induced neoplasia in nonhuman primates. A review. *In* "Medical Primatology III" (E. I. Goldsmith and J. Moor-Jankowski, eds.), pp. 176–196. Karger, Basel.

Wolfe, L. G., Falk, L. A., and Deinhardt, F. (1971a). Oncogenicity of *Herpesvirus saimiri* in marmoset monkeys. *J. Natl. Cancer Inst.* **47**, 1145–1162.

Wolfe, L. G., Deinhardt, F., Theilen, G. H., Rabin, H., Kawakami, T., and Bustad, L. K. (1971b). Induction of tumors in marmoset monkeys by simian sarcoma virus, type 1 (*Lagothrix*): A preliminary report. *J. Natl. Cancer Inst.* **47**, 1115–1120.

Wolfe, L. G., Smith, R. K., and Deinhardt, K. (1972a). Simian sarcoma virus, Type 1 (*Lagothrix*): Focus assay and demonstration of nontransforming associated virus. *J. Natl. Cancer Inst.* **48**, 1905–1908.

Wolfe, L. G., Smith, R. D., Hoekstra, J., Marczynska, B., Smith, R. K., McDonald, R., Northrup, R. L., and Deinhardt, I. (1972b). Oncogenicity of feline sarcoma viruses in marmoset monkeys: Pathologic, virologic, and immunologic findings. *J. Natl. Cancer Inst.* **49,** 519–539.

Zilber, L. A. (1961). Pathogenicity of Rous sarcoma virus for rats and rabbits. *J. Natl. Cancer Inst.* **26,** 1295–1305.

Zilber, L. A. (1964). Some data on the interation of Rous sarcoma virus with mammalian cells. *Natl. Cancer Inst. Monogr.* **17,** 261–271.

# Chapter Seven

# Biohazards Associated with Laboratory Animals

## PETER J. GERONE

## I. INTRODUCTION

All too often safety programs fail because too little consideration is given to the hazards associated with laboratory animals. From the point of view of biohazards the tendency is to concentrate on the laboratory as distinct

**225**

LABORATORY SAFETY: THEORY AND PRACTICE
Copyright © 1980 by Academic Press, Inc.
ISBN 0-12-269980-7

from the animal room. Safety devices and equipment used in the laboratory are a more visible demonstration of the so-called safety consciousness of the investigator. The animal rooms are remote and, therefore, more apt to be forgotten in a safety program. The purpose of this chapter is to define the hazards in the animal room area and to discuss measures that can be taken to minimize the risks.

Before one can deal with the hazards related to laboratory animals there must be some recognition of the problem. Most people do not look on animals as a potential hazard, or if they do, their concern is only with bites and scratches. This attitude probably reflects their early experience with pets and domestic animals. One usually learns before reaching adulthood that dogs bite, cats scratch, horses and cows kick, and wild animals, with few exceptions, cannot be trusted. This experience is reinforced when people come to work in the laboratory and soon they learn to respect certain species that are more prone to bite. The possibility of animals serving as a source of human disease seems remote, especially if the animals are not experimentally infected. The truth, however, is that there are many serious hazards associated with laboratory animals. This is supported by many published findings. In a study published by Pike (1976) it appears that animals and their ectoparasites may account for about 17% of laboratory-acquired infections.

The hazards involving laboratory animals can affect personnel, experiments, other laboratory animals, or even domestic animals. Laboratory workers can either be injured or contract infectious diseases. Experiments can be ruined or results can be badly misinterpreted if diseases are transmitted among the laboratory animals. Breeding colonies have been destroyed because diseases were inadvertently introduced to the breeding stock as a result of laboratory contamination. Finally the possibility always exists that organisms used in the laboratory may become a source of infection to domestic animals. Although each of these hazards can have important consequences and will be elaborated on later, the emphasis of this chapter will be on infectious diseases.

## II. HAZARDS TO PERSONNEL

As already mentioned, the possibility of injury to laboratory personnel resulting from the aggressive nature of certain laboratory animals is usually the first hazard to be recognized. Bites and scratches can be painful and sometimes more or less incapacitating. The most serious aspect of these injuries is the possibility of secondary infections. There are other injuries too that are not as readily anticipated because they do not involve animals directly, such as the cuts, scratches, and bruises that might result from handling animal room equipment or from falling on wet floors. There is also the possibility of electrical shocks, particularly in wet areas of the animal room, burns from the cage cleaning equipment, and irritations or allergic reactions resulting from contact with a variety of chemicals used in the animal room.

Some laboratory workers may become allergic to laboratory animals (Patterson, 1964). The allergic response can manifest itself as rhinitis, conjunctivitis, asthma, or atopic dermatitis. The hypersensitivity reaction is usually a response to the animal dander.

The infectious hazard of working with laboratory animals is of paramount importance and is discussed in detail in subsequent sections.

## III. EXPERIMENTAL HAZARD

Inadequate control of the infectious diseases in the animal room can wreak havoc with the objectives of the experiment. This can take many forms. Secondary infections may spread to animals that would not have been infected from the intended experimental procedures. If these are not recognized as such, the results of experimental work can be seriously misread. The transmission of infectious disease from animal to animal may also result in dual infections. This could either have the effect of exacerbating the infection under study or the opposite effect of causing an interference in the disease process. In either case the results could be disastrous.

Although disease transmission among laboratory animals is most likely to happen within an animal room, it should also be kept in mind that there could be room-to-room transmission. Animal attendants, fomites, and common air supplies could all be factors in room-to-room spread of disease.

## IV. HAZARD TO BREEDING STOCK

Laboratories that are involved in breeding experimental animals know that it is extremely important to keep breeding colonies free of disease. The more unique the breeding group is in terms of its genetic character, the more valuable it becomes, and therefore the need to protect such colonies is even more crucial. Uncontrolled infectious disease in laboratory animal experiments has been known to contaminate breeding stock. Such occurrences could wipe out years of research in a few weeks. Special precautions must be taken when researchers are studying diseases highly infectious for a given species in a laboratory that is also breeding that species.

## V. HAZARD TO DOMESTIC ANIMALS

The risk to domestic animals is merely an extension of the possibility discussed above. It is conceivable that organisms capable of infecting domestic animals may escape from the laboratory, causing epizootics that could have grave economic consequences. Breakdown in containment can result from contaminated exhaust air from animal room areas, escape of infected animals, animal attendants coming in contact with farm animals, and fomites. There is no evidence that this has ever happened, which is probably attributable to the strict regulations governing the use of animal disease agents in the laboratory. Exotic agents in particular have been worked with only under the most controlled circumstances (Callis and Cottrall, 1968), thereby avoiding this hazard.

## VI. IMPORTANT LABORATORY ZOONOSES

There are more than 150 recognized zoonotic diseases, that is, diseases of animals that are naturally transmitted to man (Steele, 1973; Knipping, 1972; Fraser, 1969; Hime, 1974). Of these there are about 30 zoonoses that can be transmitted from laboratory animals to man (Baldelli and Mantovani, 1974; Griesemer and Manning, 1973; Quist, 1972; Tobin, 1968; Loosli, 1967; Gay and Blood, 1967). In addition to these zoonoses one could add all the other human pathogens that have been experimentally inoculated into laboratory animals. It is clear from reviewing the literature that any time people come into contact with laboratory animals they run a certain degree of risk from injury or infection. Sometimes the hazard is extremely small, as in the case of working with specific pathogen-free rodents. At other times the risk can be great. Published evidence suggests that nonhuman primates, because of their close phylogenetic relationship to man, are particularly dangerous (Fiennes, 1967; Gerone, 1975; Perkins and O'Donoghue, 1969; Prier et al., 1964; Tauraso, 1973; Kalter and Heberling, 1975; Eyestone, 1968).

In Table I are listed some of the more serious and common laboratory zoonotic diseases. Table II lists other possible diseases that may be transmitted from laboratory animals to man. It is also important to note

TABLE I

Important Laboratory Animal Zoonoses

| Disease | Animals | Selected references |
|---|---|---|
| **Viral** | | |
| Lymphocytic chorio-meningitis | Rodents, primates | Hotchin, 1971; Baum et al., 1966; Armstrong et al., 1969; Bowen et al., 1975; Lewis et al., 1965; Maurer, 1964 |
| Marburg | Primates | Hennessen, 1968, 1969; Siegert, 1972; Martini and Siegert, 1971 |
| Herpes B encephalitis | Primates | Hull, 1973a; Keeble, 1960, 1968; Davidson and Hummeler, 1960; Perkins, 1968 |
| Hepatitis | Primates | Krushak, 1970; Hillis, 1961, 1963, 1967; Ruddy et al., 1967; Havens, 1967; Deinhardt, 1970; Friedmann et al., 1971 |
| Rabies | Primates, dogs | Boulger, 1966; Kaplan, 1969 |
| Yaba and Tanapox | Primates | Downie et al., 1971; Luby, 1971; Hall and McNulty, 1967; McNulty et al., 1968; Grace and Mirand, 1963, 1965 |
| **Bacterial** | | |
| Leptospirosis | Rodents, primates, dogs | White et al., 1961; Stoenner and Maclean, 1958; Turner, 1968; Fear et al., 1968 |
| Tuberculosis | Primates | Moreland, 1970; Kaufmann et al., 1975 |
| Shigellosis | Primates | Cook, 1969; Mulder, 1971; Carpenter, 1968; Schneider et al., 1960; Pinkerton, 1968; Carpenter and Cook, 1965 |
| Salmonellosis | Rodents, birds, primates, dogs | Smith, 1968; Schneider et al., 1960 |
| **Bedsonial** | | |
| Cat scratch fever | Cats | Warwick, 1964 |
| Ornithosis | Birds | Blackmore, 1968 |
| **Parasitic** | | |
| Toxoplasmosis | Cats, primates | Krick and Remington, 1978 |
| Larval migrans | Cats, dogs | Bisseru, 1968 |
| Malaria | Primates | Most, 1973; Cross et al., 1973; Geiman, 1964 |
| **Fungal** | | |
| Dermatosis | Rodents, primates cats, dogs | Gentles, 1968; Dolan et al., 1958; Povar, 1965 Rowsell, 1963; Booth, 1952 |

that on some occasions man has served as the source of infections in laboratory animals (Cappucci et al., 1972; Blackmore and Francis, 1970; Sellers et al., 1971).

To illustrate the biohazards associated with laboratory animals it may be useful to review some of the more common and serious laboratory zoonotic diseases.

**TABLE II**

**Possible Laboratory Animal Zoonoses**

| Organism/Disease | Animals | Selected references |
|---|---|---|
| Viruses | | |
| Monkeypox | Monkeys | Arita and Henderson, 1968; Casey *et al.*, 1967; Espana, 1971a,b |
| Foot and mouth | Pigs, cattle | Hyslop, 1973 |
| Calciviruses | Monkeys | Smith *et al.*, 1978 |
| Vesicular stomatitis | Cattle | |
| Newcastle disease | Birds | Brandly, 1964 |
| Louping ill | Sheep | Gordon, 1968 |
| Venezuelan equine encephalitis | Mice | Lennette and Koprowski, 1943 |
| Kyasanur Forest | Monkeys | Work and Trapido, 1957; Webb, 1969 |
| Yellow fever | Monkeys, mice | Berry and Kitchen, 1931; Low and Fairley, 1931 Theiler, 1930 |
| Other arboviruses | Rodents, monkeys | Hanson *et al.*, 1967 |
| Bacteria | | |
| Pasteurellosis | Mammals, birds | Mair, 1968; Mortelmans and Kageruka, 1969 |
| Brucellosis | Cattle | Morgan, 1968 |
| *Escherichia coli* | Primates, mice | McClure *et al.*, 1972; Higa and Ching, 1972 |
| *Klebsiella pneumonia* | Primates | |
| *Streptococcus pneumoniae* | Primates | |
| *Mycobacterium marinum* | Mice | Chappler *et al.*, 1977 |
| Rickettsiae | | |
| Q fever | Cattle, sheep | Foggie, 1968 |
| Parasites | | |
| Protozoa | Primates | Geiman, 1964; Neal, 1969 |
| Hydatidosis | Dogs | Blamire, 1968 |
| Fungi | | |
| *Arthroderma simii* | Monkey | Stockdale *et al.*, 1965 |
| *Microsporum canis* | Dogs, cats Monkeys | Gentles, 1968; Taylor *et al.*, 1973 |
| *Trichophyton mentagraphytes* | Rabbits, rats Mice | Alteras and Cojocaru, 1969 |
| *Keratinomyces ajelloi* | Mice | Refai and Ali, 1970 |
| Ectoparasites | | |
| Pediculosis | Monkey | Ronald and Wagner, 1973 |
| Mites | Monkey | Yunker, 1964 |

## A. Herpes B Encephalitis

*Herpesvirus simiae* (B virus) was first recognized as a human pathogen in 1932 when a laboratory worker contracted a fatal encephalitis after being bitten by a rhesus monkey (Sabin and Wright, 1934). Since that time there have been about 25 human cases, of which only 4 have survived the disease, and of these only 2 had no sequelae (Bryan *et al.,* 1975; Davidson and Hummeler, 1960; Breen *et al.,* 1958; Fierer *et al.,* 1973; Hull, 1973a).

Although Old World monkeys had been used in research for many years prior to the discovery of B virus, the disease went unnoticed probably because in simians it is relatively mild or asymptomatic. Frequently the only clinical evidence of disease is the occurrence of lesions on the surface of the tongue, buccal cavity, and lips (Keeble, 1960). These signs may not be apparent except on close examination. The incidence of B virus infection in the rhesus monkey is about 10% on capture, but it can increase to levels of 70% or higher before the animals reach the laboratory. It is most commonly found in monkeys belonging to the genus *Macaca* but other Old World and New World monkeys may become infected by contact with infected species. As is frequently the case with other herpesviruses (McCarthy and Tosolini, 1975a,b; Tosoloni and McCarthy, 1975), this virus probably persists in latent form (Vizoso, 1975) in the infected monkeys and is occasionally shed into the environment.

Although some infections in laboratory personnel are attributable to bites and scratches, there are several cases of B virus in which the mode of transmission is unknown (Hull, 1973b). A number of infections associated with the handling of monkey kidney cell cultures suggest that airborne infection may be involved. Experiments done by Chappell (1960) and later by Benda and Polomik (1969) have demonstrated the infectivity of airborne *Herpesvirus simiae* for laboratory animals. Considering the fact that rabbits can be infected by nontraumatic ocular instillation of B virus, the possibility of the ocular route as a portal of entry to man must not be excluded. As few as 1–10 $TCID_{50}$ of the virus dropped onto the cornea of a rabbit produces a lethal infection.

In summary, B virus is a common pathogen of rhesus monkeys but can be found in other species as well. The relative lack of clinical signs in infected monkeys and the possibility of reactivation of latent disease either spontaneously or as a result of experimental stress (Kirschstein *et al.,* 1961) makes this virus a particularly insidious threat. Considering the large number of laboratory personnel and pet owners who have had contact with rhesus monkeys during the past half century and the fact that there have been about two dozen human cases of B virus, one must conclude that the chances of human infection must be rather small. However, the high fatality rate in infected individuals makes this one of the most serious bio-hazards faced by laboratory workers.

## B. Marburg Disease

Marburg disease is a disease in humans that was associated with African Green monkeys (Martini and Siegert, 1971; Hennessen, 1968). This too was a highly lethal disease, resulting in 7 deaths among 31 cases. Although its associated with monkeys and high mortality is reminiscent of B virus, there are many more differences than similarities between these two diseases. Accordingly, it is worth some discussion to bring out the contrasting circumstances surrounding these diseases.

The source of the human infections was African Green monkeys that had been shipped from Uganda. The human cases occurred in Germany and Yugoslavia. Clinically, the disease was characterized as a severe hemorrhagic fever. Although most of the infected persons had direct contact with the monkeys or their infected tissues, a few were apparently infected by contact with diseased patients. Fortunately, since the 1967 episode there have been no further cases of Marburg disease related to contact with monkeys.

Evidence suggests that Marburg agent is not a natural pathogen of monkeys. This is deduced from the experimental work that showed that the virus is highly lethal for African Green and other species of monkeys (Simpson, 1969). Further support of this hypothesis comes from the absence of naturally acquired antibody in African Green and other monkeys (Slenczka et al., 1971). Monkeys that were experimentally infected had large quantities of virus in the blood, saliva, and urine and their infections were invariably fatal (Simpson, 1969). These results indicate that the African Green monkey served as a vehicle in the transmission of the disease from its natural host, which remains unknown, to man.

Marburg disease, therefore, is an example of how an accidental infection of a laboratory animal species can pose a serious threat to man. It is improbable that this is a natural infection of monkeys. The disease in the monkey rather than being benign and latent is acute and fatal.

## C. Lymphocytic Choriomeningitis

The virus of lymphocytic choriomeningitis (LCM) deserves special attention in any discussion of hazards associated with laboratory animals for several reasons. The virus is ubiquitous, especially among rodents, but apparently can infect virtually all laboratory animals. The infection in laboratory animals is often subclinical, and therefore goes undetected. LCM has produced many infections in laboratory personnel and it is apparently transmitted to man in a variety of ways. This virus has further tormented researchers by contaminating breeding colonies and affecting experimental results. As a matter of fact, the virus was discovered in 1933 by Armstrong and Lillie as a contaminant in experimental work involving St. Louis encephalitis virus (Armstrong and Lillie, 1934).

Hotchin (1971) has written a good review on LCM. The virus is widely spread in many rodents but infected animals are often asymptomatic. Persistent infections not only can last the lifetime of the animal, but can be congenitally transmitted for generations, particularly in mice. Special efforts must be made to detect persistent tolerant infections in laboratory animals (Skinner and Knight, 1971).

Clean strains of mice can be readily contaminated by bringing in infected mice from other laboratories or from the wild. In recent years a number of rodent colonies have been infected by passage of tumor cell lines (Lewis *et al.*, 1965; Biggar *et al.*, 1976; Baum *et al.*, 1966; Hotchin *et al.*, 1974; Armstrong *et al.*, 1969). The results of experimental work can be significantly altered by the contamination of mice in cell cultures with LCM virus. In some cases LCM has had an interfering effect, whereas in others it seems to have enhanced observed reactions.

LCM infections in man can be inapparent or they may take one of several clinical forms such as mild influenza-like disease, meningitis, or even fatal meningeal encephalitis. An outbreak of cases of LCM among medical center personnel in 1973 was reported by Hinman *et al.* (1975). A total of 48 cases was discovered in an outbreak traced to infected hamsters in which tumor cell lines were being passed. LCM virus was isolated from the hamsters as well as from the tumor cell lines. The investigators found that 70% of the individuals that came into direct contact with infected hamsters caged in one room became infected. The infection rate was only 28% among the staff that did not come into direct contact with the animals. Serological analysis showed that the infection rate was only 10% in individuals who entered the room less than one time per week, whereas it was 50% in those who entered more than once a week. One individual who became infected did not have any direct contact with the infected animals or cells, but did use an apple crate that had been in the infected animal room. This study provided evidence that LCM may not only be transmitted by direct contact, but also by aerosols and fomites. Experimental evidence suggests that the virus may also be transmitted by blood-sucking arthropods (Coggeshall, 1939; Milzer, 1942).

## D. Hepatitis

Hepatitis, like tuberculosis, is a disease that nonhuman primates pick up from their association with infected humans during the process of capture and transportation.to the laboratory. Once infected these primates then become a source of hazard to the laboratory personnel who must come in contact with them. There have been more than 200 cases of infectious hepatitis (hepatitis A) among individuals associated with nonhuman primates that were suspected of being the source of the human infection (Center for Disease Control, 1975). Many species of primates have been involved, but the chimpanzee appears to be the main culprit, especially young, newly imported animals (Deinhardt, 1970; Hillis, 1961, 1963, 1967; Friedmann et al., 1971; Mosley and Galambos, 1967). The chimpanzees usually appear to be well or show only nonspecific signs of illness. Infected animals probably shed virus for extended periods of time. The fecal-oral route is the most probable mode of transmission to man.

## VII. FACTORS AFFECTING LABORATORY ZOONOSES

The infectious hazard to personnel who work with laboratory animals depends on many factors. Each of these will be briefly discussed so that some estimate can be made regarding the overall risk involved.

## A. Animal Species

Some laboratory animals pose more of a risk than others. Generally speaking, the more aggressive the animal the greater the threat. Size is also important. Larger animals are harder to restrain, it is more difficult to cage them in isolation, and their bites and scratches are more damaging.

It is clear from examining Tables I and II that nonhuman primates can be a major source of human infections (Eyestone, 1968; Tauraso, 1973; Prier et al., 1964; Gerone, 1975; Hull, 1973b). There are many reasons for this. First, because of the close phylogenetic relationship of monkeys and apes to man, many of their pathogens are also pathogenic for man. Second, until recently, most nonhuman primates came from the wild bringing their naturally acquired diseases to the laboratory. Finally, nonhuman primates, which are susceptible to many human diseases, may acquire these infections from human contact while en route to the laboratory and then become a potential source of infection for laboratory personnel (Vickers, 1973).

## B. Source of Animal Infections

Workers can be exposed to infections of laboratory animals that are either natural infections or experimentally induced. Of these two categories, most of the problems seem to stem from the naturally acquired animal diseases. The reasons that natural infections are more hazardous are many. First, such infections are usually not suspected until a problem arises. Second, natural diseases of animals are often inapparent, so that special monitoring is necessary to detect their presence. Also, laboratory personnel usually do not take the same safety precautions when handling animals that are presumed to be uninfected; therefore the degree of exposure is greater.

On the other hand, laboratory people working with experimentally infected animals are acutely aware of the hazard. Because they are directly involved with the infectious agent, they are more knowledgeable about the biological characteristics of the organism and probably have information about portals of entry, shedding patterns, stability, etc. On the negative side is the fact that medical laboratories are often dealing with human pathogens, which, if transmitted to man, will almost certainly cause disease.

## C. Virulence and Infectivity of the Pathogen

The pathogenic characteristics of a microorganism carried by or experimentally inoculated into laboratory animals can have a significant effect on the hazard. The interaction of infectivity and severity of disease must be weighed. Some infections are easily transmitted to man, but the resulting infection may be mild or inapparent. Others are only rarely transmitted but infection results in severe disease or death. The worst combination, obviously, is those diseases that are severe and easily transmitted.

The "Classification of Etiologic Agents on the Basis of Hazard" published by the Center for Disease Control (1976) is a helpful guide in gauging the hazard of microbial agents. Class 3 and 4 agents are of particular concern. It is interesting to know that some organisms such as rabies, selected arboviruses, and smallpox are classified in a higher hazard class when inoculated into experimental animals as compared with their use in the laboratory.

The human infectious dose of agents found in laboratory animals can markedly influence the level of hazard. Those that can infect at low levels are particularly hazardous. Unfortunately, there are few data available in the published literature defining human doses of infectious agents (Wedum *et al.,* 1972; Wedum, 1964). A rough idea of the relative infectiousness of agents is usually based on laboratory experience; that is, those microorganisms that have caused many laboratory infections are probably infectious for man at low dose levels.

## D. Shedding of Microorganisms by Laboratory Animals

Infected laboratory animals can shed infectious material from their coats, urine, feces, or excretions. The degree of shedding contributes to the level of environmental contamination. Animals that shed infectious material in urine and feces are most apt to cause contamination problems, particularly in cases where large quantities (thousands of infectious doses) of organisms can be found in these waste products. The quantities of organisms given off by infected animals can vary depending on the organism, the route of infection, and the animal species involved. It is important to remember, however, that although quantities may vary, in most cases some shedding does occur. For that reason infected animals must always be considered a possible source of infection.

## E. Route of Infection

The route of infection both for the laboratory animals and humans can be a critical factor affecting the biohazard in the animal room area. As mentioned earlier, the route of inoculation can influence the shedding pattern of infected animals. For example, a virus that can cause both CNS disease and respiratory disease is more likely to be shed if inoculated into the animals intranasally or by aerosol as opposed to intracerebral inoculation. Likewise, enteric organisms given orally could almost certainly be found in the feces of infected animals, whereas the same organism inoculated by a parenteral route may produce disease in other organ systems with no shedding whatsoever.

The technical procedures involved in inoculating laboratory animals can also contribute to the hazard. Parenteral inoculations involving syringes and needles frequently result in self-inoculation (Chappler et al., 1977; Grace and Mirand, 1963). With laboratory infections in general the accidental needle puncture is considered to be one of the most frequent causes of disease (Stark, 1975; Pike, 1976; Sulkin et al., 1963). Intranasal inoculation of animals is very hazardous. There is an excellent chance that aerosols will be generated by the respiratory activity of the animals even when the animals are anesthetized. Animals exposed to aerosols can be dangerous because of residual infectious material on the head and coat (Wedum, 1964).

## VIII. CONTROL OF ANIMAL ROOM BIOHAZARDS

Most of the principles of good, safe laboratory practices also apply to the animal room area. This section will primarily describe biohazard control and containment measures that are particularly applicable to the animal housing section of a biomedical research laboratory (Phillips and Jemski, 1963; Darlow, 1967).

### A. Facilities

Facilities for housing laboratory animals, if properly designed, can contribute greatly to reducing the biohazards. Many designs have been described (Wedum, 1961, 1964; Kawamata and Yamanouchi, 1975; Gerone, 1975), but the salient feature incorporated into these designs is isolation. Animal rooms should be separate from other laboratory facilities. The degree of separation is dictated by the hazard of the infectious organisms to be used.

Ideally, access to animal rooms should be restricted. Entrance by way of air locks or change rooms improves isolation. Wherever possible the vivarium should include ancillary facilities, such as autoclaves, cage washing equipment, X ray, surgery, examining room, necropsy facilities, and sufficient storage space for food and supplies.

Rather than having a few large rooms, it is more desirable to have several small rooms that allow maximum flexibility in use and segregation of animals by species or infecting agent. Walls and floors should be sealed to make them waterproof and easily disinfected.

Rooms for infected animals should be under negative air pressure. Exhaust air should be filtered or incinerated to remove contaminants, and supply air, if recirculated, should be filtered. More specific information and criteria can be found in the "Guide for Care and Use of Laboratory Animals" (Moreland et al., 1978).

A wide variety of cages have been designed to accommodate animals of various sizes and to provide different levels of biologic containment. It would not be possible in this chapter to evaluate fully and discuss the attributes of all these cages. It should be stated, however, that the selection of appropriate cages is essential in controlling biohazards.

Cage designs in terms of containment range from completely open, such as wire mesh cages, to entirely sealed compartments in which the air is supplied and exhausted through filters (Wedum, 1964; Wedum et al., 1972; Horsfall and Bauer, 1940; Tauraso et al., 1969). Intermediate levels of containment are achieved with filter-top cages (Giddens et al., 1972; Wedum et al., 1972; Jennings and Rumpf, 1965; Kraft, 1958; Kraft et al., 1964;

Schneider and Collins (1966), Collins, 1969). The problem with cages in general, regardless of their degree of containment, is that sooner or later they must be opened to observe or feed the animals, or to remove them from the cage. At such times contamination can easily escape. For that reason the use of unidirectional, filtered airflow in animal rooms is helpful in reducing cage-to-cage transmission and infection of personnel (Wedum et al., 1972; Beall et al., 1971). The use of gnotobiotic isolators is also a feasible method of controlling biohazards (Carter, 1978). Radiation with ultraviolet light is also useful in reducing contamination that escapes from cages (Phillips et al., 1957; Wedum, 1964; Wedum et al., 1956).

## B. Personnel

Like any other aspect of laboratory safety, people working in the animal rooms have a critical role in reducing the biohazards surrounding laboratory animals. Well-trained and experienced technicians and animal care attendants are definite assets in a safety program. Training should include a complete indoctrination in the hazards that are faced in the vivarium. Preventive medicine and employee health service programs for personnel handling animals are worthwhile (Hummer, 1969; Wedum, 1964; Muchmore, 1975). Incoming employees should have physical examinations to establish their suitability for employment. Serum specimens should be collected and stored in a freezer for future reference. Personnel working with primates should be checked for tuberculosis by either skin tests or chest X rays. If there is a high risk of expsoure to such diseases as tetanus and rabies, workers should be immunized. If immunizations are available for other organisms that are experimentally inoculated into animals, these too should be given.

Personal hygiene is very important in the animal room. Frequent hand washing (Steere and Mallison, 1975) can reduce the chances of disease transmission by the oral route. It helps to serve as a reminder that an infectious hazard is present. Personnel should develop a habit of keeping hands away from their mouth, eyes, nose, and face.

It is good practice when working around animals that are potentially infected to wear surgical caps, gowns, gloves, and masks. Microbiologic respirators afford even greater protection against airborne infection. Heavy gloves are sometimes essential to protect workers from bites and scratches of some species. Under some circumstances it is advisable to wear boots that will remain in the animal room or disposable plastic shoe coverings, which can be discarded. To protect against pathogens that could infect by the ocular route, glasses or eyeshields can be worn.

## C. Animal Care Practices

The proper management and operation of an animal care facility unquestionably reduces the risks related to laboratory animals (Edward, 1969; Perkins and O'Donoghue, 1969). Procuring animals as free of disease as possible is the first step in minimizing hazards. Once the animals arrive in the laboratories they should be isolated in quarantine for a suitable period of time (Simmons, 1969). The greater the chance of zoonoses, the longer should be the quarantine period.

Clean animals should be segregated from infected animals. The mixing of species in a single room should be avoided. Animals should be regularly examined by qualified personnel. A diagnostic program to test for pathogens that are not easily detected by overt signs of illness can be valuable. Primates should be regularly tested for the presence of tuberculosis (Stunkard et al., 1971; Stones, 1969; Moreland, 1970; Hartley, 1969; Kaufmann, et al., 1975).

Animal rooms and quarters should be kept as clean and uncluttered as possible. Disinfectants should be liberally used in the proper concentration. Periodically, animal rooms should be thoroughly disinfected, using paraformaldehyde or other appropriate gaseous disinfectants. Measures should be taken to keep wild rodents and other vermin from entering the animal facilities. Insect control is also necessary, but the selection of insecticides should be carefully made to avoid toxic reactions in animals.

Every effort should be made to reduce aerosols around animals. Although the extent of airborne infections associated with laboratory animals is difficult to document, there is agreement among experts that it may be a common source of infection. We know from cross-infection studies (Kruse and Wedum, 1970; Krichheimer et al., 1961) and air-sampling studies that organisms are present in animal room environments (McGarrity and Dion, 1978; McGarrity et al., 1969). Activities such as dry sweeping of floors or stirring of contaminated bedding greatly contribute to airborne spread of organisms. It is also conceivable that organisms could be aerosolized by playing high-pressure streams of water on contaminated urine and feces.

There is evidence to suggest that stress in laboratory animals may cause a reactivation of latent infections. With this in mind it would be advisable to keep stress down to a minimum. This is accomplished by providing ideal conditions for the caged animals and disturbing them as little as possible.

Final disposal of carcasses and animal room wastes requires special attention. The best solution is to use an incinerator (Barbeito and Shapiro, 1977; Rosenhaft, 1974; Barbeito and Gremillion, 1968). Waste and dead animals should be placed in sealed containers or plastic bags the outer surfaces of which are disinfected and then carried to the incinerator. In the absence of an incinerator laboratories sometimes sterilize contaminated waste by autoclaving and then dispose such wastes in the same manner used for other garbage.

## D. Animal Handling

There are proper procedures for handling unanesthetized small laboratory animals so that the chances of getting bitten or scratched are minimized (Short, 1967). Methods of restraining animals are learned by observing experienced animal handlers. Animals that are appropriately restrained are not only less apt to bite and scratch, but less likely to cause self-inoculation with needles and syringes. Many laboratory accidents have occurred subsequent to sudden movements of animals in response to needle punctures. Restraint of larger animals should be accomplished with tranquilizers and sedatives.

Any procedure involving inoculation of animals with a syringe and needle should be approached cautiously. The person inoculating should be assisted so that his hands are free to do the inoculating. An alternative is to use suitable restraining devices that are either improvised or purchased commercially.

## IX. ESTIMATION OF RISK

One of the most rewarding approaches to biohazard control in the animal room is through experimental work. It is difficult to develop a safety program when there is little information regarding the extent of hazard. This gap can be filled by an experimental approach. The transmissibility of disease-producing agents can be estimated by cross-infection studies (Kruse and Wedum, 1970; Kirchheimer et al., 1961; Owen and Buker, 1956). These can be done at the intracage or intercage levels. Clearly, organisms that are easily transmitted from animal to animal are more of a threat to man and his experiment. Rigorous testing of the overall integrity of the animal room can be done using organisms that are highly transmissible from animal to animal. Such experiments would give some estimate of the chances of organisms escaping cages or animal rooms.

Laboratories should be encouraged to test disinfectants for their ability to inactivate pathogens that will be used on experimental animals. All too often disinfectants are selected because of their availability, and usually optimal concentrations have not been determined.

The shedding of microorganisms by infected animals is one of the easier tasks that can be experimentally estimated. It is important to know how much is shed, where it comes from, and how long it lasts. The stability of the pathogens in the environment is also useful information.

## X. FINAL COMMENT

Safety in the animal room area of infectious disease laboratories depends of many interrelated factors. As we have seen, the hazard varies with the infectious agent, the species of animal used, the kinds of facilities and equipment available, and finally the workers themselves. Because so many factors are involved it would appear that the task of maintaining a safe operation in a vivarium is hopelessly complex. It is complex, but the problem is solvable. It is a simple matter of recognizing the hazards and then instituting commonsense practices that will either reduce the source of contamination or suitably contain it. Current technology is available to deal with most of the problems that laboratory animals present. The first task is to recognize and estimate the magnitude of the biohazard.

## REFERENCES

Alteras, I., and Cojocaru, I. (1969). Human infection by *Trichophyton mentagrophytes* from rabbits. *Mykosen* **12**, 543–544.

Arita, I., and Henderson, D. A. (1968). Smallpox and monkeypox in non-human primates. *Bull. W. H. O.* **39**, 277–283.

Armstrong, C., and Lillie, R. D. (1934). Experimental lymphocytic choriomeningitis of monkeys and mice produced by a virus encountered in studies of the 1933 St. Louis encephalitis epidemic. *Public Health J.* **49**, 1010–1027.

Armstrong, D., Fortner, J. G., Rowe, W. P., and Parker, J. C. (1969). Meningitis due to lymphocytic choriomeningitis virus endemic in a hamster colony. *J. Am. Med. Assoc.* **209**, 265–268.

Baldelli, R., and Mantovani, E. (1974). Zoonosi trasmesse da animali da laboratorio. *Nuovi Ann. Ig. Microbiol.* **25**, 1–82.

Barbeito, M. S., and Gremillion, G. G. (1968). Microbiological safety of an industrial refuse incinerator. *App. Microbiol.* **16**, 291–295.

Barbeito, M. S., and Shapiro, M. (1977). Microbiological safety evaluation of a solid and liquid pathological incinerator. *J. Med. Primatol.* **6**, 264–273.

Baum, S. G., Lewis, A. M., Jr., Rowe, W. P., and Huebner, R. J. (1966). Epidemic nonmeningitic lymphocytic-choriomeningitis-virus infection: An outbreak in a population of laboratory personnel. *N. Engl. J. Med.* **274**, 934–936.

Beall, J. R., Torning, F. E., and Runkle, R. S. (1971). A laminar flow system for animal maintenance. *Lab. Anim. Sci.* **21**, 206–212.

Benda, R., and Polomik, F. (1969). Cause of air-borne infection caused by B virus. I. Lethal inhalation dose of B virus for rabbits. *J. Hyg., Epidemiol., Microbiol., Immunol.* **13**, 24–30.

Berry, G. P., and Kitchen, S. F. (1931). Yellow fever accidentally contracted in the laboratory: A study of seven cases. *Am. J. Trop. Med.* **11**, 365–434.

Biggar, R. J., Deibel, R., and Woodall, J. P. (1976). Implications, monitoring, and control of accidental transmission of lymphocytic choriomeningitis virus within the hamster. *Cancer Res.* **36,** 551-553.

Bisseru, B. (1968). Toxocara infections. *In* "Some Diseases of Animals Communicable to Man" (O. Graham-Jones, ed.), pp. 61-70. Pergamon, Oxford.

Blackmore, D. K. (1968). Some observations on ornithosis. Part II. The disease in cage and aviary birds. *In* "Some Diseases of Animals Communicable to Man" (O. Graham-Jones, ed.), pp. 151-155. Pergamon, Oxford.

Blackmore, D. K., and Francis, R. A. (1970). The apparent transmission of Staphylococci of human origin to laboratory animals. *J. Comp. Pathol.* **80,** 645-651.

Blamire, R. V. (1968). Hydatidosis. *In* "Some Diseases of Animals Communicable to Man" (O. Graham-Jones, ed.), pp. 93-96. Pergamon, Oxford.

Booth, B. H. (1952). Mouse ringworm. *Arch. Dermatol. Syph.* **66,** 65.

Boulger, L. R. (1966). Natural rabies in a laboratory monkey. *Lancet* **1,** 941-943.

Bowen, G. S., Calisher, C. H., Winkler, W. G., Kraus, A. L., Fowler, E. H., Garman, R. H., Fraser, D. W., and Hinman, A. R. (1975). Laboratory studies of a lymphocytic choriomeningitis virus outbreak in man and laboratory animals. *Am. J. Epidemiol.* **102,** 233-240.

Brandly, C. A. (1964). The occupational hazard of Newcastle disease to man. *Lab. Anim. Care* **14,** 433-440.

Breen, G. E., Lamb, S. G., and Otaki, A. T. (1958). Monkey bite encephalomyelitis: Report of a case—with recovery. *Br. Med. J.* **2,** 22-23.

Bryan, B. L., Espana, C. D., Emmons, R. W., Vijayan, N., and Hoeprich, P. D. (1975). Recovery from encephalomyelitis caused by *Herpesvirus simiae.* Report of a case. *Arch. Intern. Med.* **135,** 868-870.

Callis, J. J., and Cottrall, J. E. (1968). Methods for containment of animal pathogens at the Plum Island Animal Disease Laboratory. *In* "Methods in Virology" (K. Maramorosch and H. Koprowski, eds.), Vol. 4, pp. 456-480. Academic Press, New York.

Cappucci, D. T., O'Shea, J. L., and Smith, G. D. (1972). An epidemiologic account of tuberculosis transmitted from man to monkey. *Am. Rev. Respir. Dis.* **106,** 819-823.

Carpenter, K. P. (1968). Human and simian shigellosis. *In* "Some Diseases of Animals Communicable to Man" (O. Graham-Jones, ed.), pp. 35-46. Pergamon, Oxford.

Carpenter, K. P., and Cook, E. R. N. (1965). An attempt to find shigellae in wild primates. *J. Comp. Pathol. Ther.* **75,** 201-204.

Carter, P. B. (1978). How gnotobiotic isolators can assist biohazard control. *Lab. Anim.* **7,** 36-39.

Casey, H. W., Woodruff, J. M., and Butcher, W. I. (1967). Electron microscopy of a benign epidermal pox disease of rhesus monkeys. *Am. J. Pathol.* **51,** 431-446.

Center for Disease Control (1975). Nonhuman primate-associated hepatitis. *Morbid. Mortal. Week. Rep.* **24,** 115.

Center for Disease Control (1976). "Classification of Etiologic Agents on the Basis of Hazard." U.S. Dept. of Health, Education and Welfare, Public Health Service, Atlanta, Georgia.

Chappell, W. A. (1960). Animal infectivity of aerosols of monkey B virus. *Ann. N. Y. Acad. Sci.* **85,** 931-934.

Chappler, R., Hoke, A. W., and Borchardt, K. A. (1977). Primary inoculation with *Mycobacterium marinum. Arch. Dermatol.* **113,** 380.

Coggeshall, L. T. (1939). The transmission of lymphocytic choriomeningitis by mosquitoes. *Science* **89,** 515-516.

Collins, G. R. (1969). The design and use of filter cages for cats and rabbits. *Lab. Anim. Care* **19**, 659-661.

Cook, R. (1969). Simian shigellosis. *Lab. Anim. Handb.* **4**, 83-91.

Cross, J. H., Hsu-Kuo, M., and Lien, J. C. (1973). Accidental human infection with *Plasmodium cynomolgi bastianellii. Southeast Asian J. Trop. Med. Public Health* **4**, 481-483.

Darlow, H. M. (1967). Safety in the animal house. *Lab. Anim.* **1**, 35-42.

Davidson, W. L., and Hummeler, K. (1960). B virus infection in Man. *Ann. N. Y. Acad. Sci.* **85**, 970-979.

Deinhardt, F. (1970). Hepatitis in subhuman primates and hazards to man. *In* "Infections and Immunosuppression in Subhuman Primates" (H. Balner and W. I. B. Beveridge, eds.), pp. 55-63. Williams & Wilkins, Baltimore, Maryland.

Dolan, M. M., Kligman, A. M., Kobylinski, P. G., and Motsavage, M. A. (1958). Ringworm epizootics in laboratory mice and rats: Experimental and accidental transmission of infection. *J. Invest. Dermatol.* **30**, 23-25.

Downie, A. W., Taylor-Robinson, C. H., and Caunt, A. E. (1971). Tanapox: A new disease caused by a pox virus. *Br. Med. J.* **1**, 363-368.

Edward, A. G. (1969). Minimizing hazards to man in the laboratory animal facility. *J. Am. Vet. Med. Assoc.* **155**, 2158-2159.

Espana, C. (1971a). A pox disease of monkeys transmissible to man. *In* "Medical Primatology" (E. I. Goldsmith and J. Moor-Jankowski, eds.), pp. 694-708. Karger, Basel.

Espana, C. (1971b). Review of some outbreaks of viral disease in captive nonhuman primates. *Lab. Anim. Sci.* **21**, 1023-1031.

Eyestone, W. H. (1968). Recent developments of zoonoses associated with nonhuman primates. *J. Am. Vet. Med. Assoc.* **153**, 1767-1770.

Fear, F. A., Pinkerton, M. E., Cline, J. A., Kriewaldt, F., and Kalter, S. S. (1968). A Leptospirosis outbreak in a baboon (*Papio* sp.) colony. *Lab. Anim. Care* **18**, 22-28.

Fiennes, R. (1967). "Zoonoses of Primates." Weidenfeld & Nicolson, London.

Fierer, J., Baceley, P., and Braude, A. I. (1973). Herpes B virus encephalomyelitis presenting as ophthalmic zoster. *Ann. Intern. Med.* **79**, 225-228.

Foggie, A. (1968). Q fever and other rickettsioses of animals communicable to man. *In* "Some Diseases of Animals Communicable to Man" (O. Graham-Jones, ed.), pp. 247-254. Pergamon, Oxford.

Fraser, P. K. (1969). Diseases communicable from animals to man. *Ann. Occup. Hyg.* **12**, 109-119.

Friedmann, C. T. H., Dinnes, M. R., Bernstein, J. F., and Heidbreder, G. A. (1971). Chimpanzee-associated infectious hepatitis among personnel at an animal hospital. *J. Am. Vet. Med. Assoc.* **159**, 541-545.

Gay, W. I., and Blood, B. D. (1967). Zoonoses problems and their control in laboratory animal colonies. *In* "Husbandry of Laboratory Animals" (M. L. Conalty, ed.), pp. 327-341. Academic Press, New York.

Geiman, Q. M. (1964). Shigellosis, amebiasis and simian malaria. *Lab. Anim. Care* **14**, 441-454.

Gentles, J. C. (1968). Ringworm. *In* "Some Diseases of Animals Communicable to Man" (O. Graham-Jones, ed.), pp. 209-220. Pergamon, Oxford.

Gerone, P. (1975). Biohazards of experimentally infected primates. *In* "Proceedings of the National Cancer Institute Symposium on Biohazards and Zoonotic Problems of Primate Procurement, Quarantine and Research" (M. L. Simmons, ed.), Cancer Res. Saf. Monogr. Ser., Vol. 2, DHEW Publ. No. 76-890, pp. 53-65. USDHEW, Washington, D.C.

Giddens, W. E., Jr., Wolf, N. S., Carlos, A. D., Boyd, S. J., Penfold, T. W., and Dolowy,

W. C. (1972). Effectiveness of filter caging in the prevention of viral diseases of cats. *J. Am. Vet. Med. Assoc.* **161**, 591–594.

Gordon, W. S. (1968). Louping ill in animals and man. *In* "Some Diseases of Animals Communicable to Man" (O. Graham-Jones, ed.), pp. 119–124. Pergamon, Oxford.

Grace, J. T., and Mirand, E. A. (1963). Human susceptibility to a simian tumor virus. *Ann. N. Y. Acad. Sci.* **108**, 1123–1128.

Grace, J. T., and Mirand, E. A. (1965). Yaba virus infection in humans. *Exp. Med. Surg.* **23**, 213–216.

Griesemer, R. A., and Manning, J. S. (1973). Animal facilities. *In* "Biohazards in Biological Research" (A. Hellman, M. N. Oxman, and R. Pollack, eds.), pp. 316–326. Cold Spring Harbor Lab., Cold Spring Harbor, New York.

Hall, A. S., and McNulty, W. P., Jr. (1967). A contagious pox disease in monkeys. *J. Am. Vet. Med. Assoc.* **151**, 833–838.

Hanson, R. P., Sulkin, S. E., Buescher, E. L., Hammon, W. McD., McKinney, R. W., and Work, T. H. (1967). Arbovirus infections of laboratory workers. *Science* **158**, 1283–1286.

Hartley, E. G. (1969). Tuberculin testing of monkeys. *Lab. Anim. Handb.* **4**, 27–31.

Havens, W. P., Jr. (1967). Changing trends in infectious hepatitis. *Med. Clin. North Am.* **51**, 653–660.

Hennessen, W. (1968). A hemorrhagic disease transmitted from monkeys to man. *Natl. Cancer Inst., Monogr.* **29**, 161–171.

Hennessen, W. (1969). Epidemiology of Marburg virus disease. *Lab. Anim. Handb.* **4**, 137–142.

Higa, H. H., and Ching, G. Q. L. (1972). Virulent *Escherichia coli* from a mouse. *Hawaii Med. J.* **31**, 389–390.

Hillis, W. D. (1961). An outbreak of infectious hepatitis among chimpanzee handlers at a United States Air Force base. *Am. J. Hyg.* **73**, 316–323.

Hillis, W. D. (1963). Viral hepatitis associated with sub-human primates. *Transfusion* **3**, 445–454.

Hillis, W. D. (1967). Etiology of viral hepatitis. *Johns Hopkins Med. J.* **120**, 176–185.

Hime, J. M. (1974). Patterns of some diseases transmissible from animals to man—a review. *Community Health* **6**, 161–167.

Hinman, A. R., Fraser, D. W., Douglas, R. G., Bowen, G. S., Kraus, A. L., Winkler, W. G., and Rhodes, W. W. (1975). Outbreak of lymphocytic choriomeningitis virus infections in medical center personnel. *Am. J. Epidemiol.* **101**, 103–110.

Horsfall, F. K., Jr., and Bauer, J. H. (1940). Individual isolation of infected animals in a single room. *J. Bact.* **40**, 569–580.

Hotchin, J. (1971). The contamination of laboratory animals with lymphocytic choriomeningitis virus. *Am. J. Pathol.* **64**, 747–769.

Hotchin, J., Sikora, E., Kinch, W., Hinman, A., and Woodall, J. (1974). Lymphocytic choriomeningitis in a hamster colony causes infection of hospital personnel. *Science* **185**, 1173–1174.

Hull, R. N. (1973a). The simian herpesviruses. *In* "The Herpesviruses" (A. S. Kaplan, ed.), pp. 389–426. Academic Press, New York.

Hull, R. N. (1973b). Biohazards associated with simian viruses. *In* "Biohazards in Biological Research" (A. Hellman, M. N. Oxman, and R. Pollack, eds.), pp. 3–38. Cold Spring Harbor Lab., Cold Spring Harbor, New York.

Hummer, R. L. (1969). Preventive medicine concepts for animal handlers and investigators using animals. *J. Am. Vet. Med. Assoc.* **155**, 2162–2170.

Hyslop, N. St. G. (1973). Transmission of the virus of foot and mouth between animals and man. *Bull. W. H. O.* **49**, 577–585.

Jennings, L. F., and Rumpf, R. M. (1965). Control of epizootic diarrhea in infant mice. *Lab. Anim. Care* **15**, 386–391.

Kalter, S. S., and Heberling, R. L. (1975). Biohazards and simian viruses. *Bibl. Haematol.* **40**, 759–769.

Kaplan, C. (1969). Rabies in non-human primates. *Lab. Anim. Handb.* **4**, 117–118.

Kaufmann, A. F., Moultrop, J. I., and Moore, R. M., Jr. (1975). A perspective of simian tuberculosis in the United States—1972. *J. Med. Primatol.* **4**, 278–286.

Kawamata, J., and Yamanouchi, T. (1975). Protection against biohazards in animal laboratories. *Bibl. Haematol.* **40**, 779–782.

Keeble, S. A. (1960). B virus infection in monkeys. *Ann. N. Y. Acad. Sci.* **85**, 960–969.

Keeble, S. A. (1968). B virus infection in man and monkey. *In* "Some Diseases of Animals Communicable to Man" (O. Graham-Jones ed.), pp. 183–188. Pergamon, Oxford.

Kirchheimer, W. F., Jemski, J. V., and Phillips, G. B. (1961). Cross-infection among experimental animals by organisms infectious for man. *Proc. Anim. Care Panel* **11**, 83–92.

Kirschstein, R., van Hossier, G. L., Jr., and Li, C. P. (1961). Virus-B infection of the central nervous system of monkeys used for the poliomyelitis vaccine safety test. *Am. J. Pathol.* **33**, 119–125.

Knipping, P. A. (1972). Animal diseases transmissible to man. *J. Sch. Health* **42**, 385–388.

Kraft, L. M. (1958). Observations on control of natural history of epidemic diarrhea of infant mice (EDIM). *Yale J. Biol. Med.* **31**, 121–137.

Kraft, L. M., Pardy, R. F., Pardy, D. A., and Zwickel, H. (1964). Practical control of diarrheal disease in a commerical mouse colony. *Lab. Anim. Care* **14**, 16–19.

Krick, J. A., and Remington, J. S. (1978). Toxoplasmosis in the adult—an overview. *N. Engl. J. Med.* **298**, 550–553.

Kruse, R. H., and Wedum, A. G. (1970). Cross infection with eighteen pathogens among caged laboratory animals. *Lab. Anim. Care* **20**, 541–560.

Krushak, D. H. (1970). Application of preventive health measures to curtail chimpanzee-associated infectious hepatitis in handlers. *Lab. Anim. Care* **20**, 52–56.

Lennette, E. H., and Koprowski, H. (1943). Human infection with Venezuelan equine encephalomyelitis virus. A report of eight cases of infection acquired in the laboratory. *J. Am. Med. Assoc.* **123**, 1088–1095.

Lewis, A. M., Rowe, W. P., Turner, H. C., and Huebner, R. J. (1965). Lymphocytic choriomeningitis virus in hamster tumor: Spread to hamsters and humans. *Science* **150**, 363–364.

Loosli, R. (1967). Zoonoses in common laboratory animals. *In* "Husbandry of Laboratory Animals" (M. L. Conalty, ed.), pp. 307–325. Academic Press, New York.

Low, C. G., and Fairley, N. H. (1931). Observations on laboratory and hospital infections with yellow fever in England. *Br. Med. J.* **1**, 125–128.

Luby, J. P. (1971). Another zoonosis from monkeys: Tanapox. *Ann. Intern. Med.* **75**, 800–801.

McCarthy, K., and Tosolini, F. A. (1975a). A review of primate herpes viruses. *Proc. R. Soc. Med.* **68**, 145–150.

McCarthy, K., and Tosolini, F. A. (1975b). Hazards from simian herpes viruses: Reactivation of skin lesions with virus shedding. *Lancet* **1**, 649–650.

McClure, H. M., Strozier, L. M., and Keeling, M. E. (1972). Enteropathogenic *Escherichia coli* infection in anthropoid apes. *J. Am. Vet. Med. Assoc.* **161**, 687–689.

McGarrity, G. J., and Dion, A. S. (1978). Detection of airborne polyoma virus. *J. Hyg.* **81**, 9–13.

McGarrity, G. J., Coriell, L. L., Schaedler, R. W., Mandle, R. J., and Greene, A. E. (1969).

Medical applications of dust-free rooms. III. Use in an animal care laboratory. *Appl. Microbiol.* **18**, 142–146.

McNulty, W. P., Jr., Lobitz, W. C., Jr., and Hu, F. (1968). A pox disease in monkeys transmitted to man. Clinical and histological features. *Arch. Dermatol.* **97**, 286–293.

Mair, N. S. (1968). Some Pasteurella infections in man. *In* "Some Diseases of Animals Communicable to Man" (O. Graham-Jones, ed.), pp. 47–54. Pergamon, Oxford.

Martini, G. A., and Siegert, R. (1971). "Marburg Virus Disease." Springer-Verlag, Berlin and New York.

Maurer, F. D. (1964). Lymphocytic choriomeningitis. *Lab. Anim. Care* **14**, 415–419.

Milzer, A. (1942). Studies on the transmission of lymphocytic choriomeningitis virus by arthropods. *J. Infect. Dis.* **70**, 152–172.

Moreland, A. F. (1970). Tuberculosis in New World primates. *Lab. Anim. Care* **20**, 262–264.

Moreland, A. F., Barkley, W. E., Bottiglieri, N. G., Lang, C. M., Manning, P. J., Newberne, P. M., Pakes, S. P., Ringler, D. H., and Weisbroth, S. H. (1978). "Guide for the Care and Use of Laboratory Animals," DHEW Publ. No. (NIH) 78-23. USDHEW, Washington, D.C.

Morgan, W. J. B. (1968). Brucellosis in animals. *In* "Some Diseases of Animals Communicable to Man" (O. Graham-Jones, ed.), pp. 263–273. Pergamon, Oxford.

Mortelmans, J., and Kageruka, P. (1969). *Pasteurella pseudotuberculosis* infections in monkeys and man. *Lab. Anim. Handb.* **4**, 95–98.

Mosley, J. W., and Galambos, J. T. (1967). Chimpanzee-associated hepatitis; An outbreak in Oklahoma. *J. Am. Med. Assoc.* **199**, 695–697.

Most, H. (1973). *Plasmodium cynomolgi* malaria: Accidental human infection. *Am. J. Trop. Med. Hyg.* **22**, 157–158.

Muchmore, E. (1975). Health program for people in close contact with laboratory primates. *In* "Proceedings of the National Cancer Institute Symposium on Biohazards and Zoonotic Problems of Primate Procurement, Quarantine and Research" (M. L. Simmons, ed.), Cancer Res. Saf. Monogr. Ser., Vol. 2, DHEW Publ. No. 76-890, pp. 81–96. USDHEW, Washington, D.C.

Mulder, J. B. (1971). Shigellosis in nonhuman primates: A review. *Lab. Anim. Sci.* **21**, 734–738.

Neal, R. A. (1969). Intestinal protozoa of simians as a handling hazard. *Lab. Anim. Handb.* **4**, 77–81.

Owen, C. R., and Buker, E. O. (1956). Factors involved in the transmission of *Pasteurella tularensis* from inoculated animals to healthy cagemates. *J. Infect. Dis.* **99**, 227–233.

Patterson, R (1964). The problem of allergy to laboratory animals. *Lab. Anim. Care* **14**, 466–469.

Perkins, F. T. (1968). Precautions against B virus in man. *In* "Some Diseases of Animals Communicable to Man" (O. Graham-Jones, ed.), pp. 189–194. Pergamon, Oxford.

Perkins, F. T., and O'Donoghue, P. N. (1969). Proposed procedure for the housing and handling of simians to decrease the transmission of disease to man. *Lab. Anim. Handb.* **4**, 255–266.

Phillips, G. B., and Jemski, J. V. (1963). Biological safety in the animal laboratory. *Lab. Anim. Care* **13**, 13–20.

Phillips, G. B., Reitman, M., Mullican, C. L., and Gardner, G. D., Jr. (1957). Applications of germicidal ultraviolet in infectious disease laboratories. III. The use of ultraviolet barriers on animal cage racks. *Proc. Anim. Care Panel* **7**, 235–244.

Pike, R. M. (1976). Laboratory-associated infections: Summary and analysis of 3921 cases. *Health Lab. Sci.* **13**, 105–114.

Pinkerton, M. E. (1968). Shigellosis in the baboon (Papio sp.). *Lab. Anim. Care* **18**, 11–21.

Povar, M. L. (1965). Ringworm (*Trichophyton mentagrophytes*) infection in a colony of albino Norway rats. *Lab. Anim. Care* **15**, 264–265.

Prier, J. E., Sauer, R. M., and Fegley, H. C. (1964). Zoonoses associated with captive monkeys. *Lab. Anim. Care* **14**, 48–57.

Quist, K. D. (1972). Viral zoonoses of importance to laboratory animal personnel. *J. Am. Vet. Med. Assoc.* **161**, 1572–1577.

Refai, M., and Ali, A. H. (1970). Laboratory acquired infection with *Keratinomyces ajelloi*. *Mykosen* **13**, 317–318.

Ronald, N. C., and Wagner, J. E. (1973). Pediculosis of spider monkeys: A case report with zoonotic implications. *Lab. Anim. Sci.* **23**, 872–875.

Rosenhaft, M. E. (1974). An evaluation of requirements for pathological incineration facilities. *Lab. Anim. Sci.* **24**, 905–909.

Rowsell, H. C. (1963). Mycotic infections of animals transmissible to man. *Am. J. Med. Sci.* **245**, 333–344.

Ruddy, S., Jr., Mosley, J. W., and Held, J. R. (1967). Chimpanzee-associated viral hepatitis in 1963. *Am. J. Epidemiol.* **86**, 634–640.

Sabin, A. B., and Wright, A. M. (1934). Acute ascending myelitis following a monkey bite, with the isolation of a virus capable of reproducing the disease. *J. Exp. Med.* **59**, 115–136.

Schneider, H. A., and Collins, J. R. (1966). Successful prevention of infantile diarrhea of mice during an epizootic by means fo a new filter cage unopened from birth to weaning. *Lab. Anim. Care* **16**, 60–71.

Schneider, N. J., Prather, E. C., Lewis, A. L., Scatterday, J. E., and Hardy, A. V. (1960). Enteric bacteriological findings on monkeys newly received at Okatie Farms during 1955. *Ann. N.Y. Acad. Sci.* **85**, 935–941.

Sellers, R. F., Herniman, K. A. J., and Mann, J. A. (1971). Transfer of foot-and-mouth disease virus in the nose of man from infected to non-infected animals. *Vet. Rec.* **89**, 447–449.

Short, D. J. (1967). Handling, sexing and palpating laboratory animals. *In* "Husbandry of Laboratory Animals" (M. L. Conalty ed.), pp. 3–15. Academic Press, New York.

Siegert, R. (1972). Marburg virus. *Virol. Monogr.* **11**, 98–153.

Simmons, M. L. (1969). Public health aspects of quarantine programs in animals used in biomedical research. *J. Am. Vet. Med. Assoc.* **155**, 2160–2161.

Simpson, D. I. H. (1969). Marburg virus disease: Experimental infection of monkeys. *Lab. Anim. Handb.* **4**, 149–154.

Skinner, H. H., and Knight, E. H. (1971). Monitoring mouse stocks for lymphocytic choriomeningitis virus—a human pathogen. *Lab. Anim.* **5**, 73–87.

Slenczka, W., Wolff, G., and Siegert, R. (1971). A critical study of monkey sera for the presence of antibody against the Marburg virus. *Am. J. Epidemiol.* **93**, 496–505.

Smith, A. W., Prato, C., and Skilling, D. E. (1978). Calciviruses infecting monkeys and possibly man. *Am. J. Vet. Res.* **39**, 287–289.

Smith, H. W. (1968). Salmonella infection in animals. *In* "Some Diseases of Animals Communicable to Man" (O. Graham-Jones ed.), pp. 3–14. Pergamon, Oxford.

Stark, A. (1975). Policy and procedural guidelines for the health and safety in virus laboratories *J., Am. Ind. Hyg. Assoc.* **36**, 234–240.

Steele, J. H. (1973). The zoonoses: an epidemiologist's viewpoint. *Prog. Clin. Pathol.* **5**, 239–286.

Steere, A. C., and Mallison, G. F. (1975). Handwashing practices for the prevention of nosocomial infections. *Ann. Intern. Med.* **83**, 683–690.

Stockdale, P. M., Mackenzie, D. W. R., and Austwick, P. K. C. (1965). *Arthroderma simii* sp. Nov., the perfect state of *Trichophyton simii* (Pinoy) Comb. NOV. *Sabouraudia* **4**, 112–123.

Stoenner, H. G., and Maclean, D. (1958). Leptospirosis (*ballum*) contracted from Swiss albino mice. *AMA Arch. Intern. Med.* **101**, 606–610.

Stones, P. B. (1969). Incidence of tuberculosis in *Macaca mulatta* and *Cercopithecus aethiops* monkeys with special reference to the tuberculin test. *Lab. Anim. Handb.* **4**, 11–17.

Stunkard, J. A., Szatalowicz, F. T., and Sudduth, H. C. (1971). A review and evaluation of tuberculin testing procedures used for *Macaca* species. *Am. J. Vet. Res.* **32**, 1873–1878.

Sulkin, S. E., Long, E. R., Pike, R. M., Sigel, M. M., Smith, C. E., and Wedum, A. G. (1963). Laboratory infections and accidents. *In* "Diagnostic Procedures and Reagents" (A. H. Harris and M. B. Coleman, eds.), 4th ed., pp. 89–104. Am. Public Health Assoc., New York.

Tauraso, N. M. (1973). Review of recent epizootics in nonhuman primate colonies and their relation to man. *Lab. Anim. Sci.* **23**, 201–210.

Tauraso, N. M., Norris, G. F., Sorg, T. J., Cook, R. O., Myers, M. L., and Trimmer, R. (1969). Negative-pressure isolator for work with hazardous infectious agents in monkeys. *Appl. Microbiol.* **18**, 294–297.

Taylor, R. L., Cadigan, F. C., Jr., and Chaicumpa, V. (1973). Infections among Thai gibbons and humans caused by atypical *Microsporum canis*. *Lab Anim. Sci.* **23**, 226–231.

Theiler, M. (1930). Studies on action of yellow fever in mice. *Ann. Trop. Med. Parasitol.* **24**, 249–272.

Tobin, J. O'H. (1968). Viruses transmissible from laboratory animals to man. *Lab. Anim.* **2**, 19–28.

Tosolini, F. A., and McCarthy, K. (1975). Herpes virus latency in the nervous system. *Proc. R. Soc. Med.* **68**, 150.

Turner, L. H. (1968). Leptospirosis. *In* "Some Diseases of Animals Communicable to Man" (O. Graham-Jones, ed.), pp. 231–245. Pergamon, Oxford.

Vickers, J. H. (1973). Infectious diseases of primates related to capture and transportation. *Am. J. Phys. Anthropol.* **38**, 511–513.

Vizoso, A. D. (1975). Recovery of herpes simiae (B virus) from both primary and latent infections in rhesus monkeys. *Br. J. Exp. Pathol.* **56**, 485–488.

Warwick, W. J. (1964). Cat scratch disease. *Lab. Anim. Care* **14**, 420–432.

Webb, H. E. (1969). Kyasanur Forest disease virus infection in monkeys. *Lab. Anim. Handb.* **4**, 131–134.

Wedum, A. G. (1961). Control of airborne infection. *Bacteriol. Rev.* **25**, 210–216.

Wedum, A. G. (1964). Laboratory safety in research with infectious aerosols. *Public Health Rep.* **79**, 619–633.

Wedum, A. G., Hanel, E., Jr., and Phillips, G. B. (1956). Ultraviolet sterilization in microbiological laboratories. *Public Health Rep.* **71**, 331–336.

Wedum, A. G., Barkley, W. E., and Hellman, A. (1972). Handling of infectious agents. *J. Am. Vet. Med. Assoc.* **161**, 1557–1567.

White, F. H., Stoliker, H. E., and Galton, M. M. (1961). Detection of leptospires in naturally infected dogs using fluorescein-labeled antibody. *Am. J. Vet. Res.* **22**, 650–654.

Work, T. H., and Trapido, H. (1957). Kyasanur Forest disease: A new virus disease in India. *J. Am. Vet. Med. Assoc.* **161**, 1572–1577.

Yunker, C. E. (1964). Infections of laboratory animals potentially dangerous to man: Ectoparasites and other arthropods, with emphasis on mites. *Lab. Anim. Care* **14**, 455–465.

# PART THREE

# MEDICAL AND PSYCHOLOGICAL FACTORS

# Chapter Eight

# Selected Medical Problems often Associated with Laboratory Personnel

## LINDA HAEGELE

## I. INTRODUCTION

Although prevention of clinical disease may constitute the ultimate objective of medical practice, early detection combined with proper treatment represents a more practical contingency. In the general population even the early detection of illness is not always possible because of the variety of

LABORATORY SAFETY: THEORY AND PRACTICE
Copyright © 1980 by Academic Press, Inc.

diseases that can result from exposure to physical, chemical, and infectious agents in susceptible individuals and their clinical course. However, there are specific problems that have an increased probability of occurring because of the nature of the materials to which the individual is exposed. This chapter deals with the various agents that can be hazardous to laboratory personnel and the clinical programs that can best deal with illness produced by these agents. In this chapter there is no attempt to foresee every possible contingency, since the rapidly increasing body of knowledge in the field of industrial medicine would rapidly make such a review obsolete. Instead, we try to discuss the various types of illness that could occur and, in those cases where cause and effect have been established, list that information in tables with appropriate clinical tests. Finally, since carcinogenic and genetic damage to individuals has been of increasing concern recently, this area is covered in a separate chapter.

The efficacy of screening techniques in monitoring individuals at risk, as well as the early detection of patients evidencing clinical symptomatology, is influenced by identification of this population, personnel compliance, and the sensitivity of available laboratory tests. Frequent utilization of quantitative chemical assays may demonstrate evidence of toxic exposure prior to evolution of disease, when appropriate therapeutic measures may abort clinical toxicity. Table I lists available clinical tests for use with this population. Similarly, culturing techniques may detect pathogenic organisms in the environment, before laboratory personnel develop infections. Immunologic monitoring of individuals at risk is a useful protective technique, as well as a means of detecting subclinical effects. Appreciation of the multifactorial mechanisms that influence the expression of many chronic diseases will facilitate treatment and effectively reduce potentially dangerous situations.

**TABLE I**

**Chemical Agents Associated with Clinical Sequellae**

| Chemical | Principal use or source of emission | Medical sequellae | Medical surveillance |
|---|---|---|---|
| Aliphatic hydrocarbons | | | |
| Acetyline | Metallurgy, glass, chemical industry. Inhalation of gas. | Asphyxiant. Sensitization of myocardium. | Adequate ventilation. |
| Alicyclic hydrocarbons | Anesthetics, organic synthesis. Inhalation of gas. | Central nervous system (CNS) depressant. | Skin and respiratory tract irritation. Renal and liver function. |

TABLE I (*cont.*)

| Chemical | Principal use or source of emission | Medical sequellae | Medical surveillance |
|---|---|---|---|
| Gasoline | Fuel, solvent. Inhalation, percutaneous absorption. | Mucosal irritation. CNS depressant. Chemical pneumonitis. | Periodic examination. |
| Kerosene | Fuel, solvent. Inhalation of vapor. | Mucosal irritation. CNS depressant. | Neurologic examination, CBC, urinary phenol. |
| Naphtha | Organic solvents. Inhalation of gas. | Asphyxiant. | Adequate ventilation. |
| Natural gas | Heating, fuel. Inhalation of gas. | Asphyxiant. | Adequate ventilation. |
| Turpentine | Volatile base. Inhalation of vapor, percutaneous absorption. | Mucosal irritation. CNS depressant. Chronic nephritis. | Hepatic, renal, respiratory function. |
| Alcohols | | | |
| Allyl alcohol | Resins, plastics. Inhalation of vapor, percutaneous absorption. | Mucosal irritation. Pulmonary edema. Hepatic and renal damage. | Hepatic, renal, respiratory function. |
| Amyl alcohol | Chemical, textile, petroleum industry. Inhalation of vapor, percutaneous absorption. | Mucosal irritation. Narcotic. | Skin, respiratory irritation. |
| Butyl alcohol | Solvent, chemical industry. Inhalation of vapor, percutaneous absorption. | Skin irritation. Meniere's syndrome. | Skin, ocular, respiratory irritation. |
| Ethyl alcohol | Solvent, antifreeze, fuel. Inhalation of vapor, percutaneous absorption. | Tolerance. Intoxication. | Hepatic function. |

(*cont.*)

**TABLE I** (*cont.*)

| Chemical | Principal use or source of emission | Medical sequellae | Medical surveillance |
|---|---|---|---|
| Methyl alcohol | Chemical synthesis. Inhalation of vapor, percutaneous absorption. | Dermatitis. Optic nerve. | Methyl alcohol blood level. Skin, ocular, hepatic, and renal function. |
| Propyl alcohol | Pharmaceuticals, solvent. Inhalation of vapor, percutaneous absorption. | CNS depressant. | Isopropyl alcohol and acetone in blood, urine, tissues. |
| Glycols Ethylene glycol | Solvent, chemical intermediate. Antifreeze. Inhalation of vapor. | CNS depressant. Hematopoietic dysfunction. Pulmonary cardiac, renal damage. | Urinalysis for oxalic acid. |
| Ethylene glycol ethers and derivatives | Solvents. Inhalation of vapor, percutaneous absorption. | Narcosis. Pulmonary edema. Severe renal and hepatic damage. | Careful physical exam. No specific tests. |
| Ethers and expoxy compounds Bis (chloromethyl) ether | Solvent, alkylating agent. Inhalation of vapor. | Human carcinogen with 10–15 yr latency. | Periodic physical examination. CXR, sputum cytology. |
| Ethers and epoxy compounds Chloromethyl methyl ether | Methylating agent. Inhalation of vapor. | Human carcinogen. | Periodic physical examination. CXR, sputum cytology. |
| Dichloroethyl ether | Solvent; textile and paint manufacture. Inhalation of vapor, percutaneous absorption. | Conjunctival and mucosal irritation. Bronchitis. | Skin, respiratory irritation. |

**TABLE I** (*cont.*)

| Chemical | Principal use or source of emission | Medical sequellae | Medical surveillance |
|---|---|---|---|
| Dioxane | Solvent. Inhalation of vapor, percutaneous absorption. | Mucosal irritation, CNS depression, renal and hepatic damage. | Neurologic evaluation, renal and hepatic function studies. |
| Epichlorohydrin | Resins, chemical manufacure. Inhalation of vapor, percutaneous absorption. | Mucosal irritation, GI and pulmonary damage. | Periodic examination. |
| Ethylene oxide | Intermediate inorganic synthesis. Inhalation of gas. | Dermatologic and mucosal irritation, pulmonary edema, CNS depression. | Periodic examination. |
| Ethyl ether | Solvent, inhalation anesthetic, refrigerant. Inhalation of vapor. | CNS depression. High susceptibility to alcohol. | Periodic examination. Test expired breath for unmetabolized ether; blood levels. |
| Esters |  |  |  |
| Acetates | Solvents, chemical manufacture. Inhalation and ingestion. | Mucosal irritation. CNS depression. | Periodic examination. Hepatic and renal function. |
| Ethyl silicate | Metal casting, concrete. Inhalation of vapor. | Mucosal irritation. | Periodic examination. Hepatic and renal function. |
| Formatl | Solvent, chemical synthesis. Inhalation, ingestion, skin absorption. | Mucosal irritation. Narcosis. | Periodic examination. |
| Carboxylic acids and anhydrides |  |  |  |
| Acetic acid | Chemical manufacture. Inhalation of vapor. | Mucosal irritation. Bronchial pneumonia, pulmonary edema. | Periodic examination. Skin, ocular, pulmonary function. |

(*cont.*)

**TABLE I** (*cont.*)

| Chemical | Principal use or source of emission | Medical sequellae | Medical surveillance |
|---|---|---|---|
| Acetic anhydride | Acetylating agent, solvent. Inhalation of vapor. | Mucosal irritation, necrosis | Periodic examination. Skin, ocular, pulmonary function. |
| Formic acid | Reducing agent, decalcifier. Inhalation of vapor, percutaneous absorption. | Mucosal irritation. Poisoning. Shock. | Periodic examination. |
| Oxalic acid | Analytic reagent, chemical manufacture. Inhalation of mist. | Corrosive. Chronic respiratory inflammation. | Periodic examination. Urinary oxalate crystals. Blood calcium and oxalate levels. |
| Phthalic anhydride | Chemical manufacture. Inhalation of vapor. | Mucosal irritation. Chronic pulmonary disease. | History dermatologic or pulmonary allergy. Periodic examination. |
| Aldehydes and ketones Acetaldehyde | Chemical manufacture. Inhalation of vapor. | Mucosal irritation. Pulmonary edema. CNS dysfunction. | Periodic examination. Skin, ocular, pulmonary function. |
| Formaldehyde | Manufacturing, disinfectant. Inhalation of gas. | Mucosal irritation. Rare systemic intoxication. | History of allergy. Periodic examination. |
| Furfural | Solvent, chemical manufacture. Inhalation of vapor, percutaneous absorption. | Mucosal irritation, dermatitis. | History of allergy. |
| Ketones | Solvent, chemical manufacture. Inhalation of vapor, percutaneous absorption. | Dermatitis. Narcosis, peripheral neuropathy. | Periodic examination. Acetone in blood, urine, expired air. |

**TABLE I** (*cont.*)

| Chemical | Principal use or source of emission | Medical sequellae | Medical surveillance |
|---|---|---|---|
| Aliphatic haloginated hydrocarbons | | | |
| Carbon tetra-chloride | Solvent. Inhalation of vapor. | Dermatitis. CNS depression. Hepatic and renal damage. Increased hazard with ingested alcohol. | Periodic examination. Alcohol history. Hepatic and renal function studies. Expired air and blood levels. |
| Chloroform | Solvent, pharma-ceuticals. Inhalation of vapor. | Anesthetic. Hepatic damage. Cardiac arrest. | Periodic examination. Renal and hepatic function. Expired air and blood levels. |
| Chloroprene | Production of artificial rubber. Inhalation of vapor, percutaneous absorption. | Mucosal irritation. Anesthesia, respiratory paralysis. Possible carcinogen. | Periodic examination. Renal and hepatic function. |
| Dichloroethylene | Solvent, chemical manufacture. Inhalation of vapor, percutaneous absorption. | Dermatitis. Renal and hepatic damage. | Periodic examination Renal and hepatic function. Expired air level. |
| Ethyl chloride | Ethylating agent. Inhalation of gas, percutaneous absorption. | Skin irritant. Respiratory irritant. Cardiac arrest. | Periodic examination with CNS, lung, and skin evaluation. |
| Aliphatic haloginated hydrocarbons | | | |
| Fluorocarbons | Refrigerants, polymer intermediates. Inhalation of vapor. | Mucosal irritation. CNS depression. Asphyxia. | Periodic examination. |
| Methyl and ethyl bromide | Fumigant, chemical intermediate, ethylating agent. Inhalation and percutaneous absorption. | Mucosal irritation. Pulmonary, renal, and CNS damage. | Periodic examination with CNS, lung, and skin evaluation. |

(*cont.*)

**TABLE I** (*cont.*)

| Chemical | Principal use or source of emission | Medical sequellae | Medical surveillance |
|---|---|---|---|
| Methyl chloride | Methylating and chlorinating agent. Percutaneous absorption. | Ocular irritation. CNS damage. | Periodic examination with CNS, renal, and hepatic evaluation. |
| Methylene chloride | Low temperature extractant. Inhalation of vapor, percutaneous absorption. | Dermatitis, mucosal irritation. Narcosis. Elevated carboxyhemoglobin. | Hx smoking, anemia, cardiovascular disease. Hepatic, pulmonary, CNS evaluation. |
| Tetrachloroethane | Dry cleaner, fumigant, solvent. Inhalation of vapor, percutaneous absorption. | Dermatitis. Narcosis. Hepatic, hematopoietic, renal, pulmonary damage. | Comprehensive evaluation Alcohol history. |
| Tetrachloroethylene | Solvent, chemical intermediate. Inhalation of vapor, percutaneous absorption. | Dermatitis, mucosal irritation, CNS depression, hepatic damage, death. | Skin, hepatic, and renal function. Alcohol history |
| Aliphatic halogenated hydrocarbons Vinyl chloride | Vinyl monomer, solvent, chemical intermediate. Inhalation. | Skin irritant. CNS depression. Human carcinogen. | Hepatic function. Hx alcohol, hepatic toxins. Long-term follow-up. |
| Aliphatic amines Ethanolamines | Gas absorbent, emulsifier, chemical intermediate. Inhalation of vapor, percutaneous absorption. | Skin irritant. | Evaluate skin, eyes. |
| Ethylenediamine | Solvent, emulsifier, chemical intermediate. Inhalation, percutaneous absorption. | Skin irritant. | Evaluate skin. |

**TABLE I** (*cont.*)

| Chemical | Principal use or source of emission | Medical sequellae | Medical surveillance |
|---|---|---|---|
| Cyanides and nitriles | | | |
| Acetonitrile | Extractant, solvent, chemical intermediate. Inhalation, percutaneous absorption. | Mucosal irritation. Low acute toxicity. | Examination of skin, CNS, renal, and hepatic function. |
| Acrylonitrile | Plastic industry, pesticide. Inhalation percutaneous absorption. | Mucosal irritation. CNS depression. | Examination of skin, CNS, hepatic, and renal function. Hx epilepsy. |
| Calcium cyanamide | Agricultural product, desulfurizer. Inhalation of dust. | Mucosal irritation. Vasomotor reaction, circulatory collapse. | Examination of skin, respiratory tract. Alcohol Hx. |
| Hydrogen cyanide | Fumigant, chemical synthesis. Inhalation of vapor, percutaneous absorption. | Mucosal irritation. Asphyxiant. | Periodic examination. CNS, renal, and hepatic function. Blood CN levels. |
| Isocyanates | Polyurethane manufacture. Inhalation of vapor. | Mucosal irritation. Asphyxiant. | Pulmonary function, CXR. Hx allergy. |
| Aromatic hydrocarbons | | | |
| Benzene | Fuel, solvent, chemical intermediate. Inhalation of vapor. | Mucosal irritation, dermatitis. CNS depression, aplastic anemia, leukemia | Hematologic evaluation. CNS, renal, and hepatic function. Biologic monitoring. |
| Naphthalene | Chemical intermediate. Inhalation of vapor. | Dermatitis, cataracts. Intravascular hemolysis. | Skin, hematologic, hepatic, and renal evaluation. G-6-PD. |
| Toluene | Fuel, benzene manufacture. Inhalation of vapor, percutaneous absorption. | Mucosal irritation, dermatitis. CNS depression. | Skin, CNS, hepatic, and renal evaluation. Urinary hippuric acid level. |

(*cont.*)

**TABLE I** (*cont.*)

| Chemical | Principal use or source of emission | Medical sequellae | Medical surveillance |
|---|---|---|---|
| Xylene | Solvent, chemical manufacture. Inhalation of vapor. | Mucosal irritation, dermatitis. | Skin, CNS, hepatic and renal evaluation |
| Phenols | | | |
| Cresol | Chemical intermediate. Inhalation or percutaneous absorption. | Corrosive. Systemic poisoning. Renal and hepatic damage. | Skin, pulmonary, renal, and hepatic function. |
| Cresote | Wood preservative. Skin absorption. | Dermatitis, skin cancer. Systemic poisoning. | Skin, opthalmologic, CNS evaluation. |
| Hydroquinone | Photographic developer, antioxidant. Inhalation, percutaneous absorption | Mild irritant. Oral ingestion: methemoglobinemia, respiratory failure, circulatory collapse. | Opthalmologic evaluation. |
| Phenols | | | |
| Phenol | Chemical manufacture. Inhalation, percutaneous absorption. | Corrosive. Systemic poisoning. Renal and hepatic damage. | Skin, opthalmologic, renal, and hepatic function. |
| Guinone | Chemical intermediate. Inhalation of vapor. | Ulceration, cataracts. | Skin, opthalmologic evaluation. |
| Aromatic halogenated hydrocarbons | | | |
| Benzyl chloride | Industrial manufacturing. Inhalation of vapor. | Mucosal irritant, dermatitis. Pulmonary edema. | Periodic examination. |
| Chlorodiphenyls | Electric cables. Inhalation of vapor. | Skin irritant. Hepatic damage. | Skin, pulmonary, and hepatic function. |
| Chlorinated benzenes | Solvent, chemical intermediate. Inhalation of vapor, percutaneous description. | Mucosal irritation. CNS depression. | Skin, hepatic, renal function. |

**TABLE I** (*cont.*)

| Chemical | Principal use or source of emission | Medical sequellae | Medical surveillance |
|---|---|---|---|
| Chlorinated naphthalenes | Electrical industry. Inhalation of fumes, percutaneous absorption. | Chloracne. Rare systemic poisoning. | Skin, hepatic function. |
| Aromatic amines<br>Aminodphenyl | Research only. Inhalation, percutaneous absorption. | Bladder carcinogen. | Hx exposure. Urine sediment and cytology. |
| Aniline | Dye. Inhalation, percutaneous absorption. | Corneal damage. Methemoglobinemia. | Hematologic evaluation. Methemoglobin levels. |
| Benzidine | Dye. Inhalation, percutaneous absorption. | Contact dermatitis. Urinary tract carcinogen. | Hx exposure. Urine sediment and cytology. |
| Dichlorobenzidine | Pigment manufacture. Inhalation, percutaneous absorption. | Skin allergy. Carcinogen. | Hx exposure. Periodic evaluation. Sputum and urine cytology. |
| X-Naphthylamine | Dye, chemical intermediate. Inhalation, percutaneous absorption. | Urinary tract carcinogen. | Hx exposure. Periodic evaluation. Urine sediment and cytology. |
| B-Naphthylamine | Research only. Inhalation, percutaneous absorption. | Urinary tract carcinogen. | Hx exposure. Urine sediment and cytology. |
| Nitro compounds<br>Dinitrobenzene | Dye. Inhalation, percutaneous absorption. | Methemoglobinemia. CNS depression. Anemia. | Hematologic, renal, and hepatic evaluation. Methemoglobin levels. Dinitrobenzene levels. |

(*cont.*)

**TABLE I** (*cont.*)

| Chemical | Principal use or source of emission | Medical sequellae | Medical surveillance |
|---|---|---|---|
| Dinitrophenol | Dye, chemical intermediate. Percutaneous absorption, inhalation. | Dermatitis. CNS, renal and hepatic damage. | Skin, ocular, thyroid, CNS, renal, and hepatic evaluation. Urinary levels. |
| Dinitrotoluene | Explosive, dye. Inhalation, percutaneous absorption | Methemoglobinemia. Anemia. | Hematologic, renal, and hepatic evaluation. Hx alcohol; metHgb level. |
| Nitrobenzene | Explosive, dye. Inhalation, percutaneous absorption. | Ocular irritation. Methemoglobinemia. CNS depression. | Hematologic, renal, and hepatic evaluation. Hx alcohol, metHbg level. |
| Nitroglycerine | Explosive. Inhalation, percutaneous absorption. | CNS depression. Headache, angina. | CNS, opthalmologic, hematologic evaluation. Hx alcohol. |
| Nitroparaffins | Solvents, coating. Inhalation of vapor. | Mucosal irritation. | CNS, renal, hepatic function. |
| Nitrophenol | Dyes, chemical intermediate. Inhalation, percutaneous absorption. | No specific sequellae. | Cardiovascular, renal hepatic function. |
| Picric acid | Explosvie, chemical intermediate. Percutaneous absorption. | Skin irritant. CNS, hepatic, hematologic damage. | Renal and hepatic function. |
| Tetryl | Explosive. Inhalation, percutaneous absorption. | Dermatitis. CNS, hepatic, renal, hematologic damage. | Skin, hepatic, renal, hematologic, CNS evaluation. |

**TABLE I** (*cont.*)

| Chemical | Principal use or source of emission | Medical sequellae | Medical surveillance |
|---|---|---|---|
| Trinitrotoluene | Explosive. Inhalation, percutaneous absorption. | Mucosal irritation. Hepatitis, aplastic anemia. | Allergic Hx, alcohol Hx. Hematologic, hepatic evaluation. |
| Miscellaneous organic nitrogen compounds | | | |
| Acridine | Dye. Inhalation. | Mucosal irritation. | Skin, ocular, pulmonary evaluation. Blood and urine levels. |
| Ethyleneimine | Organic synthesis. Inhalation, percutaneous absorption. | Mucosal irritation, carcinogen, systemic poisoning. | Hx exposure. Periodic evaluation. Sputum, urine cytology. |
| Hydrazine | Fuel. Inhalation, percutaneous absorption. | Corrosive; carcinogen. | Hx exposure. Periodic evaluation. Sputum, urine cytology. |
| Pyridine | Solvent, denaturant. Inhalation, percutaneous absorption. | Mucosal irritation. Narcosis. CNS, GI damage. | Skin, CNS, hepatic, renal evaluation. Metabolites in urine and blood. |
| Miscellaneous organic chemicals | | | |
| Tricresyl phosphates | Plasticizer. Inhalation, percutaneous absorption. | CNS, GI damage. | Neurologic evaluation. Serum choline esterases. |
| Carbon disulfide | Chemical synthesis. Inhalation, percutaneous absorption. | Mucosal irritation. Psychological, neurologic, cardiovascular disorders. | Neurologic, hepatic, and renal function. Levels in blood, urine, expired air. |
| Dimethyl sulfate | Methylator. Inhalation, percutaneous absorption. | Mucosal irritation. Respiratory, CNS dysfunction. | Skin, ocular, CNS, pulmonary evaluation. Sputum, urine cytology. |
| Mercaptans | Chemical synthesis. Inhalation. | Mucosal irritation. CNS toxicity. | Skin, ocular, CNS, hepatic, and renal evaluation. |

(*cont.*)

**TABLE I** (*cont.*)

| Chemical | Principal use or source of emission | Medical sequellae | Medical surveillance |
|---|---|---|---|
| Halogens | | | |
| Bromide/ hydrogen | Oxidizing agent. Inhalation. | Mucosal irritation. Pulmonary dysfunction. | Skin, ocular, pulmonary evaluation. |
| Chlorinated lime | Bleach. Inhalation, ingestion. | Corrosive. | Periodic examination. |
| Chlorine | Bleach. Inhalation. | Corrosive. Asphyxiant. | Skin, ocular, pulmonary evaluation. |
| Fluorine | Organic and inorganic synthesis. Inhalation. | Corrosive. Pulmonary damage. | Skin, ocular, pulmonary, and renal evaluation. Urinary fluoride level. |
| Hydrogen chloride | Chemical synthesis. Inhalation. | Corrosive. Pulmonary damage. | Skin, ocular, pulmonary evaluation. |
| Metallic compounds | | | |
| Aluminum | Smelting. Inhalation. | Corneal necrosis; dermatitis. Pneumoconiosis. | Skin, ocular, pulmonary evaluation. |
| Arsenic | Agriculture, pharmaceuticals, chemical industries. Inhalation. | Corrosive. Chronic poisoning. | Skin, ocular evaluation. Blood counts. Urinary arsenic levels. |
| Arsine | Chemical, smelting, refining industries. Inhalation. | Ocular damage. Fatal gas exposure. | Renal, hepatic, hematologic evaluation. |
| Antimony | Alloy manufacture. Inhalation, percutaneous absorption. | Dermatitis. Pneumonoconiosis. | Pulmonary, skin, CNS evaluation. |
| Barium | Chemical manufacture. Ingestion or inhalation. | Skin irritation. Pneumoconiosis. | Skin, ocular, pulmonary evaluation. |

TABLE I (*cont.*)

| Chemical | Principal use or source of emission | Medical sequellae | Medical surveillance |
|---|---|---|---|
| Beryllium | Atomic energy industry, metallurgy. Inhalation. | Dermatitis. Systemic toxicity. Pulmonary dysfunction. | Pulmonary function. Tissue beryllium levels. |
| Bismuth | Metallurgy. Inhalation. | No significant toxicity unless ingested. | Renal and hepatic function if ingestion occurs. |
| Boron | Metallurgy. Inhalation. | Mucosal irritation. Toxicity varies with compound. | Periodic examination. |
| Brass | Metallurgy. Inhalation of fumes. | Dermatitis. | Blood lead levels. |
| Cadmium | Metallurgy. Inhalation of fumes. | Respiratory irritant. Acute and chronic toxicity. | Pulmonary, renal, and hematologic evaluation. |
| Carbonyls | Metallurgy; organic synthesis. Inhalation, percutaneous absorption. | Skin irritation. Toxicity varies with compound. | Skin and pulmonary evaluation. Urinary nickel levels. |
| Cerium | Catalyst, deoxidizer. Inhalation of dust. | Pneumoconiosis. | Pulmonary evaluation. |
| Chromium | Metallurgy; organic synthesis. Percutaneous absorption, inhalation, ingestion. | Dermatitis. Pulmonary damage. | Pulmonary evaluation. Urinary chromate levels. |
| Cobalt | Metallurgy. Inhalation of dust. | Allergen. Pneumonoconiosis. | Skin, pulmonary evaluation. |
| Copper | Metallurgy, electrical industry. Inhalation of dust. | Skin irritants. Pulmonary, GI damage. | Skin, ocular, pulmonary evaluation. |

(*cont.*)

**TABLE I** (*cont.*)

| Chemical | Principal use or source of emission | Medical sequellae | Medical surveillance |
|---|---|---|---|
| Germanium | Metallurgy, electrical industry. Inhalation of dust. | Ocular irritation. Pulmonary irritation. | Pulmonary, hepatic, and renal evaluation. |
| Iron | Metallurgy. Inhalation of dust. | Skin irritants. Siderosis. | Pulmonary evaluation. |
| Lead | Metallurgy, chemical synthesis. Ingestion of dust. Inhalation of dust. | Anemia. Neurologic and renal dysfuntion. | Neurologic, hematologic, and renal evaluation. Blood lead levels. |
| Magnesium | Metallurgy. Inhalation of fumes. | Mucosal irritation. Metal fume fever. | Periodic examinations. |
| Manganese | Metallurgy; chemical industry. Inhalation of dust. | Mucosal irritation. Chronic poisoning. CNS dysfunction. | Frequent examination. CNS evaluation. |
| Mercury | Metallurgy. Inhalation, percutaneous absorption. | Mucosal irritation. Acute/chronic poisoning. | Skin, pulmonary, CNS, and renal evaluation. Urine mercury level. |
| Molybdenum | Metallurgy. Inhalation of dust. | Mucosal irritation. | Ocular, pulmonary evaluation. |
| Nickel | Metallurgy. Inhalation of dust. | Skin sensitization. Carcinogenesis. | Hx allergy. Skin, pulmonary evaluation. Serum, urinary and nickel levels. |
| Osmium | Metallurgy. Inhalation of vapor. | Ocular irritation. | Skin, ocular, pulmonary, and renal evaluation. |
| Phosphine | Insecticide. Inhalation of vapor. | CNS depression. Pulmonary irritation. | Pulmonary evaluation. |

**TABLE I** (*cont.*)

| Chemical | Principal use or source of emission | Medical sequellae | Medical surveillance |
|---|---|---|---|
| Phosphorus | Chemical industry. Inhalation. | Potent skin irritant. Pulmonary irritation. Chronic poisoning. | Dental, skin, pulmonary evaluation. |
| Platinum | Metallurgy. Inhalation of dust. | Respiratory irritant. | Pulmonary evaluation. |
| Selenium | Metallurgy; electrical, glass industries. Inhalation, percutaneous absorption. | Mucosal irritation. | Skin, ocular, pulmonary, renal, and hepatic evaluation. Urine selenium level. |
| Silver | Metallurgy. Inhalaation, ingestion. | Corrosive. Cumulative systemic toxicity. | Hx exposure. Skin examination. Fecal silver level. |
| Thallium | Organic reactions. Inhalation, ingestion, percutaneous absorption. | Skin irritation. Cumulative systemic toxicity. | CNS, GI, hepatic, and renal evaluations. Urine levels. |
| Thorium | Nuclear energy industry. Ingestion, inhalation, percutaneous absorption. | Reticuloendothelial, bone, pulmonary damage. | Hematologic evaluation. Urine levels. |
| Tin | Metallurgy. Inhalation of dust. | Mild skin irritant. Stannosis (penumoconiosis). | Skin, ocular, CNS, and hematologic evaluation. |
| Titanium | Metallurgy. Inhalation of dust. | Skin irritation cataracts. Pulmonary damage. | Pulmonary skin evaluation. |
| Uranium | Nuclear energy industry. Percutaneous absorption. | Renal and hepatic damage. Pulmonary damage. | Renal, hepatic, and pulmonary evaluation. |

(*cont.*)

**TABLE I** (*cont.*)

| Chemical | Principal use or source of emission | Medical sequellae | Medical surveillance |
|---|---|---|---|
| Vanadium | Metallurgy. Inhalation of dust. | Mucosal irritation. | Pulmonary evaluation. |
| Zinc chloride | Chemical industry. Inhalation, ingestion. | Corrosive. Pulmonary irritant. | Skin, pulmonary evaluation. |
| Zinc oxide | Chemical industry. Inhalation of fumes. | Dermatitis. Metal fume fever. | Frequent examination. Urine zinc level. |
| Zirconium | Metallurgy. Inhalation of dust. | Dermatitis. | Skin evaluation. |
| Miscellaneous inorganic compounds | | | |
| Ammonia | Chemical industry. Inhalation. | Mucosal irritation. | Skin, ocular, and pulmonary evaluation. |
| Calcium oxide | Building materials, chemical industry. Inhalation of dust. | Caustic. | Skin, ocular, and pulmonary evaluation. |
| Carbon monoxide | Chemical industry. Inhalation. | Carboxyhemoglobin. | Cardiovascular evaluation. Carboxyhemoglobin level. |
| Graphite | Metal and chemical industry. Inhalation of dust. | Pneumoconiosis. | Pulmonary evaluation. |
| Hydrogen peroxide | Chemical industry. Inhalation. | Mucosal irritation. Pulmonary irritant. | Skin, ocular, and pulmonary evaluation. |
| Hydrogen sulfide | Chemical synthesis. Inhalation. | Mucosal irritation. Acute respiratory failure. | Skin, ocular, and pulmonary evaluation. |
| Nitrogen oxide | Chemical industry. Inhalation. | Mucosal irritation. Pulmonary damage. | Skin, ocular, and pulmonary evaluation. |

**TABLE I** (*cont.*)

| Chemical | Principal use or source of emission | Medical sequellae | Medical surveillance |
|---|---|---|---|
| Ozone | Organic chemical industry. Inhalation. | Mucosal irritation. CNS, pulmonary damage. | Skin, CNS, pulmonary evaluation. |
| Phosgene | Organic chemical industry. Inhalation. | Mucosal irritation. Pulmonary damage. | Skin, ocular, and pulmonary evaluation. |
| Portland cement | Building industry. Inhalation of dust. | Dermatitis. | Skin and pulmonary evaluation. |
| Sodium hydroxide/potassium hydroxide | Chemical industry. Inhalation. | Corrosive. Local tissue injury. | Skin, systemic evaluation. |
| Sulfur chloride | Organic chemical industry. Inhalation. | Mucosal irritation. Pulmonary injury. | Skin, ocular, pulmonary evaluation. |
| Sulfur dioxide | Chemical industry. Inhalation, skin contact. | Severe mucosal irritation. Asphyxia. Pulmonary damage. | Skin, pulmonary evaluation. |
| Sulfuric acid | Chemical industry. Inhalation. | Severe mucosal irritation. Pulmonary damage. | Skin, pulmonary evaluation. |
| Pesticides Organophosphates | Irreversible inhibition of cholinesterase. Inhalation, percutaneous absorption, ingestion. | Dependent on extent of exposure. | Prompt therapy with atropine. Respiratory support. |
| Carbamates | Reversible inhibition of cholinesterase. | Dependent on extent of exposure. | Prompt therapy with atropine. Respiratory support. |
| Chlorinated hydrocarbons | Persists in environment. Ingestion. | Dependent on extent of exposure. | Vomiting should *not* be induced. Gastric lavage. Respiratory support. |

(*cont.*)

**TABLE I** (*cont.*)

| Chemical | Principal use or source of emission | Medical sequellae | Medical surveillance |
|---|---|---|---|
| Bipyridyls | Includes paraquat. | Lens opacilification. Gastrointestinal effects. Renal and hepatic damage. Adult respiratory distress. | Therapy with absorbents to decrease absorption. Supportive care. |
| Rodenticides | Includes coumarins. Ingestion. | Dependent on agent. | Coagulation survey. |
| Fungicides | Heterogeneous compounds. Ingestion, percutaneous absorption. | Mucosal and pulmonary damage. | Physical examination. Pulmonary function. CXR. |
| Herbicides | Heterogeneous compounds. Ingestion. Inhalation, percutaneous absorption. | Corrosives. Neurologic toxicity. | Physical examination. |
| Fumigants | Increased exposure with gaseous products. Inhalation, percutaneous absorption. | Dependent on specific agent. | Physical examination. |

## II. ACUTE AND CHRONIC CHEMICAL INTOXICATIONS

### A. Acute Poisoning

Accidental ingestion of noxious agents requires prompt assessment by a physician. Delaying the onset of treatment until the appearance of symptoms may lessen the effectiveness of therapy and worsen the prognosis. Symptomatic and supportive care represent the mainstay of clinical management, although specific antidotes are indicated for certain toxins. The revered "universal antidote" is both obsolete and ineffective and may delay initiation of appropriate therapy (Arena, 1975).

Accurate history and analysis of the toxic chemical are essential for intelligent therapy. The information available at regional poison control centers is frequently invaluable in these instances.

## B. Chronic Exposure with Toxic Accumulation

The insidious development of large storage pools of toxic substances and the susceptibility of various tissues to these substances is a constant problem confronting the laboratory worker. Heavy metal exposure exemplifies the diverse manifestations caused by chronic accumulation of toxin. Although lead poisoning is frequently a pediatric problem, industrial- and laboratory-associated exposure is prevalent and may cause sudden neurologic symptoms long after exposure has ceased because of the remobilization of lead reservoirs from bone stores by intervening illnesses. The detection of many of these compounds is facilitated by their known relationships with specific tissues. Arsenic compounds can be identified by chemical assay, since they have an affinity for keratinized tissue. The excretion of mercury and cadmium produce significant urinary levels and attendant nephrotoxicity. Other manifestations of the variable expression of organ toxicity, in addition to the nephrotic syndrome or interstitial nephritis, would be the Fanconi syndrome or acute tubular necrosis caused by several metallic elements. Silicon, which has documented pulmonary effects, has recently been implicated in cases of chronic glomerular and tubular nephropathy, miming the pathologic alterations produced by heavy metals (Saldanha *et al.,* 1975). Serial chemical determinations for offending agents should be routine in an exposed population and the possibility of disease induction by these agents must be given serious consideration when patients present with renal or neurologic disease. Fortunately, various chelating agents, including BAL, Versene, and penicillamine, provide an effective therapy and have enabled rehabilitation of numerous patients.

Organic chemicals may provoke toxicity after minimal exposure, and extensive tissue necrosis may characterize the initial host response. The propensity of various insecticides to cause diffuse systemic lesions is illustrated by the neurologic and hepatic impairment produced by the organochlorine pesticide chlordecone. Because of lack of an effective antidote, the physician is required to rely on symptomatic treatment. Cholestyramine may eventually provide a viable therapeutic alternative (Cohn *et al.,* 1978).

Poisoning with halogenated hydrocarbons, such as carbon tetrachloride, produces a dramatic syndrome of hepatic, pulmonary, renal, and neurologic collapse that is invariably fatal. Since chronic contact with miniscule amounts may ultimately induce hepatic and renal dysfunction manifested by vague systemic symptoms, serial assays of liver enzymes and creatinine are advised for exposed personnel.

Benzene and other aromatic compounds can evoke significant myelosuppression resulting in aplastic anemia. Frequent peripheral blood counts are necessary in all individuals who have been exposed to these compounds. Latent toxicity may be expressed by the evolution of acute leukemia, perhaps years after initial contact with minimal amounts of benzene.

## C. Gases and Aerosols

The production of diffuse pulmonary necrosis by excessive concentrations of irritant fumes is usually initially characterized by labored breathing followed by progressive respiratory failure. In view of the extensive injury produced by such accidents, protective maneuvers and safety precautions must be employed when such fumes are generated. Quite often even intensive supportive care cannot modify the clinical damage, and fatalities may occur.

As has been described in an earlier chapter, aerosols can be a prime source of laboratory infection. This is compounded by the fact that a combination of organic and inorganic material in an aerosol may produce a pathogenic potential in an otherwise innocuous agent. Heavy metals and volatile solvents demonstrate a propensity for aerosol dissemination. Dried material contaminated with infective agents or spores may induce or exacerbate pulmonary disease and nonspecific dusts may sensitize susceptible individuals.

Concomitant administration of several agents via aerosols and other routes may produce significant synergistic effect. Often the expression of clinical disease precedes recognition of the precipitating event. Therefore the physician, safety officer, and laboratory director must become involved in a meticulous analysis of the laboratory environment to determine the causes and to design appropriate protective and therapeutic measures.

Although the most common occurrences of carbon monoxide poisoning result from the high concentrations utilized in suicide attempts or associated with fires, chronic exposure to low concentrations may evoke an insidious toxicity. A national survey of American blood donors has illustrated the pervasive character of the problem. Carboxyhemoglobin concentrations have been adversely affected by geographical location, meteorological conditions, and occupation. Significantly, active cigarette smokers exhibited pronounced elevations in blood levels (Stewart *et al.*, 1974). The danger from carbon monoxide is due to its affinity for hemoglobin, with the resultant displacement of oxygen. Moderate exposure may be accompanied by neuropsychiatric aberrations, including personality deterioration and memory impairment (Winter and Miller, 1976). Treatment involves the termination of exposure, the administration of oxygen, the detection of unsuspected sources of carbon monoxide, and the cessation of cigarette smoking.

Halothane-induced hepatic necrosis and methoxyflurane renal impairment exemplify the toxic potential of anesthetic gases in surgical patients (Churchill *et al.*, 1974). Sustained contact with these agents is an occupational hazard of human and veterinary anesthesiologists. This population is now experiencing an alarming increase in hepatic and renal disease, as well as significant fetal abnormalities. Until effective exhaust systems or improved agents are available, anesthesiologists should be carefully monitored for evidence of organic toxicity. Pregnant individuals should avoid contact with these compounds.

## III. HYPERSENSITIVITY PNEUMONITIS: A MODEL OF SENSITIZATION TO ORGANIC MATERIAL

### A. Chest Roentgenograms as a Screening Technique

Serial chest roentgenograms have long been advocated as an effective means of detecting early pulmonary disease in relatively asymptomatic individuals. Recent appreciation of the potential for induction of somatic and genetic damage by irradiation has placed this practice under critical review. Evaluation of one large series of chest roentgenograms suggests that lateral films should be restricted to those individuals suspected of having pulmonary disease and possibly to older age groups, but should not be taken of patients in the 20- to 39-year-old age group (Sagel *et al.*, 1974).

Unfortunately, chronic diffuse infiltrative lung diseases frequently are associated with normal roentgenograms. Analysis of one study detected normal chest films in 10% of patients with biopsy-proved desquamative interstitial pneumonia, sarcoidosis, and allergic alveolitis. Pulmonary function impairment included reduced vital capacity in 57% and diminished single-breath diffusing capacity in 71% of the patients studied. These tests may be efficient screening devices in patients with breathing difficulties. The inability to detect many lesions by X ray reflects the minute isolated foci of disease involvement. Early pulmonary biopsy may be indicated in symptomatic individuals with normal chest roentgenograms and evidence of impaired gas exchange in order to delineate the underlying pathologic process (Epler *et al.*, 1978).

### B. Etiologic Possibilities and Immunologic Pathogenesis

Better knowledge of immunologic mechanisms has permitted specific definition of several previously indeterminant pneumonitis and detection of initiating antigens within the immediate environment. Farmer's lung, a reaction to spores of thermoactinomyces and *Micropolyspora faeni* in moldy hay, is a classic example of a hypersensitivity pneumonitis or extrinsic allergic alveolitis. Patients characteristically develop symptoms of cough, fever, chills, and dyspnea 4–6 hr after exposure, with resolution after 12 hr. Chest roentgenograms may display infiltrates, and gel diffusion techniques may demonstrate circulating antibodies to organic material. The clinical reaction is typically reproduced by reexposure of vulnerable individuals to moldy hay or extracts of the causative organisms. Although avoidance of the antigen is recommended, symptomatic relief after exposure may be possible. Since pulmonary fibrosis may ensue after recurrent episodes, these allergic manifestations must not be ignored (Fink, 1974).

Several inhaled sera and microorganisms have been implicated, and additional inciting antigens are constantly being recognized. Air conditioning and other ventilation equipment may harbor large concentrations of spores and provide culture media for the growth of these organisms. Routine examination of filters and coolant material may detect contamination prior to the production of pathologically significant colonies or the appearance of clinical disease (Markinkovich and Hill, 1975).

Pigeon breeder's disease represents an interesting variant of hypersensitivity pneumonitis, in that immunoglobulin antibody activity against pigeon serum has been demonstrated in serum from affected individuals. Arthus-type cutaneous reactivity may be passively transferred to the skin of normal volunteers using heated serum from patients (Patterson *et al.,* 1976). Laboratory personnel exposed to birds or avian protein may be at risk for similar illnesses.

Spores from the mold aspergillus exhibited the unique capacity to produce several distinctive forms of pulmonary hypersensitivity reactions. Immunoglobulin E-mediated bronchoconstriction or allergic bronchopulmonary aspergillosis may occur in patients with atopic asthma, whereas extrinsic allergic alveolitis usually affects nonatopic patients. Immunologic evaluation of some patients has implicated lymphocyte-mediated tissue inflammation in the pathogenesis of some forms of hypersensitivity pneumonitis (Yocum *et al.,* 1976). This variability of disease expression invoked by one agent further complicates the diagnosis and effective therapeutic design.

## IV. CARCINOGENS

Prolonged exposure to numerous physical agents, chemicals, and infections may ultimately induce neoplasia, especially in individuals who have inherited the genetic makeup to develop malignancy. The propensity for solar irradiation to produce skin cancer, for aniline dyes to elicit bladder transitional cell carcinoma, and for Epstein-Barr virus to be associated with lymphoma is constantly modified by the host's inherent susceptibility to neoplastic transformation. Although this infinite variety of environmental and individual factors precludes detailed discussion in this review, the documented carcinogenic potential of numerous agents must influence the surveillance techniques utilized by physicians and dictate protective precautions employed by industries and laboratories.

The synergistic effects of multiple-agent exposure must be emphasized. Frequently, the cessation of cigarette smoking significantly diminishes the risk or extent of clinical response to other irritants, especially in susceptible patients.

## V. INFECTIONS AND INFESTATIONS

### A. Bacterial Infections

Infection in individuals with intact immune systems is best controlled by effective protective techniques and appropriate immunization. The importance of adequate tetanus prophylaxis is illustrative of this concept, since the fatality rate in this disease approaches 50%. Active immunization is achieved with alum-precipitated or fluid toxoid, and in lieu of recent injury, boosters should be administered no more than once every 10 years. Injuries that may expose patients to tetanus must be carefully evaluated by a physician who is aware of current recommendations for treatment (Weinstein, 1973). Toxoid booster administration within the year is frequently adequate prophylaxis, but additional toxoid is required if none has been given for a period longer than one year prior to the injury. Patients who have never been actively immunized pose specific problems, and effective therapy is imperative when expsoure to tetanus occurs.

Although the prevention and recognition of gross bacterial contamination are ultimately the responsibility of the laboratory supervisor, all personnel should be aware of the ability of organisms to invade superficial skin abrasions or to induce nonspecific dermatitides. Since neglect of early insignificant infections may lead to extensive illness or sepsis, even minor trauma should be reported and appropriate safety precautions should be instituted.

### B. Viral Infections

The number and diversity of viral illnesses prohibit comprehensive review and illustrate the variability of clinical presentation and diagnostic complexity. Treatment of viral disease remains supportive, in the absence of definitive chemotherapy.

The prophylaxis of type B hepatitis in personnel subjected to accidental needle-stick exposure may be feasible and routinely practical. The Veterans Administration Cooperative Study demonstrated the superiority of hepatitis B immune globulin (HBIG) over immune serum globulin (ISG) in prevention of type B hepatitis after needle-stick exposure to hepatitis B surface antigen (HBsAg)-positive donors. Clinical evidence of disease occurred in 1.4% of HBIG and in 5.9% of ISG recipients, and seroconversion was detected in 5.6% of HBIG and 20.7% of ISG recipients. Minimal side effects were experienced by both groups (Seeff et al., 1978). Hepatitis B immune globulin will probably supplant immune serum globulin as the preferred modality in hepatitis B prophylaxis.

Characteristically, viruses produce clinical disease after a latent period of several days, weeks, or months, necessitated by either viral replication or induction of viral expression. An expanding group of viral illnesses with distinctive protracted incubation intervals is now recognized. These viruses have the capacity to induce illness many years after initial infection. Although the current list of documented slow virus diseases is brief, this mechanism of pathogenesis may be implicated in several progressively debilitating diseases of undetermined etiology. This contingency should be appreciated and appropriate precautions should be exercised whenever laboratory personnel are exposed to material from patients with similar illnesses.

## C. Parasitic Infestations

Although parasitic infestation often connotes inadequate sanitation or exotic travel experiences, many diarrheal illnesses in technologically sophisticated societies are caused by infiltrating protozoans. *Giardia lamblia* is typically disseminated by fecal–oral transmission (Wolfe, 1978). Contamination of a municipal water supply by Giardia with confirmation of water infectivity, recently reported by Shaw and associates (1977), emphasizes the immediacy of the problem.

Amebiasis, infection with *Entamoeba histolytica*, illustrates the variable pathogenicity of protozoan disease. It exists as a harmless parasite unless a concurrent bacterial infection causes gastrointestinal diseases or extraintestinal lesions (Krogstad et al., 1978). Recent epidemiologic studies have revealed vast deficiencies in laboratory diagnostic competence and indicate that amebiasis is frequently misdiagnosed and, subsequently, inappropriately treated (Krogstad et al., 1978).

Parasitic involvement now parallels population mobility and "tropical" organisms have been identified in patients from northern cities. Intestinal nematodes, usually associated with rural impoverished areas, may insidiously infiltrate laboratory personnel performing diagnostic procedures

(Blumenthal, 1977). This progressive domestication of exotic organisms may eventually produce significant disease among "urban" populations.

## D. Laboratory Animals as Disease Reservoirs

Veterinary immunization and quarantine procedures have appreciably reduced the incidence of clinical illness directly attributable to infected animals. Whenever possible, wild or vicious animals should be restricted, and minor trauma inflicted by animals should receive first aid. Even superficial bites by laboratory rodents should be examined and evaluated for tetanus prophylaxis in selected instances.

Atopic personnel may experience allergic symptoms after contact with animal dander and dusts and should avoid prolonged exposure. Insect stings that may produce minor annoyance in most persons may elicit anaphylaxis in susceptible individuals.

Unsuspected contamination of excreta and secretions by infective organisms may ultimately evoke disease in animal attendants. The propensity of *Leptospirosis* species causing significant illness in humans is now appreciated, as is the increasing enzootic prevalence of infection (Andrew and Marrocco, 1977; Fraser *et al.*, 1973). Although wild rodents constitute the primary disease reservoir, most humans contact the disease from canine pets (Feigin *et al.*, 1973). Effective immunization of dogs has become surprisingly lax despite veterinarian recommendations for booster doses at six-month intervals in endemic areas (Reif and Marshak, 1973). Careful surveillance of canine populations and meticulous monitoring of personnel at risk are essential to restrict the increasing incidence of human leptospirosis.

Laboratory animals may provide sanctuary for insect vectors of various diseases. Rocky Mountain spotted fever may be confused with a variety of febrile exanthems, and fatal cases typically do not document a history of antecedant tick bite or rash (Hattwick *et al.*, 1976). Increasing incidence of this disease in different geographic areas with atypical clinical presentation, often resembling leptosprosis, has recently been reported (Rampal *et al.*, 1978). Several cases of laboratory-acquired Rocky Mountain spotted fever have been attributed to exposure to an aerosol containing infectious rickettsiae (Oster *et al.*, 1977). This potential for pleomorphic clinical expression and diverse modes of transmission emphasizes the impediments to accurate diagnosis and to effective therapy that may characterize similar illnesses.

## VI. THE PREGNANT LABORATORY WORKER

The pregnant woman and her fetus are uniquely susceptible to the effects of physical and chemical agents and infectious organisms with their attendant teratogenicity and carcinogenicity for the fetus. Appreciation of the role of irradiation in induction of chromosomal abnormalities has resulted in strict criteria limiting fetal exposure. Unfortunately, more insidious environmental hazards continue to endanger the fetuses of unsuspecting individuals.

Although the likelihood of inhalation anesthetics resulting in toxic responses in adults is recognized, their teratogenic potential has recently been described. Spontaneous abortions and congenital anomalies are significantly increased in female anesthesiologists. Their children develop unusual malignancies. Recent studies have shown a 25% increase in fetal abnormalities in children of unexposed wives of male anesthetists. Pregnant women should avoid all contact with anesthetic gases and all individuals chronically exposed to high concentrations of gas should be cognizant of the importance of adequate venting and control techniques (Corbett, 1974).

Congenital infections rarely elicit clinical signs in the mother, but may provoke devastating responses in the fetus. Toxoplasmosis provides an illustrative example of a benign adult infection with dangerous fetal consequences. Approximately 50% of Americans exhibit a chronic asymptomatic form of *Toxoplasma gondii* infection and the disease affects over 3000 infants annually. The protozoan is responsible for a feline zoonosis and cats represent both the definitive host and an important disease reservoir. Since toxoplasmosis acquired by the mother *during* pregnancy usually elicits impressive neurologic damage in the child, serial serologic evaluations to ascertain the onset of maternal infection are crucial to predict fetal involvement. Difficulties in interpretation impede the accuracy of prenatal diagnosis and the limited therapeutic options include induced abortion (Krick and Remington, 1978).

Pregnant women, those contemplating pregnancy, and men in their reproductive years must be alert to avoid any unnecessary or suspicious agents. Pregnant women should be carefully monitored by their physicians to reduce the chance of having a child who either will have a defect or will subsequently fall victim to the long-term effects of a toxic or carcinogenic agent.

## VII. ACCIDENTS AND INJURIES BY PHYSICAL AGENTS

Industrial accidents remain the primary mechanism responsible for significant morbidity in the employed population. Effective first aid may diminish the incidence of complications, but prophylaxis with sufficient safety precautions represents the ultimate objective.

Electrical injury is best prevented by adequate wiring. Laboratories should be critically inspected to detect substandard or dangerous conditions. Electrocution represents a medical emergency and prompt evaluation and treatment by a physician is essential.

Smoke inhalation during fires may facilitate dissemination of noxious substances to the pulmonary parenchyma. Combustion and thermal degradation of plastic polymers may produce serious systemic toxicity in workers exposed to small fires (Dyer and Esch, 1976).

Since many laboratories are located in close proximity to hospital facilities and physicians, prompt medical care is usually available in emergency situations. As courses in cardiopulmonary resuscitation become more popular, more individuals are conversant with simple principles which may ultimately save more patients subjected to major trauma or medical emergencies. Conversely, ignorant intervention by untrained humanitarians may be extremely deleterious and delay initiation of appropriate therapy.

## REFERENCES

Andrew, E. D., and Marracco, G. R. (1977). Leptospirosis in New England. *J. Am. Med. Assoc.* **233**, 2027–2028.

Arena, J. M. (1975). Poisoning: General treatment and prevention. *J. Am. Med. Assoc.* **233**, 358–363.

Blumenthal, D. S. (1977). Current concepts: Intestinal nematodes in the United States. *N. Engl. J. Med.* **297**, 1437–1439.

Churchill, D., Knaack, J., Chirito, E., Barre, P., Cole, C., Muekrcke, R., and Gault, M. H. (1974). Persisting renal insufficiency after methoxyflurane anesthesia. *Am. J. Med.* **56**, 575–582.

Cohn, W. J., Boylan, J. J., Blanke, R. V., Fariss, M. W., Howell, J. R., and Guzelian, P. S. (1978). Treatment of chlordecone (Kepone) toxicity with cholestyramine. *N. Engl. J. Med.* **298**, 243–248.

Corbett, T. H. (1974). Inhalation anesthesia: An occupational hazard. *Hosp. Pract.* November, pp. 81–88.

Dyer, R. F., and Esch, V. H. (1976). Polyvinyl chloride toxicity in fires. *J. Am. Med. Assoc.* **235**, 393–397.

Epler, G. R., McLoud, T. C., Gaensler, E. A., Mikus, J. P., and Carrington, C. B. (1978). Normal chest roentgenograms in chronic diffuse infiltrative lung disease. *N. Engl. J. Med.* **298**, 934–939.

Feigin, R. D., Labes, L. A., Anderson, D., and Pickering, L. (1973). Human leptospirosis from immunized dogs. *Ann. Intern. Med.* **79**, 777–785.

Fink, J. N. (1974). Hypersensitivity pneumonitis: A case of mistaken identity. *Hosp. Pract.* March, pp. 119–124.

Fraser, D. W., Glosser, J. W., Francis, D. P., Phillips, C. J., Feeley, J. C., and Sulzer, C. R. (1973). Leptospirosis caused by serotype *Fort-Bragg. Ann. Intern. Med.* **79**, 786–789.

Hattwick, M. A. W., O'Brien, R. J., and Hanson, B. F. (1976). Rock Mountain spotted fever: Epidemiology of an increasing problem. *Ann. Intern. Med.* **84**, 732–739.

Krick, J. A., and Remington, J. S. (1978). Current concepts in parasitology: Toxoplasmosis in the adult—an overview. *N. Engl. J. Med.* **298**, 550–553.

Krogstad, D. J., Spencer, H. C., and Healy, G. R. (1978). Current concepts in parasitology: Amebiasis. *N. Engl. J. Med.* **298**, 262–265.

Krogstad, D. J., Spencer, H. C., Healty, G. R., Gleason, N. N., Sexton, D. J., and Herron, C. A. (1978). Amebiasis: Epidemiologic studies in the United States, 1971–1974. *Ann. Intern. Med.* **88**, 89–97.

Markinkovich, V. A., and Hill, A. (1975). Hypersensitivity alveolitis. *J. Am. Med. Assoc.* **231**, 944–947.

Oster, C. N., Burke, D. S., Kenyon, R. H., Ascher, M. S., Harber, P., and Pedersen, C. E. (1977). Laboratory-acquired Rocky Mountain spotted fever: The hazard of aerosol transmission. *N. Engl. J. Med.* **297**, 859–863.

Patterson, R., Schatz, M. Fink, J. N., DeSwarte, R. S., Roberts, M., and Cugell, D. (1976). Pigeon breeder's disease. I. Serum immunoglobulin concentrations; IgG, IgM, IgA and IgE antibodies against pigeon serum. *Am. J. Med.* **60**, 144–151.

Rampal, R., Kluge, R., Cohen, V., and Feldman, R. (1978). Rocky Mountain spotted fever and jaundice. *Arch. Intern. Med.* **138**, 260–263.

Reif, J. S., and Marshak, R. R. (1973). Leptospirosis: A contemporary zoonosis. *Ann. Intern. Med.* **79**, 893–894.

Sagel, S. S., Evens, R. G., Forrest, J. V., and Bramson, R. T. (1974). Efficacy of routine screening and lateral chest radiographs in a hospital-based population. *N. Engl. J. Med.* **291**, 1001–1004.

Saldanha, L. F., Rosen, V. J., and Gonick, H. C. (1975). Silicon nephropathy. *Am. J. Med.* **59**, 95–103.

Seeff, L. B., Wright, E. C., Zimmerman, H. J., Alter, H. J., Dietz, A. A., Felsher, B. F., Finkelstein, J. D., Garcia-Pont, P., Gerin, J. L., Greenlee, H. B., Hamilton, J., Holland, P. V., Kaplan, P. M., Kiernan, T., Koff, R. S., Leevy, C. M., McAuliffe, V. J., Nath, N., Purcell, R. H., Schiff, E. R., Schwartz, C. C., Tamburro, C. H., Vlahcevic, Z., Zemel, R., and Zimmon, D. S. (1978). Type B hepatitis after needle-stick exposure: Prevention with hepatitis B immune globulin. *Ann. Intern. Med.* **88**, 285–293.

Shaw, P. K., Brodsky, R. E., Lyman, D. O., Wood, B. T., Hibler, C. P., Healy, G. R., Macleod, K. I. E., Stahl, W., and Schultz, M. G. (1977). *Ann. Intern. Med.* **87**, 426–432.

Stewart, R. D., Baretta, E. D., Platte, L. R., Stewart, E. B., Kalbfleisch, J. H., Van Yserloo, B., and Rimm, A. A. (1974). Carboxyhemoglobin levels in American blood donors. *J. Am. Med. Assoc.* **229**, 1187–1195.

Weinstein, L. (1973). Current concepts: Tetanus. *N. Engl. J. Med.* **289**, 1293–1296.

Winter, P. M., and Miller, J. N. (1976). Carbon monoxide poisoning. *J. Am. Med. Assoc.* **236**, 1502–1504.

Wolfe, M. S. (1978). Current concepts in parasitology: Giardiasis. *N. Engl. J. Med.* **298**, 319–321.

Yocum, M. W., Saltzman, A. R., Strong, D. M., Donaldson, J. C., Ward, G. W., Walsh, F. M., Cobb. O. M., and Elliott, R. C. (1976). Extrinsic allergic alveolitis after *Aspergillus fumigatus* inhalation. *Am. J. Med.* **61**, 939–945.

# Chapter Nine

# Medical Aspects of Occupational Health in a Laboratory Setting

## JAMES H. ANDERSON, JR.

## I. INTRODUCTION

For many years the medical profession ignored occupational health in laboratories as well as in industry in general. The American College of Surgeons pioneered a program in the 1930s by developing guidelines for the care of traumatic injury in industry. The Occupational Health Institute of the Industrial Medical Association broadened the program by increased emphasis on employee health maintenance. However, until the last decade, the programs could have been more accurately termed "occupational illness" rather than "occupational health." This problem is highlighted by statistics of the Department of Health, Education, and Welfare in 1970 that there may be as many as 100,000 deaths and 390,000 new cases of occupational disease per year in the United States. The recent realization that occupational exposures, particularly to chemical products, contribute to chronic disease as well as to acute illness has resulted in increasing

LABORATORY SAFETY: THEORY AND PRACTICE
Copyright © 1980 by Academic Press, Inc.
All rights of reproduction in any form reserved.
ISBN 0-12-269980-7

diagnosis of, if not actual increases in, occupational illnesses. Such occupation-related new diseases as hepatic angiosarcoma from exposure to vinyl chloride, sterility among manufacturers of dibromochloropropane pesticides, and altered immune function among polybrominated biphenyl-exposed farmers are being recognized (Rom, 1979). Recognition of nonoccupational exposure-related diseases has also been increasing, such as the genetic mutations in residents of the Love Canal area of New York (EPA, 1980) and the neoplastic illnesses of children of women who were treated with diethylstilbestrol. Emphasis is again on diagnosis and cure, often with less than ideal results. The laboratory worker often faces increased potential risks because of the nature of the research job.

With the establishment of the National Institute for Occupational Safety and Health (NIOSH) created by the Occupational Safety and Health Act of 1970, emphasis (and financial support) was given to providing for research and training of occupational safety and health professionals. The medical profession has also begun a shift to the "health" and preventive medicine aspect of occupational health.

It is not the purpose of this chapter to provide either a lengthy list of proven or potential biohazards or a medical "cookbook" of signs, symptoms, diagnosis, and treatment. Many such references are available (see Reading List). This chapter will explore the concepts involved in designing an occupational health program concerned with screening, medical surveillance, and laboratory evaluation of personnel working in laboratories with actual or potential biohazards.

## II. COMPONENTS OF OCCUPATIONAL HEALTH PROGRAMS

Medical screening and surveillance programs must be individually designed to function effectively and to attain the designated goals in any laboratory or work environment. Three major factors are involved in all medical programs: personnel, risk assessment, and the occupational health program itself. Obviously, there are close interrelationships among these factors.

## III. PERSONNEL

Laboratory personnel are the key factor in the success of any program. As has been stressed thoughout this book the individual has the prime responsibility for his safety. In terms of both preventive medicine and postexposure care, the individual worker must be cooperative, noncomplacent, and totally honest. It is incumbent upon both the supervisory echelons and the occupational health professionals to encourage these qualities in all personnel. Periodic examinations can be performed only if the laboratory worker keeps such appointments. Appropriate health recommendations must be followed. If the laboratory worker ignores a developing, potential health hazard (such as a faulty piece of equipment), he endangers not only himself but his fellow workers as well. Honesty is the most important characteristic. Failure to give truthful answers on preemployment or periodic examinations defeats the purpose. Not reporting accidents or exposures at the time of occurrence may delay prophylactic or therapeutic treatment as well as placing other workers or family members in danger, as in the case of exposure to infectious biologicals. The laboratory worker is thus required to play an active role in his own health and safety.

## IV. RISK ASSESSMENT

Risk assessment requires the coordinated effort of laboratory workers, administrators, safety officials, and occupational health professionals. The hazards in a research laboratory setting are usually more complex and variable than in an industrial setting and less prone to be recognized in some areas of research. The more specific the assessment of hazards, the better the medical surveillance. For example, a general medical examination and "normal" laboratory blood studies would not detect mild exposures to many organophosphates which could be detected by red blood cell cholinesterase assays or even more specific assays for individual compounds.

Many laboratory environments involve multiple risks. A worker in a nonhuman primate laboratory involved with hepatitis research is potentially or actually exposed to animal bites, animal dander, and simian herpes virus as well as hepatitis virus.

In addition to hazard recognition, all of the individuals having contact with a laboratory must be identified. Electricians, janitors, salesmen, etc. all have to be considered in risk assessment, and in some cases access may have to be limited to individuals participating in a segment of medical surveillance such as an immunization program.

When all the risks have been identified, it is then necessary to establish a record system with all personnel involved detailing the risks to which each is exposed. From this, the medical surveillance plan for each individual can be designed. Obviously, this is an ideal task for utilization of computer capabilities. In addition to a listing of all examinations and tests, the computer can coordinate testing to avoid unnecessary duplication of studies required by different health hazards.

## V. OCCUPATIONAL HEALTH PROGRAM

The occupational health program can operate successfully in a wide range of situations; however, there are certain medical standards that must be a part of all programs. These standards are best illustrated by those adopted and disseminated by the Occupational Health and Safety Programs Accreditation Commission (OHSPAC, 1977).

### A. Standards

#### 1. Policy

There shall be a written policy for the medical section of the occupational health program. The medical policy should be incorporated into the establishment's overall occupational health program policy. The policy should be reviewed and subject to revision periodically.

#### 2. Staff

The staff responsible for the direction and operation of the occupational medical program must be professionally qualified, adequate in number, and have sufficient time and authority to design and implement the medical functions of the occupational health program set forth in the written policy.

#### 3. Facilities

The medical facilities shall be of adequate quality, design, size, location, and state of readiness in which to perform the functions of the occupational health program effectively and safely.

#### 4. Equipment

There shall be sufficient equipment of good quality, properly used and maintained to effectively and safely meet the needs of the occupational medical program.

#### 5. Preplacement Assessment

The preplacement medical assessment shall be sufficiently inclusive to aid in suitable placement. It may also provide a data base for medical care, health maintenance, and medical research.

#### 6. Periodic and Special Medical Assessments

Periodic and special medical assessments shall be conducted at appropriate intervals and be sufficiently inclusive to assure that the employee's health is compatible with his job, the safety of others, and free from the adverse effects of materials or processes to which he is exposed. They should also help the employee maintain the optimum good health necessary for effective performance and minimum sick absence.

## 7. Diagnosis and Treatment of Occupational Injury or Illness

The diagnosis and treatment of occupational injury or illness shall be prompt, adequate, and directed toward rehabilitation.

## 8. Diagnosis and Treatment of Nonoccupational Injury or Illness

The diagnosis and treatment of nonoccupational injury or illness should be prompt, adequate, and of high quality.

## 9. Special Preventive Programs

Programs to prevent adverse health effects on employees shall be developed and effectively implemented.

## 10. Health Counseling and Education

There shall be a health counseling and education program to promote both general health maintenance and safe, healthful work practices.

## 11. Medical Records

An adequate medical record shall be maintained for every employee, accurately documented and readily accessible. The legal and ethical bounds of confidentiality will be strictly maintained. Additional appropriate records and summaries required for good occupational medical practice will be accurately maintained and kept current.

## 12. Disaster Health Services

There shall be written plans and provisions for the proper and timely care of casualties arising from disasters. The plans and provisions shall be periodically reviewed and tested.

## B. Discussion

The written occupational health policy should clearly reflect the planning and the programs to be established. The goals and methods to attain them should be clear and expressed in nontechnical language when possible so that all workers can understand the medical regulations and procedures, thus encouraging them to be active participants in the health care program.

The staff of the occupational medical program will vary according to the needs and size of the laboratory. Overall direction of the program should be under the supervision of a qualified physician, although in many smaller labs the physician may be on a part time or contract basis with a qualified nurse or other health professional as the full time on-site director of the medical program. Physicians with specialty training in Occupational Medicine (and, ideally, certification in Occupational Medicine by the American Board of Preventive Medicine) are limited in number, but physicians with interest and/or experience but with less formal training can establish high caliber programs.

The size of the program, which is determined by the number of individuals served and the financial resources budgeted, dictates the number and professional makeup of the remaining staff. In many programs the occupational health nurse deserves credit for the success of the program. The nurse (or nurses) has the initial, and often the only, contact with the worker after small accidents or exposures or for surveillance programs such as blood or urine samples. Special relationships often develop which increase the effectiveness of medical surveillance by increasing employee cooperation.

Large programs have several physicians, some with training in other fields important to occupational medicine such as nuclear medicine, dermatology, surgery, and emergency room medicine. The nursing staff may also have diverse backgrounds. In addition to physicians and nurses, the larger programs may have audiologists, physiotherapists, nuclear medicine technicians, and respiratory therapists as well as laboratory and X-ray technologists. Physicians' assistants and nurse practitioners are also becoming more common as medical programs expand.

Secretarial support must be sufficient, and medical records specialists with training in electronic data processing will become necessities as occupational health programs expand and as requirements and federal regulations concerning medical records proliferate.

The facilities of the medical program, like the staff, are dependent on the number of personnel to be cared for and the financial commitment. The facilities must also relate to the services to be performed. Small programs may rely on off-site facilities of a contract physician while larger programs may have complete hospital units on site. Requirements may also be dictated by the type of laboratory involved. The United States Army Medical Research Institute of Infectious Diseases laboratories have a complete hospital with both clinical diagnostic laboratories and four patient beds capable of P4 level containment for caring for patients with high hazard infections (or exposures) such as Lassa fever, Ebola hemorrhagic fever, or plague.

The occupational health program equipment is dependent upon the type of medical screening and surveillance performed and the magnitude of the program. Larger facilities may maintain nuclear medicine capabilities for bioassays, spirometers for pulmonary function studies, etc. Smaller laboratories may be totally dependent on off-site commercial sources for all medical diagnostic work. Obviously, first aid and appropriate emergency equipment are mandatory in occupational health programs.

The preplacement or preemployment medical examination should be a comprehensive evaluation of the worker's medical history and physical condition. Special aspects of the prospective applicant's job will dictate special or nonroutine questions and examinations as well as clinical laboratory studies. Workers who will be exposed, for example, to certain organic solvents will require careful evaluation of hematologic and hepatic systems while silica- and asbestos-exposed workers may have primary attention focused on the respiratory system.

The examination should serve the purpose of suitable placement of the worker, protecting him and the laboratory from increased risk due to preexisting physical or mental conditions.

The preplacement examination will also provide a data base both for monitoring the effect of the laboratory environment on the individual and for conducting occupational medicine research on larger groups. Health maintenance is dependent on accurate assessment of pre-work exposure base lines.

One point is often confusing or misunderstood and needs to be emphasized. The occupational health physician is an employee of the laboratory and information from the history and physical examinations performed are not subject to normal physician–patient confidentiality. The physician should discuss all significant medical findings with the patient whether occupationally related or not, although the legal requirements to do so have long been debated. The physician should approach his job not with the idea of hiring or firing, but of recommending the laboratory environment in which the worker can function with safety to himself and others. It is hoped that this will encourage a similar cooperative and honest attitude in the prospective laboratory worker and result in better health care and safety in the laboratory.

The periodic medical surveillance examination should be carefully designed to assess adequately all physical and mental parameters that might potentially be affected by the laboratory environment. The design of the examination is thus dependent on a thorough evaluation of the risks that might be encountered in the laboratory. Some examinations have been partially dictated by government regulations (e.g., benzene workers, radiation workers) while others must be designed by a medical director who has training in occupational medicine and a thorough understanding of the work performed by the patient. The frequency of the periodic exam is dependent on the health of the individual and the hazards to which he is exposed. The time span may range from as long as three years for personnel in nonhazardous areas to daily or weekly for certain high-hazard laboratory exposures. Often complete exams are scheduled on a longer term basis with selected specific laboratory studies done periodically during the interval. For example, certain bioassays may be performed on a monthly basis in individuals exposed to certain isotopes, with a complete physical exam performed yearly.

Special examinations may be indicated in a variety of circumstances. Laboratory workers who are changing positions or terminating their employment should have repeat exams. It is especially important for terminal exams to document health status so that the relationship of any future medical problems to prior work exposures can be fairly assessed. Special exams may be indicated when new hazards are introduced into the laboratory. Discovery of a potential or actual exposure may dictate a series of special exams. Laboratory workers who become sick should be evaluated in order to determine if the illness is occupationally related. Special exams are also indicated on an employee's return to work.

At this point it should again be stressed that examinations by themselves do little good if appropriate actions based on the findings are not taken. Often abnormal findings may require the transfer of the individual to another job area or modifications in laboratory environment or procedures in order to prevent additional exposures.

The goal in occupational medical programs is rapid and accurate diagnosis and treatment. The diagnosis of occupational injury or trauma is normally a readily apparent cause and effect situation, although treatment may range from simple to complex. The diagnosis of occupational illness in the laboratory worker may be exceedingly difficult, both for the physician and for the patient. A close examination of the diagnostic process well illustrates some of the pitfalls of medical and laboratory screening.

Occupational diseases are by definition pathologic reactions of the individual to his work environment. To diagnose correctly an occupational illness, the physician must thoroughly examine both the patient and his laboratory environment. Occupational illnesses, although often characteristic, do not usually present with pathognomonic signs or symptoms. Diagnosis requires careful investigation of the patient, the clinical laboratory tests available, and the work situation. The lung diseases from asbestos or cotton dust, the anemias from heavy metal intoxication or benzene exposure and the neurologic impairment from pesticides do not yield pathognomonic clinical or laboratory evidence producing a diagnosis. The physical findings, medical history, and laboratory tests must be viewed with a knowledge of environmental exposures in order to make an accurate diagnosis. While some diseases such as silicosis produce specific pathologic changes that can be diagnosed by biopsy or autopsy examination, these are not normally the first patient–physician interaction.

The physician (and patient as well) must be cautious in interpretation of physical findings and laboratory tests. For example, a laboratory worker may present with symptoms of headache, constipation, and abdominal pain which are characteristically seen in, but not pathognomonic of lead intoxication. A laboratory report of a blood lead concentration of 0.10 mg/100 gm would seem to be specific laboratory evidence of acute lead intoxication and indeed the diagnosis could be correct. However, a patient with a long history of chronic exposure to lead (without intoxication) and acute intestinal obstruction might present with the identical symptoms and laboratory results (Zenz, 1975).

The philosophy and ethics of the role of occupational medicine in nonoccupational diseases and provision of routine health care to workers and their families are appropriately addressed elsewhere. However, the fact remains that a diagnosis must be made in order to determine if an illness is indeed occupational. Again, symptoms, physical findings, and laboratory screens may not distinguish etiology. Toxic hepatitis from trichloroethylene cannot be distinguished from alcohol-induced hepatitis except by knowledge of environmental exposure. Cigarette-induced chronic bronchitis cannot be distinguished from chronic bronchitis resulting from occupation-related, inhaled irritants unless the physician carefully includes the work environment in his evaluation of the patient. For this reason the occupational health physician, because he is oriented to this concept, should see all laboratory personnel who develop illnesses.

## VI. OCCUPATIONAL HEALTH PROGRAM EXAMPLE

The following brief description of an occupational health program is not given as an all-inclusive "cookbook" to use in any laboratory, but as a generalized example to illustrate the need for a professional, organized approach to the complex problems of medical surveillance and laboratory screening.

As discussed earlier, identification of all personnel to be entered into the occupational health program must be done. Assessment of all actual or potential hazards in the laboratory environment must be completed and correlated with the personnel in the program. The health professionals can then design the types of examinations and laboratory studies for each category of worker and each hazard.

As an example, after correlating personnel and risks, the laboratory workers might be placed in the following categories (an individual may be included in more than one group).

1. All employees.
2. Solvent workers.
3. Organophosphate workers (including pesticides).
4. Heavy metal workers.
5. Radiation workers.
6. Microwave-laser workers.
7. High hazard virus workers.
8. High risk tuberculosis exposure.
9. Animal handlers.
10. High noise workers
11. Security workers.
12. Women workers.

The occupational health professionals might decide their program will be based on the following medical examinations and laboratory screening tests:

A. Complete medical examination.
B. Partial medical examination.
C. Special history (e.g., radiation treatment as child).
D. Multiphasic blood screen.
E. Complete blood count (CBC).
F. Urinalysis.
G. Liver function tests.
H. Kidney function tests.
I. RBC cholinesterase.
J. Pregnancy test.
K. Ophthalmologic examination.
L. Audiogram.
M. Psychological evaluation.
N. Chest X ray.
O. TB skin test.
P. Viral antibody levels.
Q. Urine for metals.

The task now required is identifying who will be screened with which test(s) and how often.

All employees would initially require a complete medical exam, including the special medical procedures listed, a complete laboratory screen, and selected special procedures in order to establish a medical data base for each individual. Thereafter, each employee might be required to repeat these examinations every three years. Solvent workers would have the initial complete exam but would be required to have a CBC, a urinalysis, and liver and kidney function tests every six months. Pesticide workers might be required to have these same exams and an RBC cholinesterase every month in addition. Laser workers may be required to have an eye exam every six months while workers with radioactive iodine might be scheduled for eye exams every three years. This hypothetical program could be charted as follows (see Table I):

While it is apparent that computerized management of the occupational health program is an almost mandatory requirement in a large laboratory or research center, rigid structuring of health programs is not sufficient. Abnormal findings may dictate new or more frequent exams. Accidents may introduce new variables. The occupational health program in the biohazard laboratory setting must be dynamic and capable of responding to the changing needs of the individuals the program serves.

## VII. CONCLUSION

Medical screening and surveillance programs determine the physical and mental abilities of a worker to perform his job without unnecessary risks or harm to himself or others and to monitor the effects of the individual's exposure to specific biological, chemical, radiological, and physical hazards. The occupational health program ideally should detect subclinical or initial effects of undetected or accidental exposures to potentially hazardous agents. Preplacement and periodic examinations are the essential tools of the medical surveillance program, but of equal importance is the identification and assessment of actual or potential hazards in the work environment. It is mandatory to note, however, that physical examinations by themselves are of little importance unless appropriate actions result from the findings. These actions may include medical treatment of the worker, modifications of personal or job habits or work environment, or transfer of the worker to another task.

To be successful, a well-designed medical surveillance program requires a great expenditure of initial effort, thought, and commitment by all personnel involved as well as periodic maintenance and updating. The health and welfare of each worker is dependent on the success of the program.

**TABLE I**

**A Hypothetical Occupational Health Program** [a]

| Personnel category | Exam or Test | | | | | | |
|---|---|---|---|---|---|---|---|
| | A. Complete exam | B. Partial exam | C. Special history | D. Multiphasic blood screen | E. Complete CBC | F. Urinalysis | G. Liver function studies |
| 1. All employees | 36 | | | 36 | 36 | 36 | |
| 2. Solvent workers | 12 | 6 | 6 | 12 | 6 | 12 | 12 |
| 3. Organo-phosphate workers | 12 | | 12 | 12 | 6 | 6 | 6 |
| 4. Heavy metal workers | 12 | | 12 | 12 | 6 | 12 | 12 |
| 5. Radiation workers | 36 | | 36 | 36 | 12 | | |
| 6. Micro-wave-laser workers | 36 | | 36 | 36 | 12 | | |
| 7. High haz-ard virus workers | 36 | | | 36 | 36 | | |
| 8. High risk TB expo-sure | 36 | 12 | | | | | |
| 9. Animal handlers | 36 | | | | | | |
| 10. High noise workers | 36 | | | | | | |
| 11. Security workers | 36 | | | 36 | | | |
| 12. Women workers | | > 40   12 | yrs | | | | |

| | | | | Exam or Test | | | | | |
|---|---|---|---|---|---|---|---|---|---|
| H. Kidney function studies | I. RBC cholinesterase | J. Pregnancy test | K. Ophthalmologic exam | L. Audiogram | M. Physchological evaluation | N. Chest X-ray | O. TB skin test | P. Viral antibody levels | Q. Urine for metals (lead, mercury) |
| | | | | | | | 36 | | |
| 12 | | | | | | | | | |
| 6 | 1 | | | | | | | | |
| 12 | | | | | | | | | 6 |
| | | | 36 | | | | | | |
| | | | 12 | | | | | | |
| | | | | | | | | 12 | |
| | | | | | | 24 | 6 | | |
| | | | | 12 | | | | | |
| | | | 12 | 12 | 36 | | | | |
| | | PRN | | | | | | | |

[a] Exams performed at intervals listed (months).

# VIII. APPENDIX

OCCUPATIONAL HEALTH ASSESSMENT

I. IDENTIFICATION

Name _____ Department (or work area) _____
SSN _____
Home address _____ Job Title _____
_____ Phone _____
Home phone _____ In emergency notify _____
Date of birth _____ _____

II. HAZARD IDENTIFICATION (check all applicable boxes)

1. Infectious organisms (specify)

   /‾/ A. Bacteria _____    /‾/  Immunized?

   /‾/ B. Viruses _____    /‾/  Immunized?

   /‾/ C. Rickettsiae _____    /‾/  Immunized?

   /‾/ D. Fungi _____

   /‾/ E. Tuberculosis _____

   /‾/ F. Protozoa _____

   /‾/ G. Metazoa _____

   /‾/ H. Other _____

2. Chemical Agents (specify)

   /‾/ A. Poisons _____

   /‾/ B. Toxins _____    /‾/  Immunized?

   /‾/ C. Solvents _____

   /‾/ D. Explosives _____

   /‾/ E. Carcinogens _____

   /‾/ F. Pesticides _____

   /‾/ G. Benzene _____

   /‾/ H. Other _____

3. Radiation

   /‾/ A. External source (x-ray, gamma and/or beta emitters in
         sealed sources)_____
         _____

/_/ B. Free isotopes (volatile and non-volatile)

_____

_____

/_/ C. High energy source (neutron generators)

_____

_____

/_/ D. Microwave (radar, heating devices)

_____

_____

/_/ E. Laser _____

_____

_____

/_/ F. Ultraviolet _____

_____

4. Eye

    /_/ A. Chemical (potential splashes, explosions)

    /_/ B. Metal working or wood working machinery

    /_/ C. Other

5. Noise

    /_/ A. Constant (greater than 85 decibels)

    /_/ B. Intermittent (greater than 100 decibels)

6. Aerosols

    /_/ A. Asbestos

    /_/ B. Silica

    /_/ C. Beryllium

    /_/ D. Agricultural products (Bagass, etc)

    /_/ E. Animal danders, cage bedding, etc.

    /_/ F. Birds, bird droppings

    /_/ G. Power sanding, grinding

    /_/ H. Other

7. Physical

    /_/ A. Falling objects

    /_/ B. Danger of heavy objects falling on feet

    /_/ C. Heavy equipment operator

/_/  D. Motor vehicle operator

/_/  E. High height exposure

/_/  F. High ambient temperature (greater than 90°)

/_/  G. Low ambient temperature (less than 45°)

/_/  H. High or low atmospheric pressure

/_/  I. Electrical hazards

/_/  J. Danger of hand injury

/_/  K. Other

8.  Special

/_/  A. Female _____

/_/  B. Pregnant _____

/_/  C. Chronic health problem (specify) _____

/_/  D. Alcohol program _____

/_/  E. Drug program _____

/_/  F. Please list any work condition(s) or exposure(s) you
        consider to be a health hazard not specified on this
        form.

_____

_____

_____

_____


_____  _____

Employee's signature             Date


_____  _____

Department Head                  Date


THIS FORM MAY BE UPDATED AT ANY TIME.

THIS FORM WILL BE REVIEWED YEARLY.

# REFERENCES

Environmental Protection Agency. (1980). Public News Release, 18 May 1980. Environmental Protection Agency, Washington, D. C.

Occupational Health and Safety Programs Accreditation Commission. (1977). OHSPAC Medical Standards for Occupational Health Programs. *J. Occup. Med.* **19**, 629–634.

Rom, W. N. (1979). Medicine re-enters the workplace. *N. Engl. J. Med.* **300**, 672–673.

Zenz, C. (ed.). (1975). "Occupational Medicine: Principles and Practical Applications." Yearbook Medical Publishers, Inc., Chicago.

# RECOMMENDED READING

Atherley, G. R. C. (1978). "Occupational Health and Safety Concepts." Applied Science Publishers, LTD., London.

Clayton, G. D., and Clayton, F. E. (eds.). (1978). "Patty's Industrial Hygiene and Toxicology. Vol. 1, General Principles," 3rd ed. Wiley, New York.

Cralley, L. V. and Cralley, L. J. (eds.). (1979). "Patty's Industrial Hygiene and Toxicology. Vol. 3, Theory and Rationale of Industrial Hygiene Practice," 3rd ed. Wiley, New York.

Fassett, D. W. and Irish, D. D. (eds.). (1979). "Patty's Industrial Hygiene and Toxicology. Vol. 2, Toxicology," 3rd ed. Wiley, New York.

Hammer, W. (1976). "Occupational Safety Management and Engineering." Prentice-Hall, Englewood Cliffs, New Jersey.

Hunter, D. (1978). "The Diseases of Occupations," 6th ed. Little, Brown, Boston.

International Technical Information Institute. (1975). "Toxic and Hazardous Industrial Chemicals Safety Manual." International Technical Information Institute, Tokyo.

Key, M. M., Henschel, A. F., Butler, J., Ligo, R. N., Tabershaw, I. R., and Ede, L. (eds.). (1977). "Occupational Diseases: A Guide to Their Recognition," DHEW Pub. No. 77–181. U.S. Public Health Serv., Center for Disease Control, Natl. Inst. Occup. Health and Safety, Washington, D.C.

Occupational Health and Safety Programs Accreditation Commission. (1976). "Standards, Interpretations and Audit Criteria for Performance of Occupational Health Programs." American Industrial Hygiene Association, Akron, Ohio.

Sax, N. I. (1979). "Dangerous Properties of Industrial Materials," 5th ed. Van Nostrand Reinhold, New York.

Zenz, C. (ed.). (1975). "Occupational Medicine: Principles and Practical Applications," Yearbook Medical Publishers, Inc., Chicago.

# Chapter Ten
# Genetic Monitoring
### KATHRYN E. FUSCALDO

## I. INTRODUCTION

As indicated in Chapters 2 and 8, a variety of foreign substances have been shown to be teratogenic, mutagenic, and/or carcinogenic on acute or chronic exposure. New developments in laboratory technology have made it more feasible to evaluate these substances both directly and indirectly in man. However, even with the availability of new and improved evaluative procedures, it remains a formidable undertaking to identify potential risks in the environment and to develop protective measures to ensure the safety of all segments of the population now and the integrity of our environment for future generations. An excellent review of this field appears in the report of the Subcommittee on Environmental Mutagenesis (Mehlman, et al., 1977).

In addition to developing the technology to identify accurately potential mutagens among the many substances with which we come in contact, it is equally important to develop methods of detecting those individuals who have responded to known mutagens early enough to be able to institute preventative measures or to treat them effectively. This chapter will review some aspects of potential damage to the genetic makeup of an individual from exposure to toxic substances.

Two major types of genetic damage can result from natural or manu-

**299**

LABORATORY SAFETY: THEORY AND PRACTICE

factured agents including chemicals and radiation. Each type involves damage at a different level. One type involves damage to the chromosomes that carry the genetic information and the second type, called point mutations, involves small errors or mutations within the genetic material itself.

This damage can occur either in the genetic material of body cells (i.e., somatic effects) or in the reproductive cells (i.e., germinal effects). Several test systems have been developed to assay for chromosomal and point mutations in somatic cells, but little is currently available to test for mammalian germinal mutations (Mehlman, 1977; Brewen, 1976; Kucerova, 1976).

The mutagenic effect of substances on chromosomes can be assessed by *in vitro* (Sobels and Vogel, 1976) as well as *in vivo* test systems (Stetka and Wolff, 1976), using yeasts, molds, drosophila (the small fruit fly), and human and mammalian cells (Mehlman, 1977). The *in vivo* assays utilizing mammalian systems have been designed to test for both somatic and germinal effects. Tests are available to assay for specific types of defects, such as dominant lethals, translocations, and the loss of the X chromosome. The nature of these tests will be discussed later in this chapter. In order to determine whether or not a specific chemical substance has a mutagenic effect, it is also necessary to evaluate the metabolites of that substance and to assess the effect in terms of the doses required to produce the effect. The evaluative procedures have usually been broken down into two broad categories. The first involves screening systems that make use of either *in vitro* microbial tests or *in vivo* drosophila systems and cytogenetic analysis of mammalian bone marrow. The second is an intensive, complete test of a suspect mutagen that uses both the tests on microorganisms with metabolic activation *in vitro* and *in vivo* or tests on drosophila. In addition, cytogenetic analysis of mammalian bone marrow and cultured human lymphocytes are done, as well as the mammalian dominant lethal test (Lang and Adler, 1977; Mark, 1977). This latter test is designed to detect chromosomal damage in male germ cells. Sperm-carrying dominant lethals are able to fertilize an egg cell, but the zygote is not viable.

## II. ECONOMIC IMPACT

The determination of whether to test a potential mutagen by the screening or complete test systems depends on several factors. They include the extent to which the substance is used by the public or the number of people who will come in contact with the chemical; the economic and medical significance of the substance; and previous information on its mutagenicity, carcinogenicity, or teratogenicity. Taking into account such factors, one would normally subject industrial chemicals, organophosphate medicines, and drugs used on a very limited number of patients to only screening procedures. The more definitive tests would be reserved for such compounds as drugs, chemicals in general use, food additives, and pesticides. However, if a compound or its metabolites is shown to be positive in the screening tests, then its genetic risk should be determined. Any compound (in average dose) that increases the level of spontaneous mutations more than 0.1% should be the subject of stringent regulation.

Few individuals will deny the necessity of continually attempting to improve the human condition, that iš, of "making progress." However, in the course of industrial and technological progress, we have produced serious side effects. The environment that surrounds and interacts with each of us has been seriously compromised. The price of progress has been high for all of us in terms of pollution of the air, water, and additions to our food supply at all stages of production. For some the price has been even higher as they have been intimately involved with the industrial processes that have produced our "progress." More recently, considerable concern has been expressed about the potential hazard to man of exposure to low levels of radiation from a variety of sources both accidental and deliberate, such as medical treatment and industrial exposure. The economic and social impact of past shortsightedness has only recently been recognized and, more important, addressed in meaningful terms. The Toxic Substances Act, recently passed by Congress, makes it mandatory that all chemicals, including those used for many years, be evaluated for toxicity (Mehlman, 1977). Toxicity is defined here in its broad sense to include mutagens, carcinogens, and teratogens.

The argument that such a testing program is not economically feasible should only serve as a challenge to develop newer methodologies and technology to deal with the problem. It is necessary to pursue this goal, as alternative proposals either not to test substances for toxicity or to return to only natural substances are untenable.

## III. TEST SYSTEM FOR DETERMINATION OF MUTAGENICITY

It is not our purpose here to review in depth a highly technical field, but rather to provide an overview of the available technology, the type of results anticipated from these tests, and the interpretation of the results. Table I lists 10 test systems that provide data as to the chromosomal effects of a potential mutagen, nine test systems designed to evaluate the effects of the mutagen on the genetic material in the form of point mutations, and

**TABLE I**    Test Systems for Mutagenicity[a]

| | | | Chromosomal effects | | | | | | | |
|---|---|---|---|---|---|---|---|---|---|---|
| | | | in vitro | | in vivo | | | | | |
| | Yeast | Droso-phila | Human cells | Mam-malian cells | Mam-malian somatic | Mam-malian germinal | Human somatic | Domin-ant lethal | Trans-location | X-chro-mosome loss |
| **I. Alkylating agents** | | | | | | | | | | |
| A. Azirdines | $\frac{2}{3}$ | $\frac{11}{35}$ | $\frac{16}{83}$ | $\frac{20}{82}$ | $\frac{1}{1}$ | $\frac{11}{54}$ | $\frac{6}{11}$ | $\frac{19}{95}$ | $\frac{3}{8}$ | $\frac{1}{1}$ |
| B. Triazines | $\frac{5}{5}$ | $\frac{2}{13}$ | $\frac{7}{10}$ | $\frac{9}{15}$ | 0 | $\frac{1}{10}$ | $\frac{1}{1}$ | $\frac{1}{25}$ | $\frac{1}{3}$ | $\frac{1}{1}$ |
| C. Nitrogen, sulfur, and oxide mustards | $\frac{9}{15}$ | $\frac{24}{33}$ | $\frac{15}{29}$ | $\frac{20}{44}$ | 0 | $\frac{2}{22}$ | $\frac{8}{24}$ | $\frac{12}{30}$ | $\frac{2}{6}$ | 0 |
| D. Phosphoric acid esters | $\frac{8}{14}$ | $\frac{2}{2}$ | $\frac{23}{62}$ | $\frac{29}{45}$ | 0 | $\frac{2}{2}$ | $\frac{16}{36}$ | $\frac{9}{22}$ | 0 | 0 |
| E. Epoxides | $\frac{1}{1}$ | $\frac{3}{5}$ | $\frac{3}{3}$ | $\frac{5}{5}$ | 0 | $\frac{2}{3}$ | 0 | $\frac{6}{7}$ | $\frac{1}{1}$ | 0 |
| F. Lactones | $\frac{1}{1}$ | 0 | $\frac{4}{6}$ | $\frac{5}{7}$ | 0 | $\frac{1}{1}$ | $\frac{2}{2}$ | $\frac{2}{2}$ | 0 | 0 |
| G. Aldehydes | $\frac{2}{2}$ | $\frac{3}{9}$ | $\frac{4}{5}$ | $\frac{6}{13}$ | 0 | $\frac{1}{1}$ | 0 | $\frac{3}{3}$ | 0 | 0 |
| H. Alkylsulfates | $\frac{1}{3}$ | $\frac{2}{2}$ | $\frac{1}{1}$ | $\frac{1}{1}$ | 0 | $\frac{1}{2}$ | $\frac{1}{1}$ | $\frac{2}{5}$ | 0 | 0 |
| I. Alkane sulfonic esters | $\frac{4}{25}$ | $\frac{11}{49}$ | $\frac{6}{14}$ | $\frac{9}{22}$ | $\frac{2}{2}$ | $\frac{8}{30}$ | $\frac{2}{6}$ | $\frac{10}{89}$ | $\frac{5}{12}$ | $\frac{1}{2}$ |
| J. Diazoalkanes | 0 | $\frac{3}{3}$ | $\frac{3}{3}$ | $\frac{3}{4}$ | 0 | 0 | 0 | 0 | 0 | 0 |
| K. Aryldialkyl-triazenes | $\frac{15}{23}$ | $\frac{3}{6}$ | 0 | 0 | 0 | 0 | 0 | 0 | 0 | 0 |
| L. Alkyl and alkane halides | $\frac{11}{11}$ | $\frac{8}{11}$ | $\frac{14}{15}$ | $\frac{20}{28}$ | 0 | $\frac{2}{2}$ | $\frac{11}{13}$ | $\frac{5}{12}$ | 0 | 0 |
| M. Azoxy- and hydrazo-alkanes | $\frac{1}{1}$ | 0 | 0 | $\frac{1}{1}$ | 0 | 0 | 0 | 0 | 0 | 0 |
| N. Sulfones | $\frac{1}{1}$ | $\frac{0}{0}$ | 0 | 0 | 0 | 0 | 0 | 0 | 0 | |
| **II. N-Nitroso Compounds** | | | | | | | | | | |
| A. Nitrosamines | $\frac{5}{7}$ | $\frac{3}{4}$ | $\frac{2}{4}$ | $\frac{2}{5}$ | 0 | $\frac{1}{1}$ | $\frac{2}{2}$ | $\frac{2}{6}$ | 0 | 0 |

[a] Reprinted by permission from Flamm *et al.*

| | | | | | | Point mutations | | | | DNA repair | |
|---|---|---|---|---|---|---|---|---|---|---|---|
| Salmon-ella | *E. coli* | Yeast | Neuro-spora | Trade-scantia | Droso-phila | Mam-malian cells in culture | Human cells in culture | Mouse | Micro-bial | Mam-malian |
| $\frac{10}{28}$ | $\frac{24}{108}$ | $\frac{2}{4}$ | $\frac{6}{14}$ | $\frac{1}{1}$ | $\frac{34}{53}$ | $\frac{3}{4}$ | $\frac{3}{3}$ | $\frac{2}{13}$ | $\frac{1}{1}$ | 0 |
| $\frac{20}{35}$ | $\frac{15}{19}$ | $\frac{1}{1}$ | $\frac{2}{3}$ | $\frac{2}{2}$ | $\frac{10}{20}$ | $\frac{1}{2}$ | $\frac{1}{1}$ | $\frac{1}{6}$ | 0 | $\frac{1}{1}$ |
| $\frac{25}{74}$ | $\frac{56}{110}$ | $\frac{3}{24}$ | $\frac{5}{42}$ | 0 | $\frac{43}{52}$ | $\frac{7}{42}$ | $\frac{7}{39}$ | $\frac{2}{2}$ | $\frac{4}{6}$ | $\frac{3}{6}$ |
| $\frac{9}{14}$ | $\frac{21}{46}$ | 0 | $\frac{1}{1}$ | $\frac{6}{6}$ | $\frac{9}{10}$ | $\frac{1}{1}$ | $\frac{1}{1}$ | 0 | $\frac{1}{1}$ | 0 |
| $\frac{10}{10}$ | $\frac{3}{6}$ | $\frac{3}{3}$ | $\frac{15}{40}$ | $\frac{1}{1}$ | $\frac{7}{9}$ | $\frac{5}{5}$ | $\frac{5}{5}$ | 0 | $\frac{1}{1}$ | 0 |
| $\frac{2}{12}$ | $\frac{5}{16}$ | $\frac{2}{6}$ | $\frac{1}{2}$ | 0 | $\frac{1}{1}$ | $\frac{1}{1}$ | $\frac{1}{1}$ | 0 | $\frac{1}{1}$ | $\frac{1}{2}$ |
| $\frac{7}{7}$ | $\frac{10}{30}$ | $\frac{1}{1}$ | $\frac{1}{9}$ | 0 | $\frac{3}{14}$ | $\frac{1}{2}$ | $\frac{1}{2}$ | 0 | 0 | 0 |
| $\frac{1}{27}$ | $\frac{3}{23}$ | $\frac{2}{14}$ | $\frac{2}{5}$ | 0 | $\frac{4}{8}$ | $\frac{5}{5}$ | $\frac{5}{5}$ | 0 | $\frac{1}{2}$ | 0 |
| $\frac{13}{58}$ | $\frac{13}{153}$ | $\frac{7}{66}$ | $\frac{2}{31}$ | $\frac{2}{9}$ | $\frac{15}{80}$ | $\frac{4}{76}$ | $\frac{4}{68}$ | $\frac{5}{21}$ | $\frac{2}{15}$ | $\frac{3}{10}$ |
| $\frac{1}{2}$ | $\frac{6}{29}$ | $\frac{3}{3}$ | $\frac{1}{2}$ | 0 | $\frac{1}{2}$ | $\frac{1}{2}$ | $\frac{1}{1}$ | 0 | 0 | 0 |
| 0 | $\frac{1}{1}$ | 0 | $\frac{2}{8}$ | $\frac{1}{1}$ | $\frac{12}{15}$ | 0 | 0 | 0 | 0 | 0 |
| $\frac{47}{61}$ | $\frac{45}{100}$ | $\frac{2}{2}$ | $\frac{6}{7}$ | $\frac{3}{8}$ | $\frac{18}{29}$ | $\frac{2}{5}$ | $\frac{2}{5}$ | 0 | $\frac{3}{1}$ | 0 |
| $\frac{2}{3}$ | 0 | 0 | 0 | 0 | $\frac{4}{5}$ | 0 | 0 | 0 | $\frac{1}{.1}$ | 0 |
| $\frac{2}{3}$ | $\frac{1}{1}$ | $\frac{1}{1}$ | 0 | 0 | 0 | 0 | 0 | 0 | $\frac{1}{1}$ | $\frac{1}{1}$ |
| $\frac{5}{34}$ | $\frac{22}{36}$ | $\frac{4}{14}$ | $\frac{4}{8}$ | $\frac{2}{2}$ | $\frac{8}{13}$ | $\frac{4}{10}$ | 0 | 0 | $\frac{2}{2}$ | $\frac{4}{6}$ |

*(cont.)*

**TABLE 1** (*cont.*)

| | Chromosomal effects | | | | | | | | | |
| | in vitro | | | | in vivo | | | | | |
| | Yeast | Droso-phila | Human cells | Mam-malian cells | Mam-malian somatic | Mam-malian germinal | Human somatic | Domin-ant lethal | Trans-location | X-chro-mosome loss |
|---|---|---|---|---|---|---|---|---|---|---|
| B. Nitrosamides | $\frac{6}{17}$ | $\frac{6}{13}$ | $\frac{1}{6}$ | $\frac{2}{14}$ | 0 | $\frac{1}{1}$ | $\frac{2}{2}$ | $\frac{1}{11}$ | 0 | 0 |
| C. Nitrosoureas | $\frac{5}{6}$ | $\frac{2}{11}$ | $\frac{3}{10}$ | $\frac{5}{18}$ | 0 | $\frac{3}{6}$ | $\frac{1}{2}$ | $\frac{3}{3}$ | 0 | 0 |
| III. Organic peroxides | 0 | 0 | $\frac{1}{1}$ | $\frac{2}{6}$ | 0 | 0 | 0 | $\frac{3}{3}$ | 0 | 0 |
| IV. Polynuclear aromatics | $\frac{9}{12}$ | $\frac{14}{33}$ | $\frac{7}{16}$ | $\frac{13}{26}$ | 0 | 0 | $\frac{5}{5}$ | $\frac{4}{8}$ | 0 | 0 |
| V. Heterocyclics | | | | | | | | | | |
| A. Benzimidaz-oles | $\frac{1}{1}$ | $\frac{1}{1}$ | $\frac{1}{1}$ | $\frac{3}{13}$ | 0 | $\frac{1}{1}$ | 0 | $\frac{2}{2}$ | 0 | 0 |
| B. Fluorenones | $\frac{9}{10}$ | $\frac{8}{11}$ | $\frac{1}{2}$ | $\frac{2}{4}$ | 0 | 0 | $\frac{2}{2}$ | $\frac{2}{2}$ | 0 | 0 |
| C. Phenothia-zines | 0 | $\frac{1}{1}$ | $\frac{4}{4}$ | $\frac{5}{6}$ | 0 | $\frac{2}{2}$ | $\frac{7}{13}$ | $\frac{2}{4}$ | 0 | 0 |
| D. Dicarboxi-mides | $\frac{2}{2}$ | $\frac{3}{5}$ | $\frac{3}{3}$ | $\frac{4}{4}$ | 0 | 0 | $\frac{1}{1}$ | $\frac{4}{7}$ | 0 | 0 |
| E. Thioxanthines | $\frac{2}{3}$ | $\frac{4}{6}$ | $\frac{2}{2}$ | $\frac{3}{3}$ | 0 | $\frac{1}{1}$ | 0 | $\frac{1}{1}$ | 0 | 0 |
| F. Acridines and quinicrines | $\frac{5}{6}$ | $\frac{6}{8}$ | $\frac{10}{11}$ | $\frac{11}{13}$ | 0 | $\frac{1}{3}$ | $\frac{2}{2}$ | $\frac{5}{15}$ | 0 | 0 |
| G. Dibenzo-*p*-dioxins | 0 | 0 | 0 | 0 | 0 | 0 | 0 | $\frac{5}{5}$ | 0 | 0 |
| H. Fluorocou-marins | 0 | 0 | $\frac{4}{4}$ | $\frac{5}{10}$ | 0 | $\frac{1}{1}$ | $\frac{1}{1}$ | $\frac{3}{5}$ | 0 | 0 |
| I. Cyclodienes | 0 | $\frac{1}{1}$ | 0 | $\frac{1}{1}$ | 0 | $\frac{1}{1}$ | $\frac{2}{2}$ | $\frac{7}{11}$ | 0 | 0 |
| J. Other | $\frac{17}{19}$ | $\frac{14}{20}$ | $\frac{50}{85}$ | $\frac{71}{120}$ | $\frac{5}{8}$ | $\frac{5}{6}$ | $\frac{33}{103}$ | $\frac{15}{19}$ | 0 | 0 |
| VI. Inorganic derivatives | | | | | | | | | | |
| A. Metal and metalloid derivatives | 0 | $\frac{18}{19}$ | $\frac{17}{19}$ | $\frac{13}{15}$ | 0 | $\frac{3}{3}$ | $\frac{12}{20}$ | $\frac{10}{11}$ | 0 | 0 |

| | Point mutations | | | | | | | | DNA repair | |
|---|---|---|---|---|---|---|---|---|---|---|
| Salmon-ella | *E. coli* | Yeast | Neuro-spora | Trade-scantia | Droso-phila | Mam-malian cells in culture | Human cells in culture | Mouse | Micro-bial | Mam-malian |
| $\frac{4}{85}$ | $\frac{9}{290}$ | $\frac{10}{67}$ | $\frac{3}{34}$ | $\frac{1}{1}$ | $\frac{4}{13}$ | $\frac{2}{69}$ | $\frac{1}{3}$ | 0 | $\frac{10}{17}$ | $\frac{4}{13}$ |
| $\frac{22}{52}$ | $\frac{3}{27}$ | $\frac{5}{12}$ | 0 | 0 | $\frac{3}{17}$ | $\frac{1}{12}$ | 0 | 0 | 0 | 0 |
| $\frac{2}{2}$ | $\frac{5}{17}$ | 0 | $\frac{3}{4}$ | 0 | $\frac{3}{3}$ | 0 | 0 | 0 | 0 | 0 |
| $\frac{23}{33}$ | $\frac{18}{39}$ | 0 | $\frac{7}{17}$ | 0 | $\frac{14}{31}$ | $\frac{16}{32}$ | $\frac{1}{1}$ | 0 | $\frac{3}{3}$ | $\frac{12}{7}$ |
| $\frac{21}{23}$ | $\frac{11}{19}$ | 0 | 0 | 0 | $\frac{7}{7}$ | 0 | 0 | 0 | 0 | 0 |
| $\frac{16}{42}$ | $\frac{7}{15}$ | 0 | $\frac{2}{3}$ | 0 | $\frac{10}{14}$ | $\frac{8}{19}$ | $\frac{3}{6}$ | 0 | $\frac{4}{5}$ | $\frac{12}{13}$ |
| $\frac{3}{5}$ | $\frac{2}{5}$ | 0 | 0 | 0 | $\frac{7}{14}$ | 0 | 0 | 0 | 0 | 0 |
| $\frac{4}{10}$ | $\frac{3}{12}$ | 0 | $\frac{1}{1}$ | 0 | $\frac{5}{7}$ | 0 | 0 | 0 | 0 | 0 |
| $\frac{18}{34}$ | $\frac{2}{4}$ | $\frac{6}{9}$ | 0 | 0 | $\frac{2}{3}$ | $\frac{4}{8}$ | 0 | $\frac{1}{1}$ | 0 | 0 |
| $\frac{22}{47}$ | $\frac{12}{135}$ | $\frac{11}{31}$ | $\frac{11}{33}$ | 0 | $\frac{8}{14}$ | $\frac{5}{31}$ | $\frac{2}{2}$ | 0 | $\frac{5}{1}$ | $\frac{2}{2}$ |
| $\frac{1}{2}$ | $\frac{2}{2}$ | 0 | 0 | 0 | 0 | 0 | 0 | 0 | 0 | 0 |
| $\frac{6}{8}$ | $\frac{8}{15}$ | 0 | $\frac{6}{13}$ | 0 | $\frac{1}{2}$ | 0 | 0 | 0 | 0 | 0 |
| $\frac{1}{1}$ | $\frac{3}{3}$ | 0 | 0 | 0 | 0 | 0 | 0 | 0 | 0 | 0 |
| $\frac{60}{73}$ | $\frac{59}{134}$ | $\frac{15}{29}$ | $\frac{4}{4}$ | $\frac{2}{2}$ | $\frac{47}{74}$ | $\frac{13}{13}$ | $\frac{1}{1}$ | 0 | 0 | 0 |
| $\frac{15}{22}$ | $\frac{48}{74}$ | $\frac{2}{4}$ | $\frac{8}{8}$ | 0 | $\frac{15}{20}$ | $\frac{1}{1}$ | 0 | 0 | $\frac{52}{1}$ | 0 |

*(cont.)*

TABLE 1 (cont.)

| | | | in vitro | | in vivo | | | | | |
|---|---|---|---|---|---|---|---|---|---|---|
| | | | | | | | | Chromosomal effects | | |
| | Yeast | Droso-phila | Human cells | Mam-malian cells | Mam-malian somatic | Mam-malian germinal | Human somatic | Domin-ant lethal | Trans-location | X-chro-mosome loss |
| B. Halogens and derivatives | 0 | $\frac{9}{9}$ | $\frac{6}{6}$ | $\frac{6}{7}$ | 0 | $\frac{1}{1}$ | $\frac{1}{2}$ | $\frac{5}{5}$ | 0 | 0 |
| C. Sulfur and nitrogen oxides and derivatives | $\frac{1}{1}$ | $\frac{5}{5}$ | $\frac{3}{3}$ | $\frac{6}{7}$ | 0 | 0 | $\frac{1}{1}$ | $\frac{1}{1}$ | 0 | 0 |
| D. Ozone | 0 | $\frac{1}{1}$ | $\frac{1}{1}$ | $\frac{1}{2}$ | 0 | 0 | $\frac{1}{1}$ | 0 | 0 | 0 |
| VII. Natural products A. Mycotoxins 1. Afla-toxin | 0 | 0 | $\frac{4}{5}$ | $\frac{5}{16}$ | 0 | $\frac{1}{1}$ | $\frac{1}{1}$ | $\frac{3}{5}$ | 0 | 0 |
| 2. Other | 0 | $\frac{1}{1}$ | $\frac{3}{6}$ | $\frac{2}{3}$ | 0 | 0 | $\frac{1}{1}$ | $\frac{1}{1}$ | 0 | 0 |
| B. Pyrrolizi-dine alkaloids | 0 | $\frac{7}{7}$ | $\frac{1}{1}$ | $\frac{6}{7}$ | 0 | 0 | 0 | $\frac{1}{9}$ | 0 | 0 |
| C. Antibiotics | 0 | $\frac{5}{12}$ | $\frac{18}{77}$ | $\frac{19}{102}$ | $\frac{2}{3}$ | $\frac{5}{6}$ | $\frac{7}{16}$ | $\frac{8}{21}$ | 0 | 0 |
| D. Xanthines | $\frac{1}{2}$ | $\frac{4}{15}$ | $\frac{12}{43}$ | $\frac{12}{71}$ | 0 | $\frac{3}{18}$ | $\frac{2}{3}$ | $\frac{3}{16}$ | 0 | 0 |
| E. Steroids | $\frac{10}{14}$ | $\frac{14}{20}$ | $\frac{36}{60}$ | $\frac{14}{19}$ | $\frac{3}{4}$ | $\frac{4}{7}$ | $\frac{37}{109}$ | $\frac{9}{9}$ | 0 | 0 |
| VIII. Halogenated ethers and halohydrins | $\frac{3}{4}$ | $\frac{1}{1}$ | $\frac{3}{6}$ | $\frac{3}{8}$ | 0 | $\frac{1}{1}$ | $\frac{1}{1}$ | $\frac{8}{8}$ | 0 | 0 |
| IX. Halogenated hydrocarbons and related derivatives A. Vinal and vinylidine derivatives | $\frac{8}{8}$ | 0 | $\frac{1}{1}$ | $\frac{3}{4}$ | 0 | 0 | $\frac{3}{4}$ | $\frac{4}{6}$ | 0 | 0 |
| B. Halogenated aromatics | $\frac{16}{16}$ | $\frac{7}{7}$ | $\frac{29}{30}$ | $\frac{33}{45}$ | 0 | $\frac{4}{5}$ | $\frac{11}{11}$ | $\frac{13}{20}$ | 0 | 0 |

| | | | | | | Point mutations | | | DNA repair | |
|---|---|---|---|---|---|---|---|---|---|---|
| Salmon-ella | E. coli | Yeast | Neuro-spora | Trade-scantia | Droso-phila | Mam-malian cells in culture | Human cells in culture | Mouse | Micro-bial | Mam-malian |
| $\frac{5}{5}$ | $\frac{18}{29}$ | 0 | $\frac{3}{3}$ | 0 | $\frac{8}{14}$ | $\frac{1}{1}$ | 0 | 0 | 0 | 0 |
| $\frac{10}{27}$ | $\frac{16}{56}$ | $\frac{2}{57}$ | $\frac{4}{4}$ | $\frac{4}{4}$ | $\frac{5}{8}$ | 0 | 0 | 0 | $\frac{1}{2}$ | 0 |
| 0 | $\frac{1}{7}$ | 0 | 0 | $\frac{1}{1}$ | $\frac{2}{2}$ | 0 | 0 | 0 | 0 | 0 |
| $\frac{2}{4}$ | $\frac{2}{4}$ | 0 | $\frac{6}{13}$ | 0 | $\frac{1}{2}$ | $\frac{1}{2}$ | 0 | 0 | $\frac{1}{2}$ | $\frac{5}{2}$ |
| $\frac{7}{7}$ | $\frac{5}{13}$ | 0 | $\frac{2}{2}$ | 0 | $\frac{1}{1}$ | 0 | 0 | 0 | $\frac{1}{1}$ | $\frac{1}{3}$ |
| 0 | 0 | 0 | 0 | 0 | $\frac{8}{10}$ | 0 | 0 | 0 | $\frac{2}{1}$ | 0 |
| $\frac{15}{38}$ | $\frac{101}{300}$ | $\frac{2}{2}$ | $\frac{3}{6}$ | 0 | $\frac{2}{2}$ | $\frac{2}{2}$ | $\frac{2}{2}$ | $\frac{1}{7}$ | $\frac{11}{5}$ | $\frac{2}{1}$ |
| $\frac{2}{5}$ | $\frac{9}{87}$ | $\frac{1}{3}$ | $\frac{2}{3}$ | $\frac{1}{1}$ | $\frac{1}{12}$ | $\frac{2}{5}$ | 0 | $\frac{1}{3}$ | $\frac{1}{1}$ | $\frac{5}{1}$ |
| $\frac{7}{42}$ | $\frac{9}{22}$ | $\frac{3}{3}$ | $\frac{1}{1}$ | 0 | $\frac{8}{10}$ | $\frac{3}{3}$ | $\frac{2}{2}$ | 0 | 0 | 0 |
| $\frac{8}{12}$ | $\frac{8}{20}$ | 0 | $\frac{3}{3}$ | $\frac{1}{1}$ | $\frac{2}{2}$ | $\frac{1}{1}$ | $\frac{1}{1}$ | 0 | 0 | 0 |
| $\frac{6}{10}$ | $\frac{13}{14}$ | 0 | 0 | 0 | $\frac{2}{3}$ | $\frac{4}{5}$ | $\frac{1}{1}$ | 0 | 0 | 0 |
| $\frac{55}{64}$ | $\frac{30}{38}$ | 0 | $\frac{3}{3}$ | $\frac{5}{5}$ | $\frac{16}{21}$ | 0 | 0 | 0 | 0 | 0 |

*(cont.)*

**TABLE 1** (*cont.*)

| | | | in vitro | | in vivo | | | | | |
| --- | --- | --- | --- | --- | --- | --- | --- | --- | --- | --- |
| | Yeast | Droso-phila | Human cells | Mam-malian cells | Mam-malian somatic | Mam-malian germinal | Human somatic | Domin-ant lethal | Trans-location | X-chro-mosome loss |
| C. Fluorocar-bons | $\frac{5}{5}$ | 0 | $\frac{2}{2}$ | $\frac{3}{3}$ | 0 | 0 | $\frac{3}{3}$ | $\frac{4}{4}$ | 0 | 0 |
| X. Nucleic acid bases and analogs | $\frac{4}{5}$ | $\frac{24}{39}$ | $\frac{38}{90}$ | $\frac{50}{183}$ | 0 | $\frac{9}{25}$ | $\frac{18}{43}$ | $\frac{18}{34}$ | $\frac{1}{1}$ | 0 |
| XI. Hydrazines, Hydroxyla-mines, Carba-mates, Hydra-zides, and ureas A. Hydraxines | 0 | 0 | $\frac{1}{1}$ | $\frac{24}{24}$ | 0 | 0 | $\frac{1}{1}$ | $\frac{3}{5}$ | 0 | 0 |
| B. Hydroxyla-mines | $\frac{8}{14}$ | $\frac{5}{7}$ | $\frac{4}{9}$ | $\frac{9}{17}$ | 0 | $\frac{1}{1}$ | $\frac{4}{4}$ | $\frac{6}{7}$ | 0 | 0 |
| C. Carbamates | $\frac{17}{20}$ | $\frac{8}{16}$ | $\frac{7}{28}$ | $\frac{10}{43}$ | $\frac{1}{1}$ | $\frac{4}{11}$ | $\frac{9}{16}$ | $\frac{6}{19}$ | 0 | 0 |
| D. Hydrazides | $\frac{1}{1}$ | $\frac{1}{3}$ | $\frac{1}{5}$ | $\frac{2}{8}$ | 0 | 0 | $\frac{3}{3}$ | $\frac{6}{11}$ | 0 | 0 |
| E. Ureas and thioureas | $\frac{19}{22}$ | $\frac{3}{12}$ | $\frac{6}{15}$ | $\frac{16}{37}$ | 0 | $\frac{1}{3}$ | $\frac{4}{5}$ | $\frac{5}{7}$ | 0 | 0 |
| XII. Aromatic amines | $\frac{12}{14}$ | $\frac{17}{27}$ | $\frac{15}{21}$ | $\frac{18}{24}$ | $\frac{2}{4}$ | $\frac{4}{6}$ | $\frac{9}{21}$ | $\frac{5}{6}$ | $\frac{1}{3}$ | 0 |
| XIII. Axo dyes | 0 | $\frac{12}{21}$ | 0 | 0 | 0 | 0 | 0 | $\frac{1}{1}$ | 0 | 0 |
| XIV. Nitro derivatives | $\frac{5}{13}$ | $\frac{4}{8}$ | 0 | $\frac{29}{44}$ | 0 | $\frac{2}{2}$ | 0 | $\frac{9}{19}$ | 0 | 0 |
| A. Nitroquio-lines and compounds | $\frac{1}{1}$ | 0 | $\frac{1}{10}$ | $\frac{2}{14}$ | $\frac{1}{1}$ | 0 | 0 | $\frac{1}{1}$ | 0 | 0 |
| B. Nitrofurans | $\frac{2}{2}$ | 0 | $\frac{6}{6}$ | $\frac{7}{9}$ | 0 | 0 | 0 | $\frac{1}{1}$ | 0 | 0 |
| C. Nitroimi-dazoles | $\frac{1}{1}$ | $\frac{1}{1}$ | $\frac{1}{4}$ | $\frac{1}{4}$ | 0 | 0 | $\frac{1}{9}$ | $\frac{1}{2}$ | 0 | 0 |
| XV. Organo-metallics A. Organo lead derivatives | 0 | $\frac{5}{5}$ | $\frac{1}{4}$ | $\frac{2}{5}$ | 0 | 0 | $\frac{1}{2}$ | $\frac{1}{1}$ | 0 | 0 |

Chromosomal effects

| | | | | Point mutations | | | | | | DNA repair | |
| Salmon-ella | E. coli | Yeast | Neuro-spora | Trade-scantia | Droso-phila | Mam-malian cells in culture | Human cells in culture | Mouse | Micro-bial | Mam-malian |
| --- | --- | --- | --- | --- | --- | --- | --- | --- | --- | --- |
| $\frac{5}{5}$ | $\frac{10}{14}$ | $\frac{2}{2}$ | $\frac{6}{10}$ | 0 | $\frac{3}{5}$ | 0 | 0 | 0 | 0 | 0 |
| $\frac{27}{67}$ | $\frac{89}{346}$ | $\frac{7}{11}$ | $\frac{10}{14}$ | $\frac{18}{3}$ | $\frac{30}{57}$ | $\frac{16}{41}$ | $\frac{10}{16}$ | $\frac{3}{8}$ | 0 | 0 |
| $\frac{2}{3}$ | $\frac{5}{6}$ | 0 | 0 | 0 | $\frac{2}{4}$ | 0 | 0 | $\frac{1}{2}$ | 0 | 0 |
| $\frac{10}{26}$ | $\frac{20}{54}$ | $\frac{3}{12}$ | $\frac{6}{28}$ | $\frac{4}{5}$ | $\frac{10}{19}$ | $\frac{15}{26}$ | $\frac{3}{5}$ | 0 | $\frac{1}{2}$ | 0 |
| $\frac{24}{49}$ | $\frac{23}{112}$ | $\frac{6}{20}$ | $\frac{4}{13}$ | $\frac{2}{2}$ | $\frac{7}{17}$ | $\frac{3}{6}$ | $\frac{2}{2}$ | $\frac{1}{7}$ | $\frac{2}{1}$ | 0 |
| $\frac{6}{9}$ | $\frac{5}{6}$ | 0 | 0 | $\frac{1}{1}$ | 0 | 0 | 0 | 0 | 0 | 0 |
| $\frac{28}{37}$ | $\frac{22}{42}$ | $\frac{12}{33}$ | $\frac{1}{1}$ | $\frac{2}{2}$ | $\frac{3}{11}$ | $\frac{2}{13}$ | 0 | 0 | 0 | 0 |
| $\frac{22}{28}$ | $\frac{42}{70}$ | $\frac{4}{4}$ | $\frac{10}{15}$ | $\frac{2}{2}$ | $\frac{27}{40}$ | $\frac{6}{6}$ | $\frac{1}{1}$ | $\frac{1}{1}$ | 0 | $\frac{11}{3}$ |
| $\frac{7}{8}$ | $\frac{19}{27}$ | 0 | $\frac{2}{4}$ | 0 | $\frac{26}{38}$ | $\frac{1}{1}$ | 0 | 0 | 0 | $\frac{1}{3}$ |
| $\frac{19}{103}$ | $\frac{31}{331}$ | $\frac{6}{45}$ | $\frac{2}{26}$ | $\frac{3}{3}$ | $\frac{8}{16}$ | $\frac{1}{65}$ | 0 | 0 | 0 | 0 |
| $\frac{2}{15}$ | $\frac{2}{27}$ | $\frac{4}{6}$ | $\frac{1}{3}$ | 0 | $\frac{1}{3}$ | 0 | 0 | 0 | $\frac{9}{6}$ | $\frac{32}{13}$ |
| $\frac{29}{32}$ | $\frac{49}{66}$ | 0 | $\frac{4}{5}$ | 0 | $\frac{4}{5}$ | $\frac{1}{2}$ | $\frac{1}{1}$ | 0 | $\frac{5}{1}$ | 0 |
| 0 | $\frac{5}{5}$ | 0 | $\frac{1}{1}$ | 0 | $\frac{1}{1}$ | 0 | 0 | 0 | 0 | 0 |
| 0 | 0 | 0 | 0 | 0 | 0 | 0 | 0 | 0 | $\frac{1}{1}$ | 0 |

(cont.)

**TABLE 1** (cont.)

| | Chromosomal effects | | | | | | | | | |
| | in vitro | | | | in vivo | | | | | |
| | Yeast | Droso-phila | Human cells | Mam-malian cells | Mam-malian somatic | Mam-malian germinal | Human somatic | Domin-ant lethal | Trans-location | X-chro-mosome loss |
|---|---|---|---|---|---|---|---|---|---|---|
| B. Organo mercury de-rivatives | $\frac{1}{1}$ | $\frac{8}{13}$ | $\frac{8}{10}$ | $\frac{8}{10}$ | $\frac{1}{1}$ | $\frac{3}{3}$ | $\frac{4}{5}$ | $\frac{1}{1}$ | 0 | 0 |
| C. Other | $\frac{32}{34}$ | $\frac{3}{3}$ | $\frac{17}{22}$ | $\frac{4}{4}$ | $\frac{3}{5}$ | 0 | $\frac{16}{33}$ | 0 | 0 | 0 |
| XVI. Miscellaneous A. Esters and anhydrides | $\frac{12}{12}$ | $\frac{7}{7}$ | $\frac{14}{21}$ | $\frac{20}{28}$ | $\frac{1}{2}$ | $\frac{1}{1}$ | $\frac{12}{17}$ | $\frac{3}{3}$ | 0 | 0 |
| B. Quaternary Ammon-ium com-pounds | $\frac{3}{3}$ | $\frac{1}{1}$ | $\frac{3}{7}$ | $\frac{6}{11}$ | 0 | $\frac{1}{1}$ | $\frac{1}{1}$ | $\frac{3}{3}$ | 0 | 0 |
| C. Quinones | $\frac{1}{1}$ | $\frac{7}{10}$ | $\frac{8}{50}$ | $\frac{8}{61}$ | $\frac{2}{3}$ | $\frac{4}{22}$ | $\frac{3}{6}$ | $\frac{4}{15}$ | 0 | 0 |
| D. N-oxides | $\frac{3}{5}$ | 0 | $\frac{2}{11}$ | $\frac{2}{14}$ | $\frac{1}{1}$ | 0 | 0 | $\frac{2}{2}$ | 0 | 0 |
| E. Sulfites | 0 | 0 | 0 | 0 | 0 | 0 | 0 | $\frac{1}{1}$ | 0 | 0 |

two tests to determine whether the presumed mutagen exerts its effect by damaging the DNA repair system. In all, a battery of 21 tests is available if one wishes to test a substance exhaustively for its ability to induce mutations. Very few chemicals are usually subjected to the entire spectrum of tests. It should be noted that not all tests are equally effective in detecting the mutagenic activity of a substance. Furthermore, some tests are designed only to determine whether a substance can induce mutations. Therefore these tests cannot provide data on how much of the substance is required to produce a particular rate of mutation in an organism like man, nor can it distinguish the type of mutational events that are induced. Flamm et al. have listed some of the factors used to determine the efficacy of a test system in assessing the risk of a particular mutagen to individuals, as (1) the genetic and metabolic similarity of the assay system to man, (2) the numbers of mutations induced and the type of mutagenic events detected by the system, and (3) feasibility extrapolation of the data from the somatic assay system to germinal effects.

| | | | Point mutations | | | | | | DNA repair | |
|---|---|---|---|---|---|---|---|---|---|---|
| Salmonella | E. coli | Yeast | Neurospora | Tradescantia | Drosophila | Mammalian cells in culture | Human cells in culture | Mouse | Microbial | Mammalian |
| $\frac{1}{1}$ | $\frac{1}{1}$ | 0 | 0 | $\frac{1}{1}$ | 0 | 0 | 0 | 0 | $\frac{2}{1}$ | 0 |
| $\frac{4}{4}$ | $\frac{4}{9}$ | 0 | 0 | 0 | $\frac{4}{5}$ | 0 | $\frac{3}{3}$ | 0 | $\frac{1}{1}$ | 0 |
| $\frac{17}{17}$ | $\frac{40}{77}$ | $\frac{2}{2}$ | 0 | $\frac{1}{1}$ | $\frac{13}{18}$ | 0 | 0 | 0 | 0 | 0 |
| $\frac{9}{10}$ | $\frac{7}{17}$ | $\frac{4}{4}$ | 0 | 0 | $\frac{2}{2}$ | $\frac{1}{1}$ | 0 | 0 | 0 | 0 |
| $\frac{2}{2}$ | $\frac{23}{98}$ | 0 | $\frac{1}{1}$ | $\frac{1}{1}$ | $\frac{15}{15}$ | $\frac{2}{2}$ | $\frac{1}{1}$ | $\frac{1}{7}$ | 0 | 0 |
| $\frac{14}{33}$ | $\frac{8}{34}$ | $\frac{8}{10}$ | $\frac{2}{6}$ | 0 | $\frac{1}{1}$ | 0 | 0 | 0 | 0 | 0 |
| 0 | $\frac{1}{2}$ | $\frac{1}{1}$ | 0 | $\frac{1}{1}$ | $\frac{1}{1}$ | 0 | 0 | 0 | 0 | 0 |

Table I provides some idea of the magnitude of the problem and the efforts expended on its solution. Over 60 classes of compounds have been identified as possible mutagens. Each class contains a large number of specific chemicals. This table lists the number of compounds in a class that have been specifically tested as the numerator and the number of papers published on the mutagenic activity of that compound in the test in question as the denominator. As an example, under the alkylating agents of the substances classified as alkane sulfonic esters, 58 have been tested for chromosomal effects, 65 for point mutations, and five for DNA repair defects. Since each compound may be tested by several assay systems, these numbers do not necessarily reflect the total number of different compounds tested. Over 830 publications have resulted from these studies, indicating a high degree of concern about the mutagenicity of these compounds.

The assessment of risk is a very complicated issue that must take into account genetic, biochemical, toxicological, economical, clinical, epidemiological, and statistical factors. It is beyond the scope of this chapter to deal with the detailed scientific, economic, and ethical considerations that must be taken into account before a substance can be rigorously controlled.

The other problem that is equally troublesome is the detection of early signs of disease in those individuals known to have been exposed to a substance that has the potential to produce genetic damage. The methods for early screening at our disposal at the present time are rather limited. A variety of biochemical tests are available to detect increased levels of known toxicological agents [i.e., lead, asbestos, polyvinyl chloride (PVC), etc.]. Whether these substances will cause an exposed individual eventually to contract cancer or give birth to a deformed child is beyond current knowledge. Attempts are being made to screen industrial workers exposed to known mutagens for the presence of abnormal chromosomes in an effort to identify high-risk individuals.

## IV. TEST SYSTEMS FOR DETERMINATION OF CARCINOGENICITY

The ability of a substance to produce effects on cellular genetic information leading to the neoplastic transformation of a cell is of immediate concern to everyone. The test systems designed to assess mutagenicity of a substance can also be used to determine carcinogenic potential. In addition, cytogenetic analysis of human chromosomes either obtained from an exposed individual or treated in culture with the suspect agent have proved to be a valuable aid in diagnosis and prognosis (Fuscaldo *et al.*, 1978; Fuscaldo *et al.*, 1980; Levan and Mitelman, 1977; Pederson, 1975). Boveri proposed in 1914 that one of the basic anomalies of malignant cells was alterations in chromosomal structure (Boveri, 1914). This observation has been consistently supported by many workers (German, 1974). The specificity of the chromosomal alterations associated with malignancy is receiving increased attention, particularly since the discovery of the Philadelphia chromosome (Ph[1]) in 1960 (Nowell and Hungerford, 1960). This abnormal chromosome has become diagnostic of chronic myelogenous leukemia. With the advent of banding techniques, additional nonrandom chromasomal changes are being found in specific human diseases such as meningiomas (Mark, 1977), Burkitt's lymphoma (Mark, 1977), acute myelogenous leukemia (Mitelman *et al.*, 1976; Pederson, 1975), and a variety of other hematological neoplasias (Mark, 1977; Levan and Mitelman, 1977) and preleukemic states (Brodsky *et al.*, 1977; Wurster-Hill *et al.*, 1976). As Rowley has stated (Rowley, 1979a), the question is no longer whether there are nonrandom chromosomal anomalies associated with human neoplasia, but what the role of these changes are in malignant transformation.

For the purpose of this review the question revolves around what the cytogeneticist can do to assess the potential of a substance to cause malignant transformation. *In vitro* testing of substances on human lymphocytes or cells in culture is one method that has proved useful. The second approach that has some merit is the cytogenetic monitoring of individuals who have been exposed to potential carcinogens or who have diagnosed neoplastic disease (Fuscaldo *et al.,* 1980; Brodsky *et al.,* 1979).

The appearance of chromosomal abnormalities such as breaks or gaps in the chromosome structure (Fig. 1), deletions and translocations (Figs. 2, 3, and 4), or the loss or addition of whole chromosomes (Figs. 5 and 6) can be indicative of chromosomal instability that would lead to or be a consequence of some malignant process.

An example of the *in vitro* testing of chemicals for their mutagenic and carcinogenic activities is the recent experience with coloring agents. Hair dyes, food dyes, and chemicals used in the graphic arts and printing have been shown to be carcinogenic in some instances. These observations led Au and Hsu (1979) to study the cytogenetic toxicity of biological stains and dyes, some of which, such as gentian violet, are used as antifungal agents in animal and poultry feed.

The second aspect, that of monitoring individuals already exposed to a potential hazard or known to have neoplastic disease, is exemplified by two studies in progress in our laboratories. One study involves the evaluation of

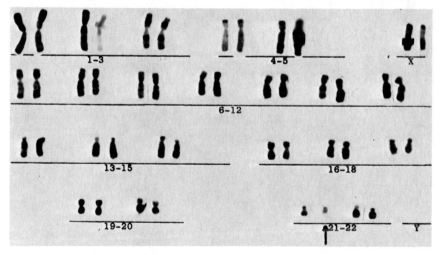

**Fig. 1.**   Examples of chromosomal breaks and gaps (arrows). Standard staining.

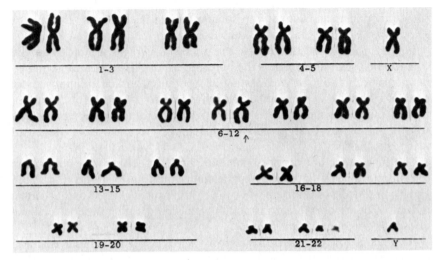

**Fig. 2.** Triradial configuration involving all of the short arm, the centromere, and a short segment of the long arm of chromosome 2 and all of chromosome 3. (46XX del 2q, triradial 2:3). Most of the long arm of chromosome 2 (2q) has been deleted. G banding.

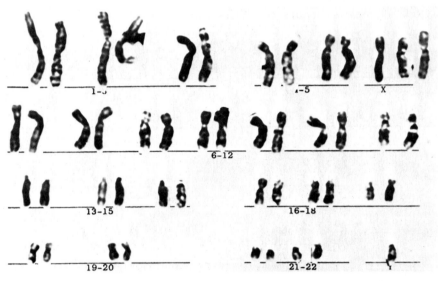

**Fig. 3.** Deletion of chromosome 3 and 12 and translocation of the two broken chromosomes [45XX -3, -12, +t (3:12)]. G banding.

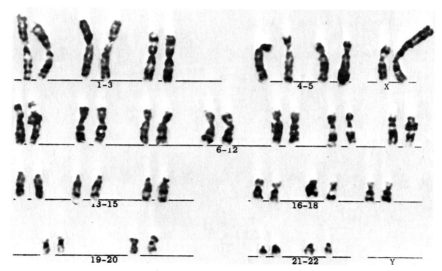

**Fig. 4.** A duplication of part of an X chromosome [46XX dup (X)]. Resulting karyotype shows additional X chromosome material. G banding.

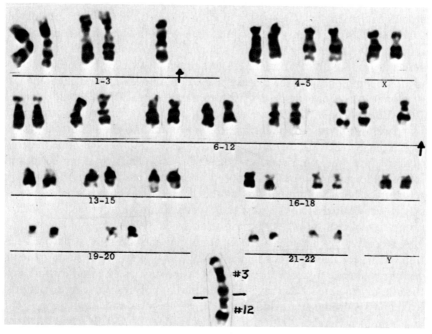

**Fig. 5.** Two additional X chromosomes—variant of Klinefelter's syndrome (48XXXY). G banding.

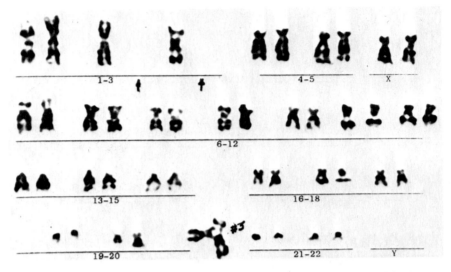

**Fig. 6.** CML patient showing two Philadelphia chromosomes in some of his dividing bone marrow cells (47XYPh[1]Ph[1]). Characteristic of advancing disease. Standard staining.

the efficacy of intensive chemotherapy based on sequential cytogenetic studies in patients diagnosed has having Ph[1] positive chronic myelocytic leukemia (CML). Twenty-six patients have been induced into hematologic remission with the drug busulfan, splenectomized, and treated with cycle-active drugs. As a part of treatment, maintenance therapy is continued during the benign phase of the disease. Based on the well-known phenomenon of clonal evolution during disease progression in CML (Fuscaldo *et al.,* 1980; Brodsky *et al.,* 1979), we are attempting to prevent or delay the onset of aggressive disease by identifying and eliminating the abnormal clones with intensive therapy. Accordingly, chromosomal surveys of bone marrow aspirates are done on a routine sequential basis. Therapy is modified and intensified when abnormal chromosomal patterns are found in the dividing cells of the bone marrow or unstimulated peripheral blood. The results of this as yet limited study suggest that survival can be improved by such a program (Brodsky *et al.,* 1979; Fuscaldo *et al.,* 1980).

A second study has identified an abnormal chromosome, a partial deletion of the long arm of chromosome 21 (21q-) (Fig. 7), in some patients with preleukemic myeloproliferative diseases, such as essential thrombocythemia, myelofibrosis and myeloid metaplasia, polycythemia vera, and preleukemic syndrome (Fuscaldo *et al.,* 1978; Zaccaria and Tura, 1978; Petit and Van Den Berghe, 1979; Van Den Berghe, *et al.,* 1979). Like the 8/21 translocation seen in acute myelogenous leukemia (Trujillo *et al.,* 1979), this abnormality is usually seen in only a small proportion of the dividing

**Fig. 7.** Deletion of part of the long arm of chromosome 21 (46XX, 21q-). R banding.

bone marrow or nonstimulated peripheral blood cells of some patients with clinical evidence of disease. A correlation appears to exist between the presence of the 21q- and a retro-virus in these patients (Fuscaldo *et al.*, 1978). These studies are being expanded to confirm the correlation and to determine, if possible, whether a causal relationship exists between these observed phenomena.

Finally, a study is under way in our laboratory to determine whether the cytogenetic analyses of human bladder tumors will provide prognostic data relative to tumor invasiveness and response to therapy. Preliminary results (Gonick *et al.*, 1980) suggest that the appearance of abnormal clones in cultured or direct preparations of tumor tissue can be correlated with poor prognosis.

The three studies summarized here offer a very limited view of the potential for human cytogenetics in clinical medicine being done in our laboratory.

The continuing interest in and importance of monitoring procedures to assess the impact of potential hazardous substances make it essential that existing procedures be improved and new ones be developed. The cost, both to the individual and to society, of the unrestrained use of and exposure to hazardous substances is too great not to take appropriate protective steps. Although the initial cost of research and development will be high, the long-term benefit will certainly more than offset those costs.

## REFERENCES

Au, W., and Hsu, T. C. (1979). Clastogenic effects of biological stains and dyes. *Mutat. Res.* **58**, 269–276.

Boveri, T. (1914). "Zur Frage der Entstehung maligner Tumoren." Fischer, Jena (Engl. transl. by M. O'G. Boveri, Williams & Wilkins, Baltimore, Maryland, 1929).

Brewen, J. G. (1976). Practical evaluation of mutagenicity data in mammals for estimating human risk. *Mutat. Res.* **41**, 15–24.

Brodsky, I., Fuscaldo, A. A., Erlick, B. J., and Fuscaldo, K. E. (1977). Effect of busulfan on oncornavirus-like activity in platelets and chromosomes in polycythemia vera and essential thrombocythemia. *J. Natl. Cancer Inst.* **59**(1), 61–67.

Brodsky, I., Fuscaldo, K. E., Kahn, S. B., and Conroy, J. F. (1979). Myeloproliferative disorders. II. CML: Clonal evolution and its role in management. *Leuk Res.* 3(6):379–393.

Flamm, W. G. and DHEW Working Group on Mutagenicity Testing (1980). Approaches to determining the mutagenic properties of chemicals: Risk to future generations. *J. Environ. Pathology and Toxicology* **1**, 301–352.

Fuscaldo, K. E., and Brodsky, I. (1978). Chronic myelocytic leukemia: predictive value of sequential surveys and their effect on survival. *Boll. Ist. Sieroter. Milan.* **57**(3), 306–321.

Fuscaldo, K. E., Erlick, B. J., and Fuscaldo, A. A. (1978). Correlation of a specific chromosomal marker, 21q- and retroviral indicators in patients with thrombocythemia. *Cancer Lett.* **6**, 51–56.

Fuscaldo, K. E., Brodsky, I., Kahn, S. B., Conroy, J. F., and Marfurt, K. (1980). Myeloproliferative Disorders: III CML: Further studies on the role of cytogenetics in diagnosis, prognosis and management. (in press). *Haematologica.*

German, J., ed. (1974). "Chromosomes and Cancer." Wiley, New York.

Gonick, P., Kalathoor, R., and Fuscaldo, K. E. (1979). Chromosomal analysis as a prognostic tool in transitional cell carcinoma: A preliminary report. Submitted for publication.

Kucerova, M. (1976). Cytogenetic analysis of human estimation of genetic risk. *Mutat. Res.* **41**, 123–130.

Lang, R., and Adler, I. D. (1977). Heritable translocation test and dominant-lethal assay in mice with methyl methanesulfonate. *Mutat. Res.* **48**(1), 75–88.

Levan, A., Levan, G., and Mitelman, F. (1977). Chromosomes and Cancer. A Review. *Hereditas,* **86**(1), 15–29.

Mark, J. (1977). Chromosomal abnormalities and their specificity in human neoplasms: An assessment of recent observations by banding techniques. *Adv. Cancer Res.* **24**, 165–22.

Mehlman, M. A. ed. (1977). "Carcinogenesis and Mutagenesis." Pathox Publishers, Inc., Park Forest South, Illinois.

Mitelman, F., Nelsson, P. G., Levan, G., and Brandt, L. (1976). Nonrandom chromosome changes in acute myeloid leukemia: Chromosome banding examination of 30 cases of diagnosis. *Int. J. Cancer* **18**, 31.

Nowell, P. C., and Hungerford, D. A. (1960). A minute chromosome in human chronic granulocytic leukemia. *Science* **132**, 1497.

Pederson, B. (1975). Possible mechanisms of pathogenesis and acute transformation in chronic myeloid leukemia. *Ser. Haematol.* [N.S.] **8**, 45.

Petit, P., and Van Den Berghe, A. (1979). A chromosomal abnormality (21q-) in primary thrombocytosis. *N. Engl. J. Med.* (Submitted for publication).

Rowley, J. D. (1977a). Nonrandom chromosomal changes in human malignant cells. *ICN-UCLA Symp. Mol. Cell. Biol.* **7**, 457–472.

Rowley, J. D. (1977b). *In* "Population Cytogenetics: Studies in Humans." (E. B. Hook and I. H. Porter, eds.), p. 187.

Sobels, F. H., and Vogel, E. (1976). The capacity of Drosophila for detecting relevant genetic damage. *Mutat. Res.* **41**, 95–106.

Stetka, D. G., and Wolff, S. (1976). Sister chromatid exchange as an assay for genetic damage induced by mutagen-carcinogens. I. In vivo test for compounds requiring metabolic activation. *Mutat. Res.* **41**(2–3), 333–342.

Trujillo, J. M., Cork, A., Ahearn, M., Youness, E. L., and McCredie, K. B. (1979). Hematologic and cytologic characterization of 8/21 translocation in acute granulocytic leukemia. *Blood* **53**(4), 695–706.

Van Den Bergh, A., Petit, P., Broeckaent-Van Orshoven, A., Louwagie, A., De Baene, H., and Verwilghen, R. (1979). Simultaneous occurrence of 5q- and 21q- in refractory anemia and thrombocytosis. *Cancer Genetics and Cytogenetics,* **1**(1), 63–68.

Wurster-Hill, D., Whang-Peng, J., and McIntyre, O. K. (1976). Cytogenetic studies in polycythemia vera. *Semin. Hematol.* **13**, 13.

Zaccaria, A., and Tura, S. (1978). A chromosomal abnormality in primary thrombocythemia. *N. Engl. J. Med.* **298**, 1422.

# Chapter Eleven

# Behavioral Factors in Laboratory Safety: Personnel Characteristics and the Modification of Unsafe Acts

JOAN C. MARTIN

LABORATORY SAFETY: THEORY AND PRACTICE
Copyright © 1980 by Academic Press, Inc.
All rights of reproduction in any form reserved.
ISBN 0-12-269980-7

## I. BEHAVIORAL FACTORS THAT RESULT
## IN ACCIDENTS

### A. Statistics

Accidental laboratory infections that resulted in death were found to occur at a higher rate (4% of such infections) than death either from motor vehicle accidents (2.7%) or from all U.S. accidents combined (1.0%) (Phillips, 1969). The accident rate in the United States in 1974 for all work settings was 8.0 for males and 1.2 for females for every 100 workers (U.S. Department of Commerce, 1976). In contrast, injury and illness rates in hospitals for that year were 9.2 per 100 employees (Bureau of Labor Statistics, 1976). It would seem that even the high degree of training and education present in laboratory and hospital workers is not proof against a high injury and accident rate.

A compendium of industrial studies found that 80–90% of all accidents were traceable to human error, with only 10–15% due to faulty machinery or job conditions. A similar result was found in laboratory settings by Phillips (1965), in which 65% were due to human error and 20% were due to equipment problems. The remaining 15% when scrutinized carefully were due to "unsafe acts" and hence fell under the rubric of human error as well. It is critical that human behavior be modified in order to reduce this percentage, since equipment modification is not the whole answer, although a distraction-free environment helps (Kennan et al., 1951).

It is a truism that the object of safety training in the laboratory is to prevent or diminish the accident rate. Individual actions must be defined so as not to *adapt* to new conditions, since the laboratory conditions themselves do not change, but to *adjust* behaviors skillfully to correspond with safe actions (McGlade, 1970).

The following factors have been found to have an effect on accident and injury rates.

### B. Age and Sex

A thesis by Davis (1977) that investigated accidents on the psychiatric units of three hospitals found that mental health staff males had two and one half times as many reported incidents as did female staff members and that older workers had significantly fewer accidents than did those who were younger. Government statistics from all work settings in 1972 found that men had approximately seven times the accident rate of women. The rate for the 17–24-year-old group was approximately twice that of the 45–64-year-old group (Vital and Health Statistics, 1976).

A study at Fort Detrick Biological Laboratories also found that individuals in the 20–29-year age group had an abnormally high accident rate. In addition, women in their laboratories were involved in significantly fewer accidents than were men (Phillips, 1965).

## C. Accident Repeaters

Whereas the concept of an "accident-prone" individual has been both defended (LeShan, 1952a; Rodstein, 1974) and challenged (Suchman and Scherzer, 1964), there is no doubt that a few individuals are involved in repeated accidents over their working lives. According to Rodstein (1974), they have had a childhood marked by early exposure to violence, severely authoritarian parents, truancy, and a history of many childhood accidents. As adults they have childless marriages or small families, a high divorce rate, frequent job changes, and demonstrable irresponsibility and aggressiveness. In addition, their impulsivity under stress probably predisposes them to accidents. LeShan (1952b) performed a study on male workers in industry, truck drivers, and drivers, in an automobile clinic, all of whom had records of two major and one or more minor accidents per year. His control group consisted of similar individuals who had been accident-free for three years. The accident repeaters were described as being social isolates in spite of a wide circle of acquaintances; they had no close emotional ties with other individuals. They were upwardly mobile individuals who had failed to rise in the social structure; they demonstrated aggression towards authority in the form of their supervisors and planned poorly and erratically for the future, often marking goals as "undecided." They blamed "fate" or "luck" for their accidents, and demonstrated impulsive, acting-out behavior with low frustration tolerance.

Too much stress should not be placed on the importance of such individuals, since one estimate finds that they account for only 0.5% of all accidents, and hence have a negligible effect on prevention.

## D. Accident-free Individuals

The "safe" counterpart of the accident repeater has been described by Phillips (1965) in the laboratory situation and by LeShan (1952b) in the industrial setting. Le Shan's group averaged only one trip to the dispensary for 13 man-years worked. These men had warm relationships with others, particularly their immediate families. A good deal of status was attached to work, and all that is part of the job, such as titles and salary. Money was important to them. They were not upwardly mobile, but wished to do well where they were in the social structure. Foremen and peer comments were quite favorable. Phillips (1965) found that at Fort Detrick the accident-free individuals had closer family ties, had more children, were less likely to have been divorced, took responsibility for their own safety in the laboratory with "defensive" work habits, and were less likely to be smokers and drinkers than were the accident-involved individuals.

## E. Temporary Emotional Stress

Gibbs (1968) found that stress has comparatively little effect upon over-learned and familiar behaviors but may impair acts that require new responses. Kerr (1964) reported a study of 400 accidents in the work setting that found that more than half of these accidents took place when the in-dividual was worried, apprehensive, or in a low emotional state. Indi-viduals who are in the throes of such stress should not be permitted to work in dangerous situations.

## F. Prolonged Emotional Disruption: The Maladjusted Individual

The percentages of individuals who seek medical help whose symptoms are actually due to behavioral problems ranges from 30% to 80%. How-ever, probably only 10–15% of the general population has serious emo-tional problems (MacIver, 1961). Such individuals should probably never be placed in a potentially dangerous situation. They can often be fairly eas-ily identified by supervisors by former work and health records and unusual behavior patterns.

Also, the socially deviant individual whose primary identity lies within a subculture rather than the cultural norm was three to five times more likely to have incurred accidental injuries over a year's time than an individual in a more conventionally oriented peer group (Suchman, 1970). It was hy-pothesized that such accidents result from a greater exposure to risk related to a departure from "safe and sane" behaviors.

Therefore, if one were to attempt to hire the laboratory worker with the least probability of being involved in an accident, one would hire a married older woman who was well adjusted, a member of the dominant cultural group, a nonsmoker and nondrinker, and the mother of more than the av-erage number of children. It is not seriously suggested that a laboratory limit itself to such a policy. Behavioral change is possible in all individuals and unsafe acts are capable of modification.

## II. TASK FACTORS RELATED TO ACCIDENTS

### A. Intrinsic Job Factors

There are some occupations that by virtue of the tasks involved pose greater risks than others. In some of them, such as skyscraper construction and tunneling, the risks are obvious. Others were originally not seen as dangerous and it was not until deaths due to injury or illness accumulated that the risk was identified. For example, Golodets *et al.* (1963) collected clinical cases in the U.S.S.R. over a 20-year period. Radiologists and radiological technicians who had worked in the early, poorly shielded laboratories presented with behaviors reminiscent of early senility: memory loss, exhaustion, emotional lability, and headaches. Apparently the symptoms, that were similar to those of arteriosclerosis, were due to the effects of X irradiation on cerebral vascularization. As an interesting aside, U.S. radiologists have the shortest life spans of any of the medical specialties. Some additional work settings that have been shown to be hazardous include textile and insulation factories (asbestos) (Newhouse *et al.*, 1972; Berry *et al.*, 1972) and rubber manufacturing (Fine and Peters, 1976). Some settings that have resulted in increased spontaneous abortions in gravid workers have included tobacco factories (Gavrilescu, 1973), plants using chlorinated hydrocarbons and gasoline (Makhametoza and Vozovaia, 1972), and operating rooms (anesthetics) (Cohen *et al.*, 1971). The possibility certainly exists that laboratory technicians and other personnel who handle viral and carcinogenic agents may also incur a greater risk to their subsequent health. It is therefore critical that individuals in such settings practice safe behaviors at all times (Yager, 1973).

### B. Task Complexity and Variability

Human factors studies done mostly in military settings have found that as tasks increase in complexity the error rate goes up proportionately. Therefore, if a particular laboratory procedure is associated with a high error or accident rate, it is probably worth examining for procedural simplification. In addition, the greater the number of ways in which a task can be performed, the greater the likelihood for error. If individuals are allowed to select procedures dependent upon the agent involved, the inevitable day will come in which the wrong procedure is used with perhaps serious consequences. It is a good idea to use the safest and/or most conservative procedure with all agents and thus reduce the number of options available.

## III. THE MODIFICATION OF UNSAFE BEHAVIOR PATTERNS

### A. Principles of Behavioral Analysis

Behavior analysis utilizes principles of both classical and instrumental (operant) conditioning (learning).

The form of learning that is applicable to most of human behavior, and the one that is of major concern to us, is the second type. An individual makes voluntary, learned responses that are governed by the environmental consequences following the behavior (i.e., rewards or punishments). It is these responses that can be modified, eliminated, or replaced by altering the consequences following the behavior (i.e., rewards or punishments). The procedure of eliminating one set of responses and substituting another set is called *behavioral modification* or *applied behavioral analysis* (Mikulas, 1972; Whaley and Malott, 1971; Malott *et al.,* 1978). Techniques for behavioral change, if used carefully, can be implemented by nonprofessionals (Ayllon and Michael, 1959). This approach is based upon the following principles:

1. An analysis of the actual behavior of the individual, rather than the individual's "thoughts" or "feelings" about the behavior.
2. Environmental events such as monetary rewards, job loss, and praise by peers may be used to modify the behavior in question.
3. Positive reward or *reinforcement* is used whenever possible. Negative reinforcement (punishment) does not result in comparable lasting change and is apt to generate anxiety and unpleasant affect as well (Estes, 1944).
4. Division of the behavior into identifiable and quantifiable segments (e.g., the number of steps to be added or eliminated to a biochemical procedure).
5. *Timing* and *consistency* are the watchwords.

The changes in the consequences following the behavior should be programed consistently and systematically. Ideally, the old behavior would never be reinforced and the new behavior would be well-defined and consistently rewarded.

### B. Modeling and Shaping Behaviors

If the new set of responses is not in the behavioral repertoire of the individual, two procedures that can be used to add them to the behavior are *shaping* and *modeling.*

Shaping consists of reinforcing responses that are already in the person's repertoire that resemble the desired behavior. Gradually, a more stringent approximation is required for reward to occur, until the individual is emitting the desired set of responses. This procedure is particularly appropriate for complex tasks and with animals, children, and individuals with learning difficulties.

Modeling is a more efficient method for use with normal adults. A knowledgeable individual models the new set of responses, which are then imitated by the person who is to learn them. The learner's errors are corrected both verbally and by example until his performance approximates some criterion level set by the modeler. Bandura (1965) and Bandura *et al.* (1969) have described these procedures extensively. There are pitfalls to be considered. These include some behavioral characteristics of both the model and the learner. For example, the model should not be of lower *status* than the person to be instructed.

## C. Employee Expectancies

Individuals model their behavior not only on the sequence of responses they observe but on what they deduce is expected of them (Clark, 1927). They pick up relevant cues from the gestures, tone of voice, and observations of the work patterns of the supervisor when he is not modeling for the employee (Greenspoon, 1955). "Do what I say and not what I do" is not an acceptable attitude if one plans to modify behavior. The anticipation of positive or negative reinforcement by the learner plays a major role in the success of the modeling sequence. If an employee perceives from his or her behavior that a supervisor is only partially convinced of the necessity for behavioral modification, he is not likely to put forth the effort to reach an acceptable performance level. He deduces that negative reinforcements will not be forthcoming for an incorrect response (Rosenthal and Jacobsen, 1969). King (1978) has described some principles in terms of the laboratory. Therefore:

1. A model must behave in a positive manner in demonstrating the new procedures.
2. The contingencies under which reinforcement occurs for the correct sequence, and punishment follows use of the old method (if punishment is to be used), must be clearly spelled out.
3. The learner should not be allowed to practice competing responses (i.e., the old method) at the same time that he is learning a new method. The reasons for this are obvious. A new response requires practice and attention before it is mastered. If at the same time an in-

dividual is allowed to continue an old, overlearned response, the new behavior will appear hopelessly clumsy by comparison. The learner will derive little personal reward from his/her attempts, since the increments in skill that are achieved will appear small by comparison with the skill at the old method. Very shortly he/she will drop back to the old method entirely, accompanying this return by verbalizations such as, "I cannot learn the new method," "It is too clumsy," or "It slows down the research."

Therefore insist that *all* use of the old method cease on the day that the new method is instituted. Do not use gradual withdrawal, since this reinforces the wrong behavior and merely postpones the inevitable day when the use of the method must stop.

## D. The Modeler

The model should not be a lower status individual than the person being instructed. He or she can be equal in status as long as a supervisor has provided him or her with the necessary authority.

The model and the immediate supervisor, if they are not the same individual, should both be convinced of the necessity for instituting the behavioral change. Otherwise, no one will learn the new method well, since they will deduce that no negative consequences will follow (Rosenthal, 1973).

Relevant to the behavior of the model, or supervisor, are the results on shared decision making in organizations obtained by Rogers and Shoemaker (1971). Briefly stated, they found that an individual's acceptance of change in working behaviors is directly related to his participation in the decision to change i.e., all workers had a vote. In situations in which it was not possible for the staff to have a voice in the decision, the change was made smoother when the reasons for change were carefully explained. It also helps to allow some choice by laboratory staff [e.g., choosing the *type* of hand pipette even if they cannot choose whether or not to use one (Schick and Hutchinson, 1974)].

## E. Contingencies Governing Change

Over the past 10–15 years a great deal of research has been performed that has investigated the efficacy of *positive* versus *negative reinforcement* (consequences) in effecting behavioral change. A combination of both may also be used if the old response is punished and the new behavior rewarded. Table I describes the experimental results of these studies that have utilized

**TABLE I**

**Effects of Positive and Negative Stimuli**

|  | Positive stimulus (reward) | Aversive stimulus (punishment) | Threat of Aversive stimulus[a] |
|---|---|---|---|
| Stimulus presented | A reward results in an *increase* in the frequency of the behavior rewarded | A punishment results in a *decrease* in the frequency of the behavior being punished, i.e., an *escape* | A threat of punishment results in a *decrease* in the frequency of the behavior being threatened, i.e., *passive avoidance* |
| Stimulus removed | Removal of the reward results in a *decrease* in the frequency of the behavior being rewarded, i.e., a return to *baseline* level of occurrence | Removal of the punishment results in an *increase* in the frequency of the behavior that had been punished, i.e., a return to *baseline* level of occurrence | Removal of the threat results in an *increase* in the frequency of the behavior previously threatened |

[a] There is also an *active* avoidance situation in which the individual is threatened with punishment if he/she *fails* to perform a certain response. In this case there is an *increase* in that behavior, e.g., "I will fire you if you do not do this my way", and a corresponding *decrease* in the frequency of occurrence when the threat is removed.

all types of organisms, from reptiles to man. An example of their applications to laboratory personnel is found in Table II.

A reward is to be preferred to aversive or threatening consequences, since its effects are longer lasting in terms of behavioral change. Punishment only suppresses the behavior for a given time period, after which the response returns unless an alternative set of responses followed by positive consequences has been programmed to replace it. Behaviors that are followed by positive consequences, on the other hand, tend to persist even when the reinforcers have been rescheduled to occur with less frequency.

## F. Schedules of Reinforcement

The simplest schedule is continuous reinforcement: One response results in one reward. This insures rapid learning initially but does not occur naturally very often and is not necessary for adult learning. A *fixed ratio* schedule requires a specified number of responses prior to the consequence.

**TABLE II**

**An Example from the Laboratory of Table I[a]**

|  | Free afternoon | Monetary fine | Threat of fine if response is *made* | Threat of fine if *fail* to make response |
|---|---|---|---|---|
| Stimulus presented | *Increase* in the frequency of instrument pipetting | *Decrease* in the frequency of mouth pipetting | *Decrease* in the frequency of mouth pipetting | *Increase* in the frequency of instrument pipetting |
| Stimulus removed | *Decrease* in the frequency of instrument pipetting | *Increase* in the frequency of mouth pipetting | *Increase* in the frequency of mouth pipetting | *Decrease* in the frequency of instrument pipetting |

[a] Respose rewarded: instrument pipetting. Response punished: mouth pipetting. Consequence: positive—free afternoon, negative—monetary fine.

This results in a high frequency of response and very stable behavior. With a *variable ratio* or *variable interval,* consequences follow in a semirandom pattern not predictable by the individual. Behavior under the control of these schedules is very resistant to extinction. A gambler operates under a variable ratio schedule and the gambler's fallacy results from this infrequent pattern of reinforcement, i.e., "Since I haven't won in several months I must be due soon." The fallacy occurs because if it is a truly random schedule, then the event is as likely to occur far in the future as soon (Latham and Dossett, 1978).

A concern often expressed by individuals who plan to use positive reinforcement to effect a change in someone's behavior can be expressed thus: "If I begin by rewarding correct behaviors with money (or praise, afternoons off, parties, etc.), then I will have to do this forever, and I do not have that much money or time to put into it." Fortunately, this problem need not occur.

Very often other reinforcers that have been associated with the original reward gradually take the place of the original reward, e.g., social praise by peers and supervisors. These *secondary* or *conditioned* reinforcers may be rewarding enough by themselves that the original reinforcement becomes no longer necessary. Verbal praise, a secondary reinforcer, is far more effective in maintaining behavior patterns in dogs than are food treats. This is often the case with middle-class children and adults, too.

Also, once the behavior has stabilized on one schedule, you can begin rewarding less often and gradually drop it out entirely. By this point, the new behavior has been so well learned that a job well done is reinforcing of itself.

The schedule changes must be carefully programmed so that no change occurs until the individual is responding well on the old schedule. At that point a schedule yielding a lower rate of reinforcement can be put into effect (Sidman, 1962).

## G. Positive Versus Negative Reinforcers

As we have stated, one method of increasing the frequency of a response is to use an effective reward. If the goal is to substitute a desired behavior for a competing, undesirable one, then reinforcement of the new response can be coupled with either punishment or the threat of punishment for the *old* behavior. The threat of punishment probably should never be used with the new response (e.g., "If you don't wear safety shoes, you will be fined"), since positive reward alone is more effective (e.g., "If you wear safety shoes, you will get an afternoon off next week"). Several combinations of positive and negative reinforcement may be used (McKelvey *et al.,* 1973).

Positively rewarding the new response without any attempt to punish the old behavior may be sufficient of itself to effect a behavioral change. The individual cannot perform both responses simultaneously, since they are incompatible. Hence the frequency of the undesirable behavior decreases if the new response is practiced. After the new behavior has become established, peer and supervisor praise plus new habit patterns take over and the old behavior is less likely to reoccur (Estes, 1972).

Punishment of the old behavior pattern alone may not be effective in establishing the new pattern, since the person may substitute another equally unacceptable mode of response for the punished one. Also, he is likely to revert to the old behavior once the punishment has ceased.

Therefore punishment should never be used alone, but only in conjunction with the reinforcement of a substitute behavior. Threat of punishment (item 3 in Table III) should *never* be used unless one plans to administer punishment the next time the incorrect behavior occurs. Otherwise, the supervisor is reinforcing the incorrect response and teaching the employee that a *threat,* but never actual punishment, is the contingency that follows the behavior.

## H. Individual Versus Group Reinforcement

The amount of time necessary to set up and to reward individual programs of modification is prohibitive, even if monitoring is done on a time-sampling rather than on a constant basis. If the laboratory has more than three to five workers, it is probably too time-consuming for one supervisor

**TABLE III**

**Examples of Positive and Negative Reinforcement**

| New response (e.g., wearing safety shoes) | | Old response (e.g., wearing street shoes) | |
|---|---|---|---|
| 1. | Positive reinforcement, e.g., verbal praise | 1. | Ignore |
| 2. | Positive reinforcement, e.g., verbal praise | 2. | Punishment, e.g., "Go home and change shoes" |
| 3. | Positive reinforcement, e.g., verbal praise | 3. | Threat of punishment, e.g., "If that occurs again, you will be sent home to change" |

to attempt individual behavioral modification procedures unless the majority of the individuals are tractable and present no problems.

When the behavior of a large group of individuals must be modified, a group reinforcement procedure might be attempted. Under this procedure the large group is broken down into manageable clusters of five or six individuals with one senior-level laboratory technician appointed to each to keep records. Individuals within the group can also choose one of their members for this task.

Reinforcement is made contingent upon group change rather than individual change. The entire group must perform correctly for a given time period stated in advance before a reward for the entire group is forthcoming. Reversions to the old method of behavior by any member of the group resets the time period to the beginning again. For example, the group is to be rewarded at the end of a two-week period with an afternoon free if all perform no mouth pipetting during that period. Sally mouth pipettes during the second week, whereupon the entire group then has another two weeks before a reinforcement can be won. Although children under such contingencies must be followed and rewarded daily, weekly and biweekly consequence are sufficient for mature adults who are able to delay gratification for longer time periods. The time periods between reinforcers can be gradually lengthened and then removed altogether after new habit patterns have become established.

This method has the added advantage of adding group peer pressure in addition to supervisor pressure on the individual. In the case of the recalcitrant person, group censure may have the desired effect of forcing him or her to modify the behavior. If the stubborn hold-outs are senior staff members, an attempt should be made to place others at or near his or her level into some of the other groups.

## I. IMMEDIATE BUT TEMPORARY BEHAVIOR CHANGE

One phenomenon that never fails to surprise beginners who attempt behavior modification should be mentioned. The fact that another individual is watching and keeping track of what one does is often sufficient of itself to effect an immediate change in behavior. This is also true if a person attempts to change his or her own behavior. The obese individual finds that merely keeping a record of what is consumed over a week is enough to cause a reduction in caloric intake. *However, these changes are only temporary.*

Therefore, a technician who starts to monitor a particular behavior change who discovers that everyone changes without effort should not necessarily expect this to continue past the first week without the addition of reinforcers. In any case, some return to baseline (backsliding) should be expected at first. As soon as the reinforcers are put into effect and the new behaviors become more automatic, the change should become relatively permanent.

## J. Steps in Changing Behavior

The following are steps that can be taken to change behavior:

1. Select the behavior to be modified.
2. Make certain that this is well enough defined so as to be distinctly different from the old behavior or errors of definition may occur.
3. Select a method of measurement. The frequency count is the usual measure employed, and if one behavior is to replace another, then an error count of the number of reversions to the old response is a reasonable measure.
4. Model the desired behavior for each group and make certain that all individuals can perform it correctly. Shaping may be appropriate with difficult or complex procedures that are not within the technicians' repertoire. Insist on more stringent criteria each day until the individual performs adequately. Acceptance of a lower level of performance is only temporary.
5. A time-sampling procedure is necessary in order to check on performance. In the case of small groups, the chosen leader should have this task. In a small laboratory the supervisor should make periodic checks on each individual.
6. Reinforcers should be chosen and the staff should be aware of them and of the contingencies governing them. If verbal reinforcers are used, these soon become self-evident and no explanation is necessary.

In the case of more tangible rewards, the method of earning them should be explained.

## K. Appropriate Reinforcers

What is an appropriate reinforcer? This is not an easily answered question, since types of laboratories and types of supervisors vary considerably. A free afternoon, which would be an excellent incentive in a tightly run laboratory, would fail miserably if the supervisor is in the habit of giving afternoons off for the asking. Verbal praise or censure might be an excellent reinforcer in a small, closely knit group that identifies strongly with the supervisor and the research problem.

A large laboratory could be split into smaller groups as described above. The groups that first achieved an error-free two weeks could be rewarded with a cook-out, beer party, or free afternoon. This schedule might be repeated over a two-month period before it is lengthened.

The list of possible reinforcers is virtually endless, but the appropriate one must be chosen by someone who knows the laboratory and personnel involved. A number of reinforcers that might prove effective in a laboratory include a choice of work projects, handling a project from beginning to completion, scheduling one's working hours for a period of time, choice of desk position or office space, gifts of lab glassware, and choice of vacation time.

The program described above that results in a reward every five or ten days is an example of a fixed-ratio schedule. After this has been in effect for a few weeks, a greater ratio might be employed (e.g., with 15–20 days intervening between reinforcements). The more stringent schedule should not be placed in effect until behavior has stabilized on the more lenient schedule and groups are obtaining the rewards regularly at a 10-day interval. When behavior on the more stringent schedule has stabilized, the supervisor could put a variable ratio schedule into effect. The supervisor would announce that from that point on, he/she would check the records at random intervals that would be known only to him or her and should be unpredictable to the personnel involved. This type of schedule results in steady, predictable behavior emission. After this period, reinforcements could probably be discontinued with no loss in efficiency.

At this point it is necessary to evaluate the effectiveness of the program. If the reinforcers are not having the desired effect, it may be necessary to change them or to add a punishment contingency for lapses into the old behavior patterns. Probably most laboratory procedures should have been placed into effect and be working fairly smoothly within one month. Progress should be checked weekly and recorded by the supervisor on a chart.

This is the only method by which accuracy is assured and the success or failure of the procedures may be assessed.

## L. Examples of Specific Behavioral Problems in the Laboratory

These examples have been brought to the author by laboratory personnel. For the solutions to be most effective it is assumed that

1. There is a clear chain of command within the laboratory so that each individual knows who is his or her immediate superior.
2. It is preferable to use rewards and some minor negative reinforcement rather than such sanctions as withholding pay, denying promotion, and in the extreme instance, firing an individual.

The simplest solution should always be tried first. For example, if it is desirable for individuals in the laboratory to wear coveralls, then the day on which these arrive the lab coats should be gathered together and thrown away, which eliminates the possibility that an individual will forget and put on a coat.

### 1. Example: Behavior Modification by a Lower Status Individual

Laboratory Y employs three middle-level technicians and one senior technician who reports to a Ph.D. laboratory chief.

Jane, one of the middle-level technicians, has attended a course on safety and has been chosen by the laboratory chief to implement the following changes, since the senior level technician had been unable to attend: (1) instrument pipetting is to replace mouth pipetting, (2) no food or drinks are to be consumed in the work area, (3) tasks that pose a risk of infection are only to be performed in one area in the laboratory.

**The Status Question.** Jane should first go to the senior-level technician and enlist his (or her) aid in setting up a couple of short meetings with the other technicians. These could be over lunch. The senior-level technician would first speak briefly on the reasons for the changes and then mention that the laboratory chief is anxious to implement them. Jane would then describe the course she has attended and set up a session in which she models the appropriate behaviors (e.g., instrument pipetting). She should observe each individual and give appropriate feedback and encouragement on his or her performance. The senior technician is also instructed in this technique.

At this point either the senior technician or the lab director, not Jane, should state that the new method is to be implemented immediately and that the old method ceases from that time. He (or she) also states that someone whom he designates will check from time to time to see if anyone is having any problems. Ideally, the record-keeping would be a job for the senior technician, but he probably will delegate it to the individual who attended the course. If this is her task, she should report violations to the senior technician. (Never allow the incorrect behavior to occur without negative reinforcement.) Needless to state, if neither the senior technician nor the lab director is cooperative, then no changes can be implemented.

If violations occur, the senior-level technician might handle them in the following fashion. A reward procedure should be tried first. He or she might state to the group that if all of them are performing the new methods adequately within three weeks, then the entire group will get an afternoon off, have a beer party at his house, or obtain some other designated reinforcement. Jane will then continue to check and report all violations as before. It is important when utilizing a reward to state a *definite* date when this will be given. The date should be distant enough that the new behavior will be well practiced and individuals will not be tempted to return to the old method once they have received the reward. It cannot, however, be too distant or it will not function as a reward. Three weeks should probably be the outside limit.

He may find it necessary to censure individuals who are maintaining the old response. This can be verbal and should be done in private, *not* in front of the individual's peers. The senior person should perform this task. The important point in this example is that a lower status individual cannot implement change without the aid and concurrence of a higher status person. This aid should also be made explicit for the greatest effectiveness.

## 2. Example: The Recalcitrant Technician

Laboratory Z employs 20 middle-level technicians, four senior-level technicians, two Ph.D. virologists, and a lab director over all personnel. One of the senior-level technicians attends a course in biohazards and returns to teach this to the other members of the laboratory. (Ideally, in a laboratory of this size more than one individual should attend the course. It places quite a burden on one individual to convey all that has been learned, especially when individuals are being asked to change old, familiar behaviors.) The senior technician first reports to the individual to whom he/she is responsible, who may set up a meeting with the senior personnel to discuss what has transpired and what changes should be instituted in the laboratory. (We will assume the laboratory director is not involved except to sanction whatever changes are necessary.) The attendee models whatever changes are appropriate and all four senior technicians learn the chosen technique with each made responsible for the individuals under him/her. One of the Ph.D's meets with all the lower level technicians and explains the changes necessary and the reasons for the changes. Alternately, he could merely introduce the senior technician who attended and in doing so state that the individual would give the explanation. The important point to be made is that the behavioral changes are sanctioned by the highest rank and are not the senior technician's idea. This is particularly critical if the attendee is an individual who is not usually listened to by the others by virtue of being younger, attached to the laboratory for a shorter time period, etc.

The changes are implemented and one senior technician soon discovers that one old (in terms of service) technician refuses to make some of the changes. He or she assumes that by virtue of length of service no negative consequences will be forthcoming. Let us assume that the recalcitrant individual dislikes the senior technician and therefore personal pleas are not likely to effect much change. The senior technician might take one of two approaches: (1) have his/her own superior invoke sanctions against the technician or (2) enlist the aid of the rest of the technicians who work with the difficult individual. The latter is done indirectly by offering some group reward if each member performs adequately for the next month (or whatever time period is set). Now the entire group will work to see that the hold-out changes his behavior. The senior technician continues to make periodic checks. The first time he/she catches the individual in a violation, he/she tells the group that the date of the reward is now extended for an additional time period, since one member is still using the old method. There is no need to name the individual, since the members will find out quickly enough. One violation will probably be all that occurs; however, this one should be anticipated, *since that type of individual is likely to test the limits.*

A slightly different solution is necessary when the hold-out is the lab director, or a Ph.D.-level scientist. In the case of the director, nothing can be accomplished by anyone in the laboratory. (We are assuming the individual is hostile to change, not merely lazy.) If the individual is a scientist, then the laboratory director or another individual at or above his/her level are the only individuals who could invoke sanctions.

## 3. Example: The Recalcitrant Laboratory Director

The problem that has been brought to the author most often by laboratory technicians, postdoctoral fellows, and junior-level scientists is that of the senior researcher or laboratory director who refuses to implement safety procedures. This reluctance is usually an unwillingness to change techniques they have been using for many years, in a situation in which the hazards are not clearly defined and the level of risk is unknown.

There are a variety of problems involved in implementing behavioral changes in a high-status individual. Obviously this cannot be handled by a technician or a lower level scientist except at a very informal educational level. Implementing the techniques with the rest of the laboratory staff can be accomplished as long as *someone* in authority sanctions this (e.g., an assistant laboratory director or higher level technician).

It may never be possible to accomplish change in the laboratory director unless the funding agency or governmental regulations demand such change. Alternatively, if the entire laboratory is utilizing safe techniques, a director may find him/herself pressured into amending old habit patterns.

The problem that arises when a laboratory director does not see the necessity for change is that laboratory staff may be slow to make changes as well. They deduce that negative consequences will not accrue to such a refusal and they are probably correct.

## 4. Example: The Lone Wolf

This example is not directly related to safety. Indeed, one study has demonstrated that individuals who work alone are *less* likely to have accidents (Keenan *et al.,* 1951). It is included here because it may be a real problem in a smaller laboratory.

In the laboratory of today, group effort and cooperation are usually required to achieve scientific progress. The individualist who cannot be integrated into the group is an anachronism. However, these people do exist, albeit in decreasing numbers, and may be superb technicians when working by themselves. What does one do if one has an individual of this type in a small laboratory where there is not enough work for such a person always to operate independently? This individual must share in the group effort at least part of the time for the lab to become cost effective. All efforts to integrate him/her have failed and the person refuses to work with anyone or insists on directing the operation instead of sharing in it, if placed with other technicians.

It is important to understand the probable environmental contingencies that have shaped this individualistic mode of behavior. Such individuals have been rewarded in the past for working alone and/or either have seldom been reinforced or have been negatively reinforced in groups. Their behavior needs to be shaped so that working with other individuals is also rewarding.

Begin by choosing a capable but noncompetitive individual who is different on as many parameters as possible: older and female if the loner is young, and male, young and higher status if the individualist is older and lower status. This can be a rough approximation. Don't worry if the chosen person is not too different. The important parameter is the *noncompetitiveness:* the motherly or fatherly figure (whatever the age) who gets along with everyone is the one for whom we are searching.

Set up a task for these two people to perform as a unit so that both get credit and praise together when the task is completed. Either make the first task rather short in duration or allow an extended time period to complete it so the loner at first only works short periods of time with the other person.

The time periods when the two work as a unit can be gradually lengthened as complaints by the individualist become fewer and the cooperative individual reports that they are working better as a team. The following points are important:

1. Be sure that the schedule of cooperative activity is clearly spelled out to both individuals before they begin. The loner should thus be assured that working with another individual is a small part of total working time. The partner should also know this so that he or she does not attempt to add more cooperative time in order "to finish the job." The loner will probably tolerate small amounts of such (to him/her) distasteful activity fairly well, especially with a sympathetic partner.

2. Do not add additional cooperative work hours with the sympathetic individual until feedback from that person and observation by another individual indicate that the loner is beginning to receive some positive reinforcement from such activity.

3. When the cooperative activity with the one individual is being well tolerated, a third person can be added to the group (retaining the original sympathetic individual) for brief time periods at first. It is hoped that the positive reactions that have been conditioned by the sympathetic individual will generalize to the new person. Cooperative time periods can then be gradually increased. The time periods with the original individual can then be gradually lessened until the loner is spending the group work periods with the new individual.

By this method coupled with praise for *group effort* by the supervisor, the loner should begin to work more easily and for longer time periods with others.

There will be practical limits as to what you can hope to accomplish with such people and they should always be left a finite time period for solitary work. It is unrealistic to expect them to become extraverted people lovers, but it should be possible to turn them into effective working partners for a portion of the work week.

## M. Consultants

Specific help with behavior modification can usually be obtained gratis within a university setting from faculty in the departments of psychology, psychiatry, or social work. A consulting fee may be anticipated if your own company is outside of such a setting. Such fees are usually not excessive, since behavior modification procedures can be implemented relatively rapidly.

## ACKNOWLEDGMENT

This chapter in modified form has been presented as part of a short course entitled "Principles of Biohazard and Injury Control in the Biomedical Laboratory," sponsored by the National Cancer Institute, National Institutes of Health, Contract #NO1-CP-4-3285.

## REFERENCES

Ayllon, T., and Michael, J. L. (1959). The psychiatric nurse as a behavioral engineer. *J. Exp. Anal. Behav.* **2**, 323-334.

Bandura, A. (1965). Behavioral modifications through modeling procedures. *In* "Research in Behavior Modification" (L. Krasner and L. P. Ullman, eds.), pp. 310-339. Holt, New York.

Bandura, A., Blanchard, E. B., and Ritter, B. (1969). Relative efficacy of desensitization and modeling approaches for inducing behavior, affective and attitudinal changes. *J. Personality & Social Psychol.* **13**, 173-199.

Berry, G., Newhouse, M. L., and Turok, M. (1972). Combined effect of asbestos exposure and smoking on mortality from lung cancer in factory workers. *Lancet* **2**, 476-479.

Bureau of Labor Statistics (1976). General, annuals and biennials. *In* "Handbook of Labor Statistics."

Clark, E. L. (1927). The value of student interviewers. *J. Personality Res.* **5**, 204-207.

Cohen, E. N., Bellville, J. W., and Brown, B. W. (1971). Anesthesia, pregnancy, and miscarriage: A study of operating room nurses and anesthetists. *Anesthesiology* **35**, 343-347.

Davis, S. H. (1977). The relationships between the incidence of accidents, locus of control, and life crisis. M.S. Thesis, University of Washington, Seattle.

Estes. W. K. (1944). An experimental study of punishment. *Psychol. Monogr.* **57**, No. 3.

Estes, W. K. (1972). Reinforcement in human behavior. *Am. Sci.* **60**, 723-729.

Fine, L. J., and Peters, J. M. (1976). Respiratory morbidity in rubber workers. I. Prevalence of respiratory symptoms and disease in curing workers. *Arch. Environ. Health* **31**, 5-9.

Gavrilescu, C. (1973). Avortul spontan repetat. *Obstetr. Ginecol.* **21**, 201-208.

Gibbs, C. B. (1968). Some practical applications of research on impairment. *Can. Res. Counc., Memo.*

Golodets, R. C., Zeigarnik, B. V., and Rubenstein, S. I. (1963). A clinical and pathopsychological description of asthenic conditions developed under long-term irradiation. *Vopr. Psikhol.* **3**, 39-47.

Greenspoon, J. (1955). The reinforcing effect of two spoken sounds on the frequency of two responses. *Am. J. Psychol.* **68**, 409–416.

Keenan, V., Kerr, W., and Sherman, W. (1951). Psychological climate and accidents in an automotive plant. *J. Appl. Psychol.* **35**, 108–111.

Kerr, W. (1964). Complementary theories of safety psychology. *In* "Accident Research: Methods and Approaches" (W. Haddon, Jr., E. A. Suchman, and D. Klein, eds.), pp. 304–308. Harper, New York.

King, E. C. K. (1978). A design for laboratory managers to encourage staff motivation. *Lab. Manage.* **7**, 45–48.

Latham, G. P., and Dossett, D. L. (1978). Designing incentive plans for unionized employees: A comparison of continuous and variable reinforcement schedules. *Personnel Psychol.* **31**(1), 47–61.

LeShan, L. L. (1952a). Dynamics of accident prone behavior. *Psychiatry* **15**, 73–80.

LeShan, L. L. (1952b). The safety prone: An approach to the accident-free person. *Psychiatry* **15**, 465–468.

McGlade, F. S. (1970). "Adjustive Behavior and Safe Performance." Thomas, Springfield, Illinois.

MacIver, J. (1961). Safety and human behavior. *In* "Behavioral Approaches to Accident Research" (H. H. Jacobs *et al.,* eds.), pp. 59–76. Assoc. Aid Crippled Children, New York.

McKelvey, R. K., Engen, T., and Peck, M. B. (1973). Performance efficiency and injury avoidance as a function of positive and negative incentives. *J. Saf. Res.* **5**, 90–96.

Makhametoza, G. M., and Vozovaia, M. A. (1972). Reproductive power and incidence of gynecological disease among female workers exposed to a combined effect of gasoline and chlorinated hydrocarbons. *Gig. Tr. Prof. Zabol.* **16**, 6–9.

Malott, R. W., Tillena, M., and Glen, S. (1978). "Behavior Analysis and Behavior Modification: An Introduction." Behaviordelia, Inc., Kalamazoo, Michigan.

Mikulas, W. L. (1972). "Behavior Modification: An Overview." Harper, New York.

Newhouse, M. L., Berry, G., Wagner, J. C., and Turok, M. E. (1972). A study of the mortality of female asbestos workers. *Br. J. Ind. Med.* **29**, 134–141.

Phillips, G. B. (1965). "Causal Factors in Microbiological Laboratory Accidents and Infections." U.S. Army Biol. Labs. No. 2, Fort Detrick, Maryland.

Phillips, G. B. (1969). Control of microbiological hazards in the laboratory. *J. Am. Ind. Hyg. Assoc.* **30**, 170–176.

Rodstein, M. (1974). Accident proneness. *J. Am. Med. Assoc.* **229**, 1495.

Rogers, E. M., and Shoemaker, F. F. (1971). Authority, innovation-decisions, and organizational change. "Communication of Innovations: A Cross-Cultural Approach," Chapter 10. Free Press, New York.

Rosenthal, R. (1973). The Pygmalion effect lives. *Psychol. Today* September, pp. 56–63.

Rosenthal, R., and Jacobsen, L. (1969). "Pygmalion in the Classroom: Self-fulfilling Prophecies and Teacher Expectations." Holt, New York.

Schick, S., and Hutchinson, B. A. (1974). A no-fuss approach to safer pipetting. *Med. Lab. Observer* May-June, pp. 115–120.

Sidman, M. (1962). Operant techniques. *In* "Experimental Foundations of Clinical Psychology" (A. J. Bachrach, ed.), Chapter 6. Basic Books, New York.

Suchman, E. A. (1970). Accidents and social deviance. *J. Health Social Behav.* **11**(1), 4–14.

Suchman, E. A., and Scherzer, A. L. (1964). Accident proneness. *In* "Accident Research: Methods and Approaches" (W. Haddon, Jr., E. A. Suchman, and D. Klein, eds.), pp. 387–389. Harper, New York.

U.S. Department of Commerce (1976). "Statistical Abstract of the United States," pp. 83, 389. Bureau of the Census, Washington, D.C.

Vital and Health Statistics (1976). "Persons Injured and Disability Days by Detailed Type and Class of Accident: United States 1971–1972," DHEW Publ. (HRA) 76-1532. U.S. Department of Health, Education and Welfare, Washington, D.C.

Whaley, D. L., and Malott, R. W. (1971). "Elementary Principles of Behavior." Appleton, New York.

Yager, J. W. (1973). Congenital malformations and environmental influence: The occupational environment of laboratory workers. *J. Occup. Med.* **15,** 724–728.

# Appendix:
# Classification of Agents*†

A. Bacterial
Class 1
All bacterial agents not included in higher classes according to "Basis for Agent Classifications"
Class 2
*Actinobacillus*—all species except *A. mallei,* which is in Class 3 *Arizona hinshawii*—all serotypes
*Bacillus anthracis*
*Bordetella*—all species
*Borrelia recurrentis, B. vincenti*
*Clostridium botulinum,*
    *Cl. chauvoei, Cl. haemolyticum,*
    *Cl. histolyticum, Cl. novyi,*
    *Cl. septicum, Cl. tetani*
*Corynebacterium diphtheriae,*
    *C. equi, C. haemolyticum,*
    *C. pseudotuberculosis,*
    *C. pyogenes, C. renale*
*Diplococcus (Streptococcus) pneumoniae*
*Erysipelothrix insidiosa*
*Escherichia coli*—all enteropathogenic serotypes
*Hameophilus ducreyi, H. influenzae*
*Herellea vaginicola*
*Klebsiella*—all species and all serotypes
*Leptospira interrogans*—all serotypes
 *Listeria,* all species
*Mima polymorpha*
*Moraxella*—all species

*Mycobacteria*—all species except those listed in Class 3
*Mycoplasma*—all species except *Mycoplasma mycoides* and *Mcyoplasma agalactiae*
*Neisseria gonorrhoeae, N. meningitidis*
*Pasteurella*—all species except those listed in Class 3
*Salmonella*—all species and all serotypes
*Shigella*—all species and all serotypes
*Sphaerophorus necrophorus*
*Staphylococcus aureus*
*Streptobacillus moniliformis*
*Streptococcus pyogenes*
*Treponema carateum, T. pallidum,* and *T. pertenue*
*Vibrio fetus, V. comma,* including biotype E1 Tor, and *V. parahemolyticus*
Class 3
*Actinobacillus mallei* (USDA permit also required for import or interstate transport.)
*Bartonella*—all species
*Brucella*—all species
*Francisella tularensis*
*Mycobacterium avium, M. bovis, M. tuberculosis*
*Pasteurella multocida* type B ("buffalo" and other foreign virulent strains, USDA permit also required for import or interstate transport)
*Pseudomonas pseudomallei* (USDA

---

* Prepared by the Office of Biosafety, Center for Disease Control, Public Health Service, Atlanta, Georgia 30333.

† This classification does not include strictly animal pathogens. A PHS permit is required to import any agent or to transfer within the United States any agent imported under permit.

Class 1 *(cont.)*
permit also required for import or interstate transport.)
*Yersenia pestis*

B. Fungal

Class 1

All fungal agents not included in higher classes according to "Basis for Agent Classifications"

Class 2

Actinomycetes (including *Nocardia* and *Actinomyces* species and *Arachnia propionica*)
*Blastomyces dermatitidis*
*Cryptococcus neoformans*
*Paracoccidioides brasiliensis*

Class 3

*Coccidioides immitis*
*Histoplasma capsulatum*
*Histoplasma capsulatum* var. *duboisii*

C. Parasitic

Class 1

All parasitic agents not included in higher classes according to "Basis for Agent Classification"

Class 2

*Endamoeba histolytica*
*Leishmania* sp.
*Naegleria gruberi*
*Toxoplasma gondii*
*Toxocara canis*
*Trichinella spiralis*
*Trypanosoma cruzi*

Class 3

*Schistosoma mansoni*

D. Viral, Rickettsial, and Chlamydial

Class 1

Class 1 includes all viral, rickettsial, and chlamydial agents not included in higher classes according to "Basis for Agent Classification." Specifically listed are:

Influenza virus A/PR8/34
Newcastle virus—strains licensed for vaccine use in U.S.
Parainfluenza virus 3, SF4 strain
(These viruses are included because the committee agreed that they are suitable for science experiments at a junior level.)

Class 2

Adenoviruses—human, all types
Cache Valley virus
Coxsackie A and B viruses
Cytomegaloviruses
Echoviruses—all types
Encephalomyocarditis virus (EMC)
Flanders virus
Hart Park virus
Hepatitis-associated antigen material
Herpes viruses—except herpesvirus simiae (monkey B virus) which is in Class 4
Cornona viruses
Influenza viruses—all types except A/PR8/34, which is in Class 1
Langat virus
Lymphogranuloma venereum agent
Measles virus
Mumps virus
Parainfluenza viruses—all types except parainfluenza virus 3, SF4 strain, which is in Class 1
Polioviruses—all types, wild and attenuated
Poxviruses—all types except Alastrim, smallpox, monkey pox, and whitepox, which, depending on experiments are in Class 3 or Class 4
Rabies virus—all strains except rabies street virus, which should be classified in Class 3 when inoculated into carnivores
Reoviruses—all types
Respiratory syncytial virus
Rhinoviruses—all types
Rubella virus
Simian viruses—all types except herpesvirus simiae (Monkey B virus) and Marburg virus, which are in Class 4
Sindbis virus
Tensaw virus
Turlock virus
Vaccinia virus
Varicella virus
Vole rickettsia
Yellow fever virus, 17D vaccine strain

Class 3

Alastrim, smallpox, monkey pox, and whitepox, when used *in vitro*

Arboviruses—all strains except those in Class 2 and 4 (arboviruses indigenous to the United States are in Class 3, except those listed in Class 2. West Nile and Semliki Forest viruses may be classified up or down, depending on the conditions of use and geographical location of the laboratory.)

Dengue virus, when used for transmission or animal inoculation experiments

Lymphocytic chorimeningitis virus (LCM)

Psittacosis-ornithosis-trachoma group of agents

Rabies street virus, when used in inoculations of carnivores (see Class 2)

Rickettsia—all species except Vole rickettsia when used for transmission or animal inoculation experiments

Vesicular stomatitis virus (USDA permit also required for import or interstate transport.)

Yellow fever virus—wild, when used *in vitro*

Class 4

Alastrim, smallpox, monkey pox, and whitepox, when used for transmission or animal inoculation experiments

Hemorrhagic fever agents, including Crimean hemorrhagic fever (Congo), Junin, and Machupo viruses, and others as yet undefined

Herpesvirus simiae (monkey B virus)

Lassa virus

Marburg virus

Tick-bone encephalitis virus complex, including Russian spring-summer encephalitis, Kyasanur forest disease, Omsk hemorrhagic fever, and Central European encephalitis viruses

Venezuelan equine encephalitis virus, epidemic strains, when used for transmission or animal inoculation experiments

Yellow fever virus—wild, when used for transmission or animal inoculation experiments

# Index

## A

Accident(s), 278–279
  behavioral factors related to
    accident-free individuals and,
      321–322
    accident repeaters, 321
    age and sex, 320–321
    prolonged emotional disruption, 322
    statistics and, 320
    temporary emotional stress, 322
  limiting damage occurring in, 8
  modification of unsafe behavior patterns
    appropriate reinforcers and,
      332–333
    behavioral analysis and, 324
    consultants and, 338
    contingencies governing change and,
      326–327
    employee expectancies and, 325–326
    immediate but temporary behavior
      change and, 331
    individual vs. group reinforcement and,
      329–330
    modeling and shaping behaviors and,
      324–325, 326
    positive vs. negative reinforcers and,
      329
    reinforcement schedules and, 327–329
    specific behavioral problems and,
      333–337
    steps in changing behavior and,
      331–332
  task factors related to
    intrinsic job factors, 323
    task complexity and variability,
      323

Accident causation, 4–5
  energy factors and control strategies in,
    6–8
  human factors in, 5–6
Acetaldehyde, 256
Acetates, 255
Acetic acid, 255
Acetic anhydride, 256
Acetonitrile, 259
2-Acetylaminofluorene (2-AAF), 47
Acetylation, 38
Acetyline, 252
Acetyl mercapturic acid, 38
Acridine(s), 263, 302–303
Acrylonitrile, 259
Adenoviruses, 120–121
Administration route of chemicals, 31
  dermal, 31–32
  ingestion, 36
  inhalation, 33
    gases and vapors and, 35–36
    particle density and, 34–35
    particle shape and, 34
    physical state of particles and, 33–34
    temperature and respiration rate and,
      35
  ocular, 32–33
Aerosols
  control of, in viral oncology research,
    211–212
  in animal rooms, 329
  intoxication by, 272–273
Aflatoxin, 304–305
Age
  accidents and, 320–321
  toxicity and, 42
Air barriers, 190

## S

## DATE DUE

| FE 5 '85 | | | |
|---|---|---|---|
| MAR 31 '87 | | | |
| | | | |
| | | | |
| | | | |
| | | | |
| | | | |
| | | | |
| | | | |
| | | | |
| | | | |
| | | | |
| | | | |
| | | | |
| | | | |
| | | | |
| | | | |
| GAYLORD | | | PRINTED IN U.S.A. |